Jamaica in the Age of Revolution

JAMAICA
in the
AGE *of* REVOLUTION

TREVOR BURNARD

PENN

UNIVERSITY OF PENNSYLVANIA PRESS

PHILADELPHIA

Published by
University of Pennsylvania Press
Philadelphia, Pennsylvania 19104-4112
www.upenn.edu/pennpress

Printed in the United States of America on acid-free paper
1 3 5 7 9 10 8 6 4 2

Library of Congress Cataloging-in-Publication Data

Names: Burnard, Trevor G. (Trevor Graeme), author.
Title: Jamaica in the age of revolution / Trevor Burnard.
Description: 1st edition. | Philadelphia : University of Pennsylvania Press, [2020] |
    Includes bibliographical references and index.
Identifiers: LCCN 2019034405 | ISBN 9780812251920 (hardcover)
Subjects: LCSH: Plantations—Jamaica—History—18th century. |
    Slavery—Jamaica—History—18th century. | Revolutions—History—
    18th century. | Jamaica—History—18th century. | Jamaica—Social
    conditions—18th century. | Jamaica—Economic conditions—18th century. |
    United States—History—Revolution, 1775–1783—Influence.
Classification: LCC F1884 .B87 2020 | DDC 972.92/033—dc23
LC record available at https://lccn.loc.gov/2019034405

Contents

Introduction

In 1774, the white residents of Jamaica were the richest group of people in the British Empire. They probably had more influence within the British imperial state than any other colonial subjects of George III.[1] On the eve of the American Revolution Jamaica was undoubtedly the jewel in the British imperial crown. It was the British colony that was most indispensable to imperial prosperity and the single colony that the British could least afford to lose to the French during the War of American Independence in the 1770s and 1780s.[2] Consequently, Britain in 1781 was prepared to sacrifice its interests in the thirteen colonies to preserve its control over Jamaica.[3]

At present, such statements do not seem as outlandish as they seemed in 1987, when I took my first academic position as a lecturer in American history at the University of the West Indies at Mona and began to explore the riches of the various archives and libraries in Jamaica. At that stage, few historians of colonial British America ever thought very much about how Jamaica was at the forefront of imperial officials' minds in the 1760s through the 1790s. Nor did many historians consider very seriously the idea that Jamaica had as much importance in America before 1776 as Virginia and Massachusetts. The advent of Atlantic history in the 1990s and 2000s, however, has widened the geographical boundaries of early America so that the British West Indies, which was part of seventeenth- and eighteenth-century British America but which did not join the United States of America after 1787, can be thought of as an early American place.

Of course, there have always been works that put British North America and the British West Indies together. Richard Dunn's magnificent account of the rise of planter societies in the seventeenth century from 1972 and Carl and Roberta Bridenbaugh's account in the same year of seventeenth-century Englishmen leaving England for the New World, which included a considerable section on the West Indies, were two such notable works.[4] In the last decade it has been relatively common for early American historians to write on West Indian topics.[5] One such notable example is Vincent Brown's meditation on mortality and cultures of death in eighteenth-century Jamaica from 2008.[6] Another is the social analysis of mid-eighteenth-century Jamaica produced by my mentor, Jack P. Greene.[7] And Richard Dunn has culminated his long research into Jamaica with a wonderful work comparing slave life on Mesopotamia plantation in western Jamaica with slavery in Tidewater Virginia.[8] Younger scholars place Jamaica within an Atlantic and early American context as a matter of course, as can be seen in books by Emily Senior, Sasha Turner, Katherine Gerbner, Aaron Graham, Daniel Livesay, Christer Petley, and Brooke Newman, all illuminating aspects of slavery within an Atlantic framework.[9] Jamaica is no longer incidental, or aberrant, within larger Atlantic patterns, but emblematic of how the Atlantic system worked. Robert DuPlessis's superb analysis of the material life and consumer habits of Atlantic peoples, for example, uses data from Jamaica to outline the fashioning of a wider Atlantic world.[10]

In addition, West Indian historians have long written about the eighteenth- and early nineteenth-century West Indies using frameworks derived from the historiography of colonial and antebellum America. My work was informed by many of these historians, including people I worked alongside in Jamaica between 1987 and 1989. The most important of these historians has been my former colleague at Mona, Barry Higman, who has delved deeply into all aspects of West Indian social and economic history during the period of abolition, mostly works on subjects from 1807 to 1834, but including some works on subjects extending back into the middle of the eighteenth century.[11] Without his works, none of us writing today would know very much at all about the underlying foundations of West Indian slave society.[12] Another former colleague at Mona, Kamau Braithwaite, is a great poet, but before he devoted himself to that occupation he was an important historian, author of a highly suggestive work on Creole society in Jamaica around the time of the eighteenth century that not only has influenced what I thought about Jamaica but also was pivotal in shaping my understanding of Chesapeake society in my first work on eighteenth-century Maryland planters.[13]

But the rich depth of local studies and quantitative histories that made the history of early America in the 1970s and 1980s so deeply exciting and which formed the social history foundation from which "the cultural history turn" of the 1990s developed was less rich when extended to eighteenth-century Jamaica.[14] The relative paucity of empirical studies of Jamaica in this period meant that there were marvelous opportunities to do work on a variety of social history topics already covered by historians elsewhere. Few scholars had exploited Jamaica's rich archives. These archives contain little of the traditional sources—letters, pamphlets, written testimony of all kinds—but wonderful material for the social historian—wills, deed, inventories, manumission records. The chapters in this book are heavily based on the now more than thirty years of work in these archives.[15]

If the quality of work on the history of Jamaica has often been very strong, the quantity of work is not as impressive as the scholarship on other parts of the Atlantic world in the eighteenth century.[16] The reasons are quite straightforward. Jamaica was an indispensable colony in the eighteenth century, but its importance faded after the end of slavery in 1834 and even more after emancipation seemed to fail and as the Morant Bay rebellion of 1865 confirmed to imperialists that Jamaica was a hopeless case.[17] The island became by the early twentieth century a marginal colony of little importance in the wider world. It gained independence, but not prosperity, in 1962. By the late 1970s, the country had fallen on hard times, and its history in the last thirty years has been one of continual disappointment—everlasting economic crisis, debilitating rates of crime, heavy outmigration to Britain and to the United States, a culture marked by shameful homophobia, and limited geopolitical relevance.[18] Even its once mighty cricketers stopped being world-beating sportsmen around 2000. Since the death of Bob Marley in 1981 no other Jamaican cultural figure has achieved worldwide fame, although the incomparable Usain Bolt and the magnificent Shelley-Ann Fraser-Pryce—the fastest man and the fastest woman in the world—and terrific writer and Booker Prize winner Marlon James continue to be Jamaicans with global name recognition. But Jamaica is hardly central to world culture, let alone to the global economy, in the way that it certainly was, for good or ill, in the period of the American Revolution.[19]

This marginalization of Jamaica has happened before. The division of the plantation regions of British America because of the American Revolution saw Jamaica disconnected from American history. Jamaica remained a diminished place within the British Empire while the plantation colonies of North America formed part of an expanding and dynamic United States. This book uses the American Revolution as a pivot for Jamaican history because it was after that event, and the separation of the thirteen colonies that came to make up the United States, that the twinned histories of the British West Indies and British North America truly came apart, despite some divergences arising as early as the Glorious Revolution of 1688–89.[20]

The American Revolution marks a historiographical fissure. It disconnected the future United States—the first nation founded because of an anticolonial revolt—from a set of colonies, including Jamaica, that stayed in the British Empire. Indeed, the historiography of the American Revolution is curious when seen from a Jamaican perspective. The dispute split plantation America in two, with the northern half (from the Chesapeake to Georgia) rebelling and the southern part (the West Indies) staying loyal to Britain. There were still some similarities in the experience of both sections of the prerevolutionary British plantation world. The American Revolution had some negative economic consequences for all British American plantation colonies. Indeed, the American Revolution was an economic disaster for much of the American South, especially for poor whites and blacks, leading to an economic depression hitherto unprecedented in American history, from which the American South has never quite recovered.[21] Yet few accounts of the consequences of the American Revolution for the new American nation dwell on the costs that ordinary Americans paid for defeating the British. An economic boom after 1792 and Alexander Hamilton's successful financial reforms made the depression of the 1780s appear a blip rather than a fundamental structural change in the relationship of the South to the North.[22]

By contrast, historians of the British Caribbean have argued that the American Revolution heralded the beginning of the end for the plantation system. The war itself, it has been argued, was economically devastating and encouraged Britons to believe that the economic value of the West Indian plantations was not sufficient to overcome the moral problems involved in the transatlantic slave trade.[23] The weight of opinion now suggests that the American Revolution did not have as severe economic and political consequences as once thought, especially once troubles in neighboring Saint-Domingue removed Jamaica's biggest competitor in international trade from the transatlantic economy. Yet it is hard to get past the historical consensus about the American Revolution in the British Caribbean marking an end to West India's privileged position in the British Empire.[24] Historians differ considerably about whether the American Revolution initiated terminal decline in the British West Indian plantation system, with the weight of scholarly opinion behind the idea that a "decline" has been overstated. The difference in tone, however, between scholarship on America in the period of the early republic and that on the British West Indies in the age of abolition is remarkable—even the idea of discussing "decline" in respect to the effect of the American Revolution on American development is generally unthinkable.[25]

It is thus a feat of imagination, as well as historical scholarship, to envision a period when Jamaica was rich (even if all that wealth went to a very small proportion of the population), geopolitically crucial, and a place of vital cultural creation, especially within its majority slave population but even within its white population. It is not always noted, for example, that Jamaica was a foundational place for the development of the gothic novel.[26] Several major practitioners derived their wealth from there or used it as a point of reference. And it was a place well suited to the gothic imagination, being the home of notoriously cruel and overmighty slave owners and a vibrant African culture that both fascinated and terrified Europeans. Because of its subsequent history, Saint-Domingue, which became Haiti in 1804 after a successful slave revolt, has occupied center place in the Western imaginary of the terrors of a slave society (and for nineteenth-century Americans and Europeans of the horror of slave insurrection, as well). But before the Haitian Revolution it was Jamaica that was the focus of fervent, if intermittent, interest by European Enlightenment philosophers, not just in Britain but in France.[27]

Jamaica embodied some of the major features that led to modernization and some of the characteristics that showed that the Enlightenment was compromised by the savagery some Enlightenment peoples embraced. The white Jamaicans I study were as much part of the Enlightenment as cosmopolitan intellectuals in Paris, London, Edinburgh or Philadelphia were. And so too, Laurent Dubois argues, were many of the black Jamaicans whom the white Jamaicans of the Enlightenment so cruelly mistreated.[28] Jamaica was not a refined place. It was a raw frontier society marked by aggressively entrepreneurial planter elites. It had extremely exploitative social relations characterized by remarkable levels of violence against enslaved people. That violence included sexual rapacity by white men against black women on a grand scale. It resulted in a social system that was fluid, chaotic, destructive, and, for most of the enslaved population, full of trauma. The highly dysfunctional character of white Jamaican society was evident in the remarkable diaries of an English immigrant, Thomas Thistlewood, who lived mostly among enslaved people in rural Jamaica for thirty-seven years, about whom I have written. I moved from writing a microhistory of Jamaican society as seen through the life of a single person to completing two books in 2015 and 2016 in which Jamaica was placed within larger contexts, first within the plantation systems of British America from the mid-seventeenth century until the early nineteenth century, and then within the context of the Greater Antilles, comparing Jamaica with the neighboring French colony of Saint-Domingue.[29] Increasingly, my attention has moved toward the period in which the American Revolution reshaped the politics of plantation societies in British America, the United States of America, and the British Empire.[30]

Studying the American Revolution from a Jamaican perspective shows that the British government had little interest in slavery disappearing, contrary to a developing literature that presumes that American rebels were united in a common cause against the British because they felt that the British were undermining slavery in the colonies.[31] Why would they even think about attacking slavery when some of their most valuable colonies, notably Jamaica, depended utterly on slavery as the foundation of their society and economy? As Katherine Paugh and Woody Holton note, a major point of dissension between the Virginian elite and the British government in the late 1760s and early 1770s was Virginians' anger that Britain kept on insisting that the Atlantic slave trade needed to continue to operate in the Chesapeake. Thomas Jefferson was so upset by what he considered British intransigence over keeping the slave trade to Virginia going that he listed Britain's rejection of a law by Virginia in 1772 to impose a prohibitory tax on Atlantic slave imports as one of the grievances that were leading Virginia to rebellion.[32] Thus, it was a rash decision for North American planters to join in a rebellion started by northern colonists whose attitudes toward the continuation of slavery as a social system were decidedly suspect by planters' standards.[33]

The plantation system in Jamaica was built on an especially vicious system of slavery. The chaos and cruelty of slavery in eighteenth-century Jamaica was a product of both accident and design. As several chapters in this book make clear, white Jamaican society possessed decidedly unattractive features. White

Jamaicans fornicated too much, drank and ate in excess, were overly devoted to the main chance and short-term thinking, and saw everything in terms of material gain. Contemporary British playwrights got it right also when they depicted Jamaican slave owners as uncouth philistines with aberrant sexual mores. Samuel Foote's 1764 farce *The Patron* featured Sir Peter Pepperpot, "a West Indian of overgrown fortune," and "an ingenuous and absurd character" given to hot-blooded violence at a moment's nature and leering sexual assaults on black women.[34]

These repellent people presided over a highly productive economic system, a precursor to the modern factory in its management of labor, its harvesting of resources, and its scale of capital investment and output. As Barry Higman concludes, planters made their money from "the manipulation of a complex, agro-industrial technology, an integrated trade network, and brutal system of labor exploitation."[35] Underpinning and sustaining Jamaican prosperity and white Jamaican control over enslaved people was a brutal form of labor exploitation. Jamaica perfected a hard-driving slave system in which large and continual declines in the enslaved labor force year after year were made up by additions to the labor force from an ever-expanding Atlantic slave trade. To call an agricultural system based on chattel slave labor "modern" is to employ a problematic and anachronistic term, but there is no better way to describe the agro-industrial system perfected in the booming years between 1760 and 1788 in Jamaica than to say it was remarkably modern, especially in its careful if callous management of human resources. Planters, supported by a dynamic merchant class in Kingston, created a plantation system in which short-term profit maximization was the main aim. The Jamaican plantation complex was a carefully calibrated and finely tuned system in which planters devoted all their attention to growing as much plantation crops as they could, no matter how hard this system of production was on overworked and underfed slaves. It was an economic system that was extremely far from being self-sufficient. This dependence on the outside world for material essentials caused considerable problems in the War for American Independence when provisions were hard to procure from Europe and North America and when the slave trade was greatly disrupted.[36]

Planters did little to make the material conditions of slavery any easier. On the contrary, they adopted more scientific methods of slave management during the eighteenth century that made plantations more efficient and slave work more difficult. Consequently, slaves were highly stressed, prone to disease, and likely to die.[37] A few prescient planters, like Simon Taylor (1740–1813), of Golden Grove estate in the Southeast of Jamaica, whose life has been sensitively chronicled in a new book by Christer Petley, realized that pushing slaves to exhaustion and thus to death was not a viable long-term strategy. To make his point, Taylor compared slaves to the hardened metal pivot pieces that held the rollers in a vertical sugar mill. He pointed out that "they [the slaves] are not steel or iron and we see neither gudgeons nor capooses last in this country." Because the plantation's productivity depended on an adequate supply of labor, he thought, slaves should not to be worked "above their ability."[38]

But most planters did not heed Taylor's warnings. Given that the Atlantic slave trade brought in new captives who could be bought to replace dead slave workers, planters preferred to work slaves as hard as they could to produce the greatest possible crops without concerning themselves overmuch about how this might have an impact on future productivity. Taylor feared that slaves "killed by overwork & harassed to Death" would result in a "Land without Negroes." He was incensed at how his overseer, John Kelly, worked slaves on Golden Grove to death without caring about replacing them. He thought Kelly worked slaves so hard that "their hearts have been broke." Kelly gave them so few rations and favored his own slaves so methodically over the slaves of his employer when allocating provisions that the Golden Grove slaves "were Starving" and a "very feeble Set indeed."[39] But Kelly did not worry about what might happen in the future. He was focused on producing the largest crop possible and thus getting a large commission. This short-term obsession with profits did indeed break slave hearts—and their bodies even more so.

Jamaica's brutal slave system worked because the men, and occasional women, who ran it were extremely powerful. Of course, even the most hegemonic powers had their limitations. Hegemony, as practitioners of subaltern history have told us, can be undermined by peasants and slaves refusing to accept their place in the world. Everyone enjoys laughing at one's superiors. Breaking wind behind the prince's back or stealing the slave master's sheep or participating in carnivalesque inversions of order at Christmastime or even taking the tenets of Christianity and interpreting them as radical proposals for a new social order are transgressions of what rulers consider the rightful relationship of ruler to ruled. But the real effects of such inversions of power in a slave system as completely dominated by white slave owners as in Jamaica in the age of revolution were very small. We should be careful not to overstate how vulnerable slave owners were to everyday slave resistance that fell short of violent revolt. One reason why it is important to study the men and women who gained so much from the labor of enslaved people in Jamaica is that it is an opportunity to study how power is exercised when the powerful had a tyrant's mandate and were unconstrained by custom, law, or (for the most part) public approbation or disapproval over whether or not they exercised their immense power over individuals with care or concern. The unremitting war by the powerful against the poor and powerless, evident in the day-to-day struggles slaves had with masters, is an important context within which to understand what enslaved people had to endure.

The chapters in this book focus on the years between Tacky's Revolt, the American Revolution, and the beginnings of parliamentary abolitionist legislation in 1788. The first three chapters look at planter culture and why planters were so committed to pushing their slaves so hard that one would have thought that slave resistance was inevitable. Indeed, it seems that slave owners doubled down in the 1760s and 1770s on overworking their slaves so that they could maximize their wealth. That they did so after Tacky's Revolt in 1760, the subject of the fourth chapter in this book, takes some explaining. My explanation is that slave owners convinced themselves that putting down this revolt meant that they could withstand even the most serious slave rebellion ever seen before 1791 in the Caribbean and in British Americas. Perhaps the group in Jamaica who were most affected by Tacky were free people of color, who found their freedom and opportunities severely constricted by legislation passed by the Jamaican Assembly after 1761. Nevertheless, the number and importance of free people of color increased in the second half of the eighteenth century, as Chapter 5 details. The next four chapters look at how Jamaica fits into wider political contexts with specific reference to two revolutions—the American Revolution and the Industrial Revolution. One chapter looks at the *Somerset* case of 1772 and another at the *Zong* case of 1783, both presided over by Britain's leading judge, Lord Mansfield, and each case being vital to the development of an abolitionist movement in Britain. They also were cases that had strong resonances in Jamaica. The last two chapters address two long-standing historiographical debates, the first on why Jamaica did not join the American Revolution and the second on the contribution of West Indian slavery to the origins of industrialization.

One theme in my work is that the major threat to white hegemony came less from enslaved people, despite the shock of a major slave rebellion in 1760, than from opponents in the metropole. White Jamaicans worked out ways in which they could keep enslaved people under control, but their means of doing this, as well as their distinctive and largely distasteful manner of living, repelled a large and influential section of British society who by the 1780s, and especially after the scandal of the *Zong*, turned toward abolitionism. They did so in part because they disliked West Indian planters. Jamaicans had their supporters, and the plantation economy of the island brought Britain great wealth. But in the long run white planters became isolated figures in the empire.[40]

For white Jamaicans, the period between the 1750s and the end of the slave trade in 1807 was a high point politically and economically in their history as a colony of Britain. They were rich, secure from attack from either internal or external enemies, influential within the highest reaches of imperial counsel, and confident that the plantation system they had perfected would enable them to become even richer in the future. There were, however, many clouds on the horizon. Tacky's Revolt was an extraordinary shock as was the break in the plantation world of British America that was initiated in the American Revolution. Enslaved people always challenged, to the extent that they were able, the power of white Jamaicans to determine their lives.[41] But the biggest threats to white Jamaicans in the mid-eighteenth century came from an unexpected source—from the small and marginal group of opponents of slavery led by Granville Sharp who from the 1760s extended their dislike of tyrannical West Indian planters, resident in Britain, to a criticism of the whole system of slavery that sustained these planters, starting with limiting slavery in Britain as in the *Somerset* case.

It was an unexpected area of opposition because Jamaica was central to the dominant British visions of empire in the protean period between the end of the Glorious Revolution and the start of the Seven Years' War. British statesmen in the age of Robert Walpole thought that maximizing labor productivity, as with slavery, and producing goods, such as sugar, at the lowest possible cost for the benefit of a growing consumer class was central to any imperial policy. Thus, the British West Indies occupied more attention and received more favorable treatment than did the northern colonies while advancing the Atlantic slave trade was a key commercial aim. The plantation system was at the center of imperial thought and practice, and until the 1760s virtually no one doubted its utility to the British nation.[42] Abolitionism, thus, was an unexpected and, to white Jamaicans, an unwelcome development. It was one reason why the leading intellectuals of the British West Indies were moved as a group to challenge the decisions enunciated in the *Somerset* case of 1772, as noted in Chapter 6 of this book. The American Revolution slowed down abolitionist sentiment, but in the 1780s abolitionism moved in a very short space of time from being a minor to a major social movement. The cause célèbre of the *Zong* in 1783 proved pivotal in making Britons aware of the enormity of the crimes committed in the Atlantic slave trade. Usually, these cases are studied from the perspective of Britain and imperial history, but both *Somerset* and especially the *Zong* can be understood better if the Jamaican context is added to existing interpretative frames.

The wealth and political importance of Jamaica in the age of the American Revolution was more pronounced than at any other time in its history. It meant that Jamaica was involved in world historical events and that its history had a global significance as well as local resonance. The biggest event in the region before the defining crisis of the French Revolution was the American Revolution.[43] Some white Jamaicans had some sympathy for the republican ideas that propelled British North Americans into rebellion, but most did not. Jamaica was instinctively loyalist and never showed any inclination to join with their American cousins in rebellion. Given that instinctive loyalty to Britain we do not need to argue that Jamaica was prevented from joining in rebellion because white Jamaicans were terrified of slave rebellion. Nevertheless, including Jamaica in histories of the American Revolution enriches that story. If we include Jamaica as part of the American Revolution, we can understand why Britain believed that most colonists did not want independence.

The history of Jamaica connects also with the history of Britain. It was a primary destination for slaves, making it central to one of the larger British mercantile trades in the eighteenth century, the Atlantic slave trade. And its exports of tropical produce to Britain altered consumption patterns, facilitated the development of ancillary industries, made many people in Britain rich, and shored up the British state through its substantial contribution to customs revenue. Of course, it was in the same century in which Jamaica was most important in the British economy that Britain began its remarkable rise to world economic dominance, achieved through the Industrial Revolution. In the half century between the Treaty of Utrecht in 1713 and the end of the Seven Years' War in 1763, West Indian planters had a privileged position within British society. No one doubted the efficacy of slavery, and plantation produce was an essential part of Britain's economy. Planters and merchants, especially from Jamaica, had the ear of sympathetic British statesmen. This book deals with Jamaica when it was at its economic peak, with the years immediately before the American Revolution probably being the high point in white prosperity on the island. Its political peak, however, came earlier, during the Seven Years' War, before the start of abolitionism and when the plantation colonies were almost universally considered the most dynamic and valuable parts of the empire.

My findings connect with a recent flourishing of work on how the wealth generated by slavery in Jamaica and the West Indies when plantation agriculture and the slave system that sustained it were at their peak in the second half of the eighteenth century and show the continuing presence and significance of slavery in British metropolitan society. Leading the way in this research has been a team of researchers at University College, London, who have been recovering patterns of slave ownership by Britons living in Britain in the eighteenth and nineteenth centuries in a well-funded project called the Legacies of British Slaveholding. Slavery reached everywhere in imperial Britain, as can be seen in the painstaking analysis of the role of slavery in the history of the ducal Bentinck family, undertaken by Sherylynne Haggerty and Susanna Seymour. They show how the tentacles of slavery in Jamaica were deeply rooted in this prominent family, even if it is a family not commonly associated with the island. It was to this island where Henry Bentinck, First Duke of Portland (1682–1726) repaired when his fortune was dissipated through unwise speculation in South Sea stock. He was a highly successful governor between 1723 and his death in 1726, his early demise prompting a grieving Jamaican legislature to name a new parish in the Northeast after him. His grandson William Henry Bentinck, Third Duke of Portland (1738–1809), was involved in Jamaican affairs as prime minister, using his high position to secure the lucrative secretaryship of the island for his grandson Charles Greville (1794–1865). Another relative, Henry William Bentinck (1765–1820), was governor of several eastern and southern Caribbean colonies from 1802 to 1820.

Perhaps not surprisingly, given their attachments to Jamaica, both the Third and Fourth Dukes of Portland were firm opponents of the abolishing of the slave trade. Their involvement in plantation slavery can be traced, as with many elite families, in the lineament of English country houses.[44]

That political influence and economic power was forged out of trauma. The period when white Jamaicans most prospered was when black Jamaicans endured especially miserable lives in a particularly vicious and all-encompassing slave system. The benefits Britain received from Jamaica in the age of the American Revolution were substantial, but they were linked always to the exploitation of enslaved men and women of African descent. Jamaica helped make Britain great, but it did so through slavery. One of the challenges historians of eighteenth-century Jamaica must grapple with is acknowledging the twin faces of the island's history—great wealth and immense poverty within an imperial system that was always complicit with slavery until the start of abolitionism as a major social movement in 1787–88. Historians increasingly stress that slavery touched the lives of most Britons and Americans in the eighteenth century. There were few areas of life in either Britain or its empire disconnected from slave owning, from the influence of slave money, and from the systemic impacts of slavery on how people in Africa, Europe, and the Americas lived because of the transformation of small islands in the Caribbean Sea into mighty engines of production and commerce. Hence, the history of Jamaica in the period of the American Revolution is relevant to the making of the modern world and in the ramifications of an Atlantic system characterized by the movement of goods, ideas, and most of all people. The following chapters explicate some themes in the history of this quintessentially American, Atlantic, imperial, and African place.

This volume also is intended to show my engagement with a body of scholarship becoming thicker and more sophisticated year by year, with the current spate of scholarship in the last five years emulating the burst of interest in West Indian slavery that was prominent in the 1960s and 1970s.[45] Readers of this book will be aware throughout the text of the extent to which I am influenced by the social history of West Indian slavery that developed in the years before I arrived at Mona in October 1987. There is no need to outline here my indebtedness to this rich scholarship in the pages that follow. But perhaps it is useful to write briefly on where my work fits within the new scholarship on Jamaica in the age of slavery, much of it written so recently that I have not been able to absorb the insights derived from this work in all the work that follows.

To my mind the central problem that shapes modern historiography on Jamaica is a moral question different from that which animated a previous generation of scholarship. For historians in the 1960s and 1970s, the principal aim was of recovery, of investigating through the mechanisms of social history means whereby "invisible" lives could be made visible. It is not surprising that so much of

the best work in this period revolved around understanding the social and economic structures of slavery, with an attention to slave demographic patterns. The
moral imperative was to show that enslaved lives were valuable and could be
chronicled. It went almost without question that the social and economic facts of
life under slavery were building blocks of a Caribbean culture that was vital and
important both in its own time and in the history of a contemporary Caribbean.
Moreover, an insistent theme in this scholarship was that as powerful as the
planter class was, slaves always resisted their dehumanization and succeeded
despite great odds in developing lives full of value, even if disparaged at the time
by the rulers and shapers of society and largely ignored by posterity, as the travails
of slaves and the inequity of slavery faded in public consciousness.[46]

That moral question has hardly disappeared, either in scholarship or in public
understandings of Jamaican history. We can listen to the lyrics of Bob Marley's
"Redemption Song" to realize that "songs of freedom" and redemption from
oppression "in this generation triumphantly" arise out of historical experience.
To this is added an even larger question—is colonialism evil? I would generally
subscribe to such a view, though I would not frame it in such a simplistic way.
That issue is not one I address directly in this work.[47] To those worthy concerns
has been added, at least in professional scholarship, an anxiety over how we
acknowledge the dysfunctional, monstrous, and catastrophic in the history of
eighteenth-century Jamaica while gesturing to ways in which the thoughts, skills,
and actions of the enslaved are entangled with transformations of the Atlantic
economy and influence the making of the American world.[48] How do we move
away from seeing slaves as abstractions underpinning planters' fortunes to seeing
them as historical actors inextricable from colonial society? Answering that question, at least for scholars coming to academic maturity in the decades following
the "cultural turn" in history writing, has meant thinking hard about the production of the sources that shape historical writing. It is difficult to escape the "planters' voice" when it was only white colonists and white imperial observers who left
records that can be examined.[49] I deal with this issue in Chapter 4, when I analyze
the silences in the archive—archival fragments, really—about the person, Wager,
who I think was the mastermind behind the series of revolts we call, probably
incorrectly, Tacky's Revolt in 1760.

Other scholars writing now have examined this intractable problem intensively. Marisa Fuentes, for example, confronts directly in her examination of the
lives of mostly enslaved women in colonial Bridgetown, primarily in the ameliorationist period of slavery after 1788, the difficulties in narrating ephemeral archival
presences through imaginative readings "across the archival grain" and the
deployment of theoretical perspectives derived from modern feminist scholarship. She employs a methodology that, she argues, "purposively subverts the over-
determining power of colonial discourses by changing the perspective of a

document's author to that of an enslaved subject." The results are illuminating, especially to those of us who use more traditional methods derived from social history that mine those sources intensively but without so self-consciously looking at "archival sources and pausing at the corruptive nature of this material."

Fuentes has not let what she initially saw as a failure—returning "empty-handed" from the archive with silence about the topics she was most concerned about—the battered bodies of vulnerable enslaved women in a cruel plantation system—deter her from making conclusions that were less about finding out details of their quotidian lives than about "the manner in which enslaved women were silenced" and "developing methodological pathways into their lives." Fuentes's project is not mine: I am too wedded to social historical techniques and to the task of increasing our empirical knowledge about the past to delve as deeply as she does into archival silences. And while everyone looking at Jamaican records confront sources "that show only terror and violence," not all of us can identify with Fuentes when she sees these sources as "a danger to the researcher who sees her own ancestors in these accounts." But the questions she raises about archival research in this period and these topics are ones that historians, not just of the Caribbean but of many places and times in history, need to consider. Do we risk "emotional strength," and is objectivity "obliterated" and research become a matter of "endurance" when encountering records that often are about exploitation and human degradation?[50]

One of the ways in which my work can be distinguished from recent work is through temporality. Although I believe there were significant continuities in the structure of plantation slavery over time, slavery was not an unchanging or unevolving institution.[51] It is not an accident that myself and Orlando Patterson, both shaped in our understanding of Jamaican slavery by looking at it in the age of the Atlantic slave trade, have each emphasized more than other scholars the wretched existence of slaves in this period. Thus, I depict slaves as existing in a world of uncertainty where, in Patterson's words, the enslaved were always vulnerable to repeated depredations that led to "significant slave dehumanization as masters sought, with considerable success, to obliterate slaves' personal histories." Patterson portrays Jamaica as a Hobbesian society (a theme taken up in Chapter 1 of this book), in which planters inflicted on slaves "a reign of terror" and "a holocaust" in a slave system that was "uniquely catastrophic." Out-of-control whites, as described in the diaries of Thistlewood, operated with "near genocidal cruelty." Patterson emphasizes that the horror of slavery was made worse by the chaos of white life, with white Jamaicans "screwing themselves stupid . . . smoking too much . . . drinking too much." That chaos made it an appalling society in which people born into slavery—the Creole slaves—could concentrate only on surviving enslavement with some degree of their psyche intact.[52]

All of us working on slavery in the period before abolitionism struggle with the realization that enslaved people's lives were miserable and stunted in ways that make it hard to see how Jamaican slaves could have led any sort of lives that held any meaning for them. We try to look at slavery in this period as having some features that bear testimony to the creativity of enslaved people's cultures, and we want to show that enslaved people cannot be defined just by how hard masters made their lives. In a recent article with economists Laura Panza and Jeffrey Williamson, I used economic data on standards of living in 1774 and 1779 to show not only that Jamaica was the most unequal society on the planet in the years before and during the American Revolution but that the standards of living for enslaved people meant they lived barely above subsistence and in appallingly bad conditions, with severe consequences for their health and well-being.[53] Vincent Brown aptly describes the treatment meted out to enslaved people in the quarter century before the American Revolution as a form of "spiritual terror," in which whites used both material and psychic tools to keep enslaved people cowed, terrified, and psychologically depressed. The large numbers of suicides that occurred within enslaved populations are an indication of just how bad the conditions of slavery were in mid-eighteenth-century Jamaica.[54]

Despite his own work, however, Brown is resistant both to Patterson's ideas of social death and to my own "unflinching" examination of enslaved life under Thomas Thistlewood's arbitrary care. "Surely," he argues in response to my work on Thistlewood and his slaves, "they must have found some way to turn the 'disorganization, instability, and chaos' of slavery into collective forms of belonging and striving, making connections when confronted with alienation and finding dignity in the face of dishonor." He argues that "rather than pathologizing slaves by allowing the condition of social death to stand for the experience of life in slavery, then, it might be more helpful to focus on what the enslaved actually made of their situation."[55]

Brown cites approvingly works on slavery that catalog enslaved people's ability to preserve and transform their culture, arguing that "the preservation of distinctive cultural forms has served as an index both of a resilient social personhood, or identity, and of resistance to slavery itself." What he suggests as an alternative is that in order to get past seeing slavery in Jamaica in nihilistic ways as a form of social death and so as to not over-celebrate the vibrancy and cultural resistance of slave communities and cultures in ways that allow the damages that slavery inflicted to be minimized, we should look less at finding an integrated and coherent ethos among the enslaved and instead concentrate "more on the particular acts of communities that allowed enslaved people to articulate idioms of belonging, similarity, and distinction." If we did, he memorably says, we can show how the enslaved "made the threat of social chaos meaningful."[56]

Jamaica looks less chaotic, especially for slaves, from the perspective of looking at slavery in the period of amelioration, when pressure from abolitionist initia-
tives from Britain opened spaces for slave agency that had been less possible in
the period of master domination and slave degradation in the first three-quarters
of the eighteenth century. Fulfillment of Brown's hopes for a meaningful study of
slave life look more possible for scholarship on Caribbean slavery after the American Revolution as planter power was tested by metropolitan opposition. Most
recent scholarship has focused on Jamaica in the late eighteenth and early nineteenth centuries. Only Katherine Gerbner and Daniel Livesay have devoted much
attention on the mid-eighteenth century, when the large integrated plantation
was fully established and when the authority of white Jamaicans over black Jamaicans was untrammeled. In both cases, the object of attention is not plantation
slavery. Gerbner traces the contradictions inherent in the concept of a "Christian
slave" in societies in which Christianity and slavery were kept conceptually distinct. Livesay pays concerted attention to a small but vital subset of the nonwhite
population (or not fully white population), the mixed-race children of wealthy
Jamaicans who went to Britain and whose treatment in law and in practice complicated ideas of race in a period when racial ideologies were forming and becoming rigid.[57]

The period of amelioration in the face of abolitionist pressure has received
considerable attention. Christer Petley, J. R. Ward, and Aaron Graham look at this
period from the perspective of the white ruling elite, an elite that still maintained
enormous power within a state that operated effectively to maintain white
supremacy and the slave system well into the nineteenth century. Indeed, all three
writers, along with Justin Roberts, who is doing work like my own on the nature
of slave management in the late eighteenth century, share my views of the dynamism and vitality of the plantation system and the planter class who ran that
system throughout the 1790s and up until the abolition of the slave trade.[58] The
plantation system in the British West Indies was in no danger of collapsing from
internal pressures in the period of the French Revolutionary and Napoleonic wars
and the Haitian Revolution.[59]

Nevertheless, abolitionism in the wake of the American Revolution made a
huge difference to enslaved people's ability to confront planter power. Several
scholars have drawn inspiration from the work of Jennifer Morgan on
seventeenth-century Barbados and especially Diana Paton's ongoing work on the
relationship between private punishment and state formation in the nineteenth
century to pay attention to the enslaved body, how that body was a site for
disciplinary practices, and how enslaved women used the interstices between the
myth of total planter control and the reality of the need for African labor to
forge lives that were not entirely determined by planter desire. Dawn Harris, for

example, examines the importance of the body as a site of disciplinary and puni-
tive practices, tracing transformations as Jamaica moved from slavery to emancipation.[60] Sasha Turner, Katherine Paugh, Audra Diptee, and Colleen Vasconcellos
explore what Turner calls "contested bodies" in the relations of enslaved women
and children with slave owners and with each other.

Their work marks a new stage in the history of slavery in Jamaica during
abolition and amelioration where there is a strong concentration on gender and
on women's and children's lived experience of slavery and especially reproduction. This emphasis contrasts with a generation of earlier scholars whose interest
was in the discovery of evidence relating to slave demography to answer empirically the question of why Jamaican slave populations did not experience natural
population increase even in pronatalist environments. This new generation of
scholars is less interested in these questions of demography and pronatalism and
is more interested in the management and politics of reproduction and how those
politics affected enslaved women's ability to contest planter ambitions. It is worthy work. But it is not work that is part of the studies I am doing, which focus on
the still necessary task of establishing what the contours of slave management
practices were that these enslaved women needed to navigate.[61]

Finally, my work in this volume is spatially focused although it insistently
connects the local specifics of Jamaica to global concerns. This linking of the local
to the global can be seen in the chapter on the *Zong* as an event usually seen
within a British context but which has an important Jamaican context (that is
where the crime occurred) that when examined from the Jamaican angle compli-
cates and enriches wider debates within imperial history. I am an Atlantic World
historian and believe that we can understand Jamaica only if it is placed in an
Atlantic context. Moreover, studies of the landscape and social and economic
structures of Jamaica make the reality of the Atlantic World in the age of the
American Revolution richer and help us understand how one part of the Atlantic
World was connected to several other parts. This aspect of my work connects to a
developing strand of scholarship that looks at the physical environment of
Jamaica and its representation in art and architecture as a means of working out
larger structures within a surprisingly diverse plantation structure.

Jack Greene's recent detailed explication of socioeconomic data describing
the people and land of Jamaica in the 1750s is very similar to how I approach
studying Jamaica, with Greene emphasizing the variety of social and physical
environments that made Jamaica more diverse than other British West Indian
islands and connected it to diverse societies in British North America.[62] One
theme of scholarship increasingly apparent is that the West Indies were a primary
place for scientific experimentation, the practice of natural history, and the
eighteenth-century enthusiasm for collecting. James Delbourgo and Deirdre
Coleman have each written stimulating books on collectors and naturalists whose

collections focused on the Atlantic World and the Caribbean—one, a major collector, Hans Sloane, the founder of the British Museum, and an upholder of the values of the early Enlightenment, and one, Henry Smeathman, less heralded and certainly idiosyncratic but especially interesting as a chronicler of a subaltern natural history that paid greater than normal attention to Africa and to Africans. Delbourgo shows Sloane as a collector of objects, an interest that developed from his position as someone implicated in the collection of people in slavery. Sloane derived much of his fortune from his marriage to a Jamaican heiress. His collecting impulse was thoroughly implicated in the reproduction of enslavement in scholarly discourse, a reproduction that did not worry him. Sloane found African chattel slavery unproblematic, reflecting the general British consensus about slavery that predominated in the years of the Seven Years' War. Smeathman, on the other hand, was the more profound scientist, and, because of his experiences in Africa, the Atlantic slave trade, and the West Indies, he developed through his study of termites an understanding of slavery and the plantation system as being a complex ecosystem that raised issues around morality, politics, and the economy of nature. What he saw, twenty years after Sloane had died at an ancient age, disturbed him so much that he became a committed and highly observant antislaver.[63]

The study of eighteenth- and nineteenth-century Jamaica has also been enriched by interdisciplinary perspectives from art historians, literary critics, and architectural historians, each of whom bring their disciplinary skills to bear on the issue of "place" and how the geography, representation, and built environment of the island informs our understanding of social, economic, and political processes. We can take a few recent works as examples. Charmaine Nelson explores the central role of geography and its racialized representation in the landscape, comparing landscape painting in the West Indies and in Canada in the early nineteenth century. She shows how vision and cartographic knowledge translated into authority, mastery, and a form of erasure. Jamaica, for example, was depicted in landscape painting as a place denuded of most of the population. Slaves were hardly there, or, if there, were depicted in ways that conformed to metropolitan and colonial understandings of contented workers. Rebellion, resistance, and contestation are absent: slaves were "hidden in plain sight," in Simon Newman's evocative phrase about runaways in the Jamaican environment.[64]

Elizabeth Bohls continues and expands such a theme in her fine-grained analysis of "the capacious genre of travel writing." She investigates closely what she calls "the bifocal capacity of aesthetics to enact imaginative intertwinement between colony and metropole at the same time obscuring the ugliness of slavery's site-specific practices." She notes the interplay between Britain and the West Indies and how in the period of abolition what pro-and antislavery writers shared was a desire to posit a break between the two places, either because the islands were home to uncivilized Africans or because they were controlled by a decadent and philistine ruling class. The aim was to make Jamaica seem radically different to the metropole and to deny what was self-evident, that the Caribbean was inextricably linked with Britain, no matter how reluctant all actors were to acknowledge such links.[65]

Bohls and Charmaine Nelson concentrate on the period of abolition. Louis P. Nelson deals with the period that connects with the subject matter of this book in his lavishly illustrated and intensively researched examination of Jamaica's built environment. He confirms some of the standard tropes of Atlantic history, that there were multiple flows and exchanges in capital, people, and things crossing the Atlantic and intruding themselves into local landscapes. He does this through an intensive examination of the design, construction, and occupation of buildings, especially housing, among the enslaved and free populations. He shows, as I do, that colonial British Jamaica cannot be detached, as later British writers wanted, from its origins as a major contributor to British wealth and well-being, as well as shaping in formative ways imperial ways of thinking. As in my scholarship, Louis Nelson interrogates how colonial life on a small tropical island seemingly on the edge of empire followed or diverged from patterns and practices elsewhere in the empire and especially in the metropolitan center.[66]

His work, like most of the new scholarship but explicitly so, concerns what I see is the overarching interest in the historiography of Jamaica in the twenty-first century, which is the nature of power in slave societies. That connects to the moral concerns noted above. Here is the real difference between the histories of eighteenth-century Jamaica written in recent years and those of the previous surge of scholarship in Caribbean history in the 1960s through the 1980s. Scholars writing in this period were hardly indifferent to power and its operations in the British West Indies, but their primary interest was in culture. Sidney Mintz may be taken as both representative of this tendency and a shining exponent of how to do it. His work was about how the Caribbean was a region with a dynamic past with a confluence of people from different cultural backgrounds who together created Creole societies. Creolization, he argued, was a tremendous creative act, mostly by the enslaved and in conjunction with the rulers; was active not passive; and led to hybridity and new forms of culture within patterns of globalization that showed that this phenomenon was much older than it is often taken to be.[67]

Contemporary scholars are hardly uninterested in culture, but their attention is more insistently on power—who had it; who exercised it and in what ways; how power was contested; and how the effects of power were manifested in ordinary life and in imperial relationships. The operation of power was part of the modernity of the Caribbean—a shared interest in the Caribbean as a crucible of modernity is what links scholarship from the late twentieth century with scholarship of the early twenty-first century. That modernization came out of the structure of the sugar plantation, an institution that C. L. R. James claimed was "the most civilizing as well as the most demoralizing influence in West Indian development." It modernized the planter class but modernized even more the enslaved population, who, James continues, "from the start lived a life that was in essence a modern life. That is their history—as far as I am able to discover, a unique history." Jamaica was one of the first places where humans were changed into anonymous units of labor and everything and everybody was commoditized.[68] Modernization came from culture—the mingling of ethnic groups under enslavement. But it was fundamentally underpinned by power, especially by the naked exercise, as seen most graphically in Tacky's Revolt, of physical violence. Slaves learned to modernize only under the threat of constant repression.

White Jamaicans survived because they mastered the real and symbolic weapons of violence. And power in the Caribbean is closely connected to trauma. The plantation landscape (and understanding that even in Jamaica the plantation was not the only site in society where power was exercised, as the island was remarkably diverse in its physical and economic settings) was a highly rationalized, highly efficient, but extraordinarily brutal way of producing money through the sacrifice of humans, as made explicit in Chapter 3, to the manufacture of tropical products for sale back in Britain. By the 1750s, this system purred along very well. It was resilient enough to survive some of, but not all, the shocks of the age of revolution. Investigating how this island worked (or did not work) in the latter half of the eighteenth century in a period when the world was beginning major convulsions, some of which reached Jamaica, gives us an insight into the operation of power in a complex but fascinating geopolitical region.

Chapter 1

Planter Politics and the Fear of Slave Revolt

Fear is all the rage nowadays. It seems to be the dominant emotion, along with its twin—terror—which dominates contemporary discourse. It is also increasingly important in Atlantic World historiography. One reason why we are interested in fear as an emotion influencing people's actions in the past comes from contemporary concerns, most notably the changed political state of the world since 9/11. As Michael Ignatieff presciently stated in 1997, the contemporary international order increasingly has rested "less on hope than on fear, less on optimism about the human capacity for good than on dread of human capacity for evil, less on a vision of man as a maker of his history than of man the wolf towards his own kind."[1] The quintessential political theorist that helped make sense of the state's relationship to the individual has been Thomas Hobbes, whose work has been fundamental to shaping our recognition of the importance of fear in modern life.[2]

Of course, the political events that shook up the world in the aftermath of 9/11 and the calamitous War on Terror have deep roots. Thinkers, including historians, have become more conscious of the importance of fear in shaping people's experience in the last twenty years. Social theorists have tried to work out why people in the postindustrial West have become both remarkably risk averse and unable to measure risk accurately.[3] Of more importance to historians of early America and the Atlantic World, the rise of what has been termed "emotions history" or, rather awkwardly, "emotionology," has focused historians' attentions on how feelings as much as facts have shaped historical understanding.[4]

Fear was an emotion of great intellectual importance in the eighteenth century. It was thought of by philosophers like Baruch Spinoza and Edmund Burke as a means of gauging a society's worries and concerns. Spinoza contrasted fear—a thing that is coming that we believe will be bad—with hope—the expectation that good things are going to happen. Significantly, he defined fear by reference to temporality—a concern about future events—rather than seeing it purely in physiological terms, as something that can be an observable biological response to an event that is sudden, dangerous, and unwelcome. Burke also connected fear to an apprehension of future events. He argued that "no passion so effectively robs the mind of all its powers of acting and reasoning as fear, for fear being an apprehension of pain or death, it operates in a manner that resembles actual pain." He identified fear as "the ruling principle of the sublime," making the crucial point that fear is not universally an unpleasant emotion, given that we enjoy experiencing other people's fears vicariously. It is no surprise that it was in the aftermath of Burke's outline of fear and the sublime that the literary genre of the gothic emerged—a genre in which West Indian planters like Matthew Lewis and William Beckford were conspicuous early exponents.[5]

Gothic writers, as Emily Senior argues, represented "the sugar islands as corrupted by disease and death." Through application of a Caribbean sensibility they "rearticulate[d] the landscape imagery of the picturesque in terms of a colonial gothic mode which drew more overtly from the contemporary medical discourse which was by now increasingly pessimistic about the possibility of human adaptation and harmony between Europeans and tropical character."[6] Most gothic works from Jamaica hailed from after the 1790s, such as Charlotte Smith's novella *The Story of Henrietta* (1800), set in the Blue Mountains of Jamaica where the terrors of the heroine's life are amplified by her atavistic fears of Jamaican obeah and possible attack from frightening others, notably highly sexualized and threatening black enslaved men. Nevertheless, the gothic is also connected to the period of the American Revolution. Horace Walpole's *The Castle of Otranto* (1764), the first gothic novel, with its concern about tainted bloodlines, was published almost contemporaneously with writings by Edward Long and others that allowed a fear of miscegenation to enter public discourse. Two major gothic writers—William Beckford and Matthew Gregory Lewis—derived their vast fortunes from West Indian plantations, though their productions largely avoided the Caribbean. Lizabeth Paravisini-Gebert argues that the gothic novel was heavily influenced by gothic themes and that with the inclusion of Caribbean themes a new set of darkness—race, landscape, erotic desire, and despair—entered and enriched the genre. She argues that the gothic was fundamental to Caribbean literature in general. The birth of this literature is "intrinsically connected to the exploration of the tensions and perversions of the political, economic, physical and psychological bond between master and slave which, especially in Haiti, had culminated in widespread destruction and violence, rape, mutilation, and untold deaths."[7]

Thus, fear has a definable historical context that has meant different things in different places at different times. It is true, of course, that fear is a universal human emotion. Judith Shklar notes of fear that "it can be said without qualification that it is as universal as it is physiological. It is a mental as well as a physical reaction and it is common to animals as well as to human beings." In short, she argues, "to be alive is to be afraid."[8] Nevertheless, fear is historically conditioned as well as being universal. As Joanna Bourke insists, fears are social phenomena, as are all emotions, and even if it is an emotion mediated through the neural or hormonal systems—"the emotion of fear is fundamentally about the body—its fleshiness and its precariousness"—fear cannot be understood without appreciating the encounters that give rise to it. The difference between fear and anxiety, she asserts, is that individuals "in a fear state" can see an immediate, subjective threat and can externalize that threat, either by action or through scapegoating, in ways that allow fearful people to believe they have some protection from what they fear. That ability to externalize threats that cause fear, and thus deal with them, is dependent on being able to act, and that in turn is dependent on power relations within historical communities. What historians need to do, she argues, is to understand how fear operates as a means of action in historical contexts rather than seeing fear merely as an emotional response. Emotions, she notes, are not "simply reports of inner states" but are things that "mediate between the individual and the social."[9] It is this aim—of seeing how fear works, in the context of how planters used fear as a weapon to frighten and control enslaved people—that is the theme of this chapter. This task is not easy. As Lucien Febvre, a pioneer in emotions history, wrote a long time ago, "reconstituting the emotional life of a given era is a task that is both extremely seductive and terribly difficult," because "even though emotional change needs to be woven into the historical context," emotions like fear enter, Joanna Bourke says, "the historical archive only to the extent to which they transcend the insularity of individual psychological experience and present the self in the public realm."[10]

Several historians have argued that fear presented itself in the public realm especially conspicuously in the early modern period. The most prominent proponent of this idea was the great French historian Jean Delumeau, whose first book of a trilogy of works on fear, guilt, and sin in late medieval and early modern Europe, published in 1978, argued that early modern Europeans were beset by a panoply of fears, like fears of plague, of storms, of the Devil, of Jews, and of witchcraft.[11] The advent of the calamitous seventeenth century only intensified such fears, especially fears of war and of transgressive religion. In seventeenth-century England and Scotland, such fears were manifested in particular in an obsession with Catholics, their threat to established order, and their propensity, it was alleged, for conspiracy and for fomenting rebellion.[12] Historians of early America and the Atlantic World, especially in the last twenty years, have taken with gusto to this thesis of societies transfixed by fears, especially by fears of the outsider, notably Amerindians but also Africans and African Americans. The throwaway comment made by John Murrin in 1990 that eighteenth-century commentators like Abbé Raynal were right in thinking that "early America was a catastrophe—a horror story, not an epic" has become a leading principle in the writing of early American history.[13]

In studies of early American slavery, it has become axiomatic that the fear of slave revolt was a paramount concern for planters in plantation societies. That fear, it is argued, began in the seventeenth century, was intensified in the eighteenth-century in the crisis that led to the American Revolution, and increased in the nineteenth-century, as what happened in Saint-Domingue after 1791, leading to the black republic of Haiti, haunted the imaginations of planters in the American South.[14] Examples abound for the eighteenth century.[15] And for the period of the early Republic, Alan Taylor has made the fear of slave revolt and slave assertiveness central to his reinterpretation of Virginian politics in the War of 1812.[16]

—

A considerable amount of evidence exists to support the contention that the fear of slave revolt was central to planters' thinking. James Madison, for example, thought that "should America and Britain come to a hostile rupture an Insurrection among the slaves may and will be promoted."[17] One of the most famous letters from colonial Virginia, by William Byrd to the Earl of Egmont on July 12, 1736, expressed Byrd's fear that as Virginia's slave numbers grew—"we have at least ten thousand men of these descendants of Ham fit to bear arms and their numbers increase every day"—"a man of desperate courage" might emerge and "with more advantage than Catiline kindle a servile war." He believed that Virginia had the power to render such a servile war unsuccessful, but "such a man might be Dreadfully mischievous before any opposition could be formed against him and tinge our rivers, wide as they are, with blood."[18] In the aftermath of Lord Mansfield's decision in the *Somerset* case of 1772, when Americans and West Indians first realized that there was opposition to slavery in England, the Virginian Arthur Lee drew on parallels with Rome, "brought to the very brink of ruin by the insurrections of their Slaves," and warned Virginians that their greater number of slaves than in Rome and annual importations of several thousand captive Africans per annum would be a "fearful odds, should they ever be excited to rebellion."[19]

Fears of slave revolt were even more intense in the Caribbean than in Virginia. Seventeenth-century governors were concerned about the growing disproportion between black and white numbers and about what this meant for island security. In Barbados, for example, Lord Willoughby wrote to Lord Clarendon after becoming governor in the 1660s that "I feare our negros will growe too hard for us."[20] By the middle of the eighteenth century, the huge numbers of blacks and very low numbers of whites made some Jamaican leaders assume that a devastating slave revolt was inevitable. Governor Edward Trelawney, an impressive mid-eighteenth-century Jamaican governor, argued that "in the Rage that Planters have for buying Negroes" and in their careless management of those slaves "sooner or later" a revolt would occur, leading "to the island's being overrun, and ruined by its own slaves." He thought, moreover, that Jamaican planters were naturally fearful as they were "not only alarm'd by every trifling Armament of the Enemy, but under the greatest Apprehensions frequently from their own Slaves."[21]

By the Seven Years' War, fears of internal and external challenges to white settlement in plantation America became more urgent. In 1757, William Burke argued that "the colonies are endangered, both from within and without; how much exposed to the assaults of a foreign enemy, and to the insurrection of their own slaves." Like Trelawney, he blamed planters for their predicament. They refused to encourage white settlement and indeed deterred it through allowing slaves to take jobs from ordinary white tradesmen. Moreover, they treated slaves so badly that it was no surprise they revolted. Planters seemed not to understand, he contended, "what every English waggoner clearly comprehended: that, if he works his horse but moderately, and feeds him well, he will Draw more profit from him in the end." Masters thought of slaves as "a sort of beast, without souls," and this ill-treatment stopped slaves from being "more honest, [and] tractable."[22]

As the American Revolution neared, white Jamaicans had become accustomed to using the supposed threat of slave revolt as an excuse for adopting policies, such as asking for substantial numbers of soldiers to be sent to Jamaica, that were intended to shore up security on the island. In 1773, the Jamaican Assembly sent a petition to the king asking for more troops, declaring that "your loyal subjects of Jamaica are constantly expos'd to massacre and desolation from an internal Enemy." They requested "on our constant apprehension of the revolt of our slaves" that they be availed of "your Majesty's gracious protection."[23] Statements like these suggest that Jamaican planters saw themselves as a group besieged, wanting metropolitan help for protection not just from the French or Spanish but also from their slaves, which they regularly spoke of as an "intestinal enemy." For Andrew O'Shaughnessy, who has written a highly influential account of the American Revolution in the British Caribbean, this fear of slave revolt was what stopped nervous West Indian whites from joining in American rebellion.[24]

This emphasis on the fear of slave revolt as an animating feature in plantation life from its establishment in the late seventeenth century to the cataclysm of the Civil War, which destroyed it in 1865, has developed from and contributed to a view of the character of colonial and antebellum planters. In this view, they are not seen as they portrayed themselves in portraits or written work—as masterful men, characterized by sometimes reckless courage and by an expansive, liberal generosity of feeling and behavior. Instead, historians depict planters as psychologically conflicted and deeply troubled people, consumed by fears and anxiety. That planters were troubled people goes back in recent historiography to the powerful depiction of eighteenth-century planters by Edmund Morgan in the 1970s. Morgan concentrated mostly on describing the rapacious, philistine, and exploitative white Virginians of the seventeenth century but added a few comments about planters in the eighteenth century that have proved enduringly influential.

For Morgan, the attractive hospitality that elite planters displayed toward ordinary white men and their automatic assumptions of authority were masks that hid deeply held prejudices and fears that white people had about the black people they owned. Morgan argued that planters had the sort of contempt for slaves that in England wealthy people had for the "inarticulate lower classes." Racism, he famously declared, "absorbed in Virginia" that fear and contempt for the poor that was part of English commitment to inequality. It was only and precisely because planters were racist toward Africans that Virginians developed a fierce commitment to republican liberty—they were able to raise up all white men into a single master class because they lumped together "Indians, mulattoes and negroes into a single pariah class."[25]

Historians have found this psychological interpretation appealing. They have interrogated planters' statements of racial and class superiority for what fears and anxieties remained hidden behind their bravado and bluster. For Rhys Isaac, the determination of planters in colonial Virginia to resist what they considered to be tyrannical British actions in the coming of the American Revolution was directly related to their anxiety over their position in society being challenged by assertive evangelicals. Evangelicals did not share planters' commitment to ideas of honor, social hierarchy, and gentility. Planters made out that evangelicals were not worthy of their notice, but, nevertheless, the more assertive planters were in public, the more conflicted and fearful they were in private. Isaac exemplified these large themes in a collective portrait of Virginia planters and an in-depth and psychologically informed study of Landon Carter, a wealthy but deeply insecure planter, beset on all sides by what he saw as rebellion against duly constituted patriarchal authority.[26]

That slave owners were consumed by anxiety is a constant trope within the historiography of the planter classes in America, especially in Virginia.[27] Kathleen M. Brown's comparison of the planter psyche to a poorly built building is typical and elegantly expressed. She argues that "for colonial gentlemen, authority was a delicate project, much like a home built upon an unstable foundation. To keep such a structure standing, the owner had to be extremely sensitive to fine cracks and imperfection, shoring up the edifice to prevent the entire home from tumbling down." It meant that just about everything worried them, from the ways they lived to the people (wives, children, slaves) they commanded. And the biggest worry of all, except for slave revolts, was whether the English elite respected their claims to gentility. There were, it seems, many signs that their claims to authority were not accepted by others, either their dependents or their superiors in London, meaning that anxiety was natural and that it represented the "tortured perfectionism of colonials who could never achieve enough of an English inflection."[28]

That white men were too lustful, too cruel, and too full of anxiety and uncertainty about their status to be anything other than fearful about their place in plantation society is commonsense but problematic. It does not fit with other depictions that stress planters' adherence to warrior virtues, and their devotion to a military ethos of stoicism and to ideas of courage that tended toward the reckless, even suicidal. White men in plantation settings were not easily scared or unusually fearful. Bravery and aggression were essential components of planter manliness. In Jamaica, for example, white men were expected to be willing to fight. William Dorrill told the English immigrant Thomas Thistlewood in 1750, "in this Country it is highly necessary for a Man to fight once or twice, to keep Cowards from putting upon him."[29] Being able to stand up for oneself was a crucial measure of masterfulness. It was a central way in which white Jamaicans saw themselves as free people. White Jamaicans were a combative people, "liable to sudden transports of anger." Given the omnipresence of slavery, it is not surprising that almost all white men had "something of a haughty Disposition." As Charles Leslie commented in 1740, a noticeable feature of the white male character in Jamaica was that white men "required Submission" from all around them. Moreover, every man insisted on being the "absolute master of himself and his actions."[30]

There is not a great deal of evidence that planters, even in Jamaica, where slave revolts were not just conspiracies but had in 1760 become realized in an actual revolt, were so paralyzed by fear that they reverted into such timidity and panic about their throats being cut by vengeful slaves that they refused to act to protect themselves. The wealthy Jamaican planter Simon Taylor (1740–1813), for example, had no misconceptions about how much his slaves feared and hated him, but he rarely voiced any explicit concern about the precariousness of his position, although, as Christer Petley argues, he was always conscious that the low number of whites on the island made this precariousness permanent. He never expressed any fear for his life, even when in 1765 his overseer on a St. Mary's property in north-central Jamaica was murdered "by a parcel of new Negroes." He remained vigilant and resolute against slave rebels, enquiring on behalf of Chaloner Arcedeckne, whose attorney he was on Golden Grove estate in St. Thomas parish in southeast Jamaica, whether any slaves on that estate had known of the rebellion in St. Mary's (they had not). That a rebellion had happened on an estate he managed, leading to the death of a white employee, did not faze him. He continued to expand his properties and remained an eager buyer of "new Negroes."[31]

White men in other parts of plantation America were equally quick to anger and ready to use violence to get what they wanted. Indeed, that planters were violent and that slaves were not, or at least were not prepared to openly display feelings of rage, was a marker of slaves' servile character and a justification for their enslavement. One legacy of the American Revolution was that it justified linking freedom with an obligation to resist. If they were not free, planters in the American South argued, they should be dead, as Patrick Henry famously was supposed to have declared: "Is life so dear, or peace so sweet as to be purchased at the price of chains and slavery? I know not what course others may take but as for me, give me liberty or give me death."[32] It was not accidental that he chose death, not slavery, as the antithesis to liberty. To choose slavery was to make a cowardly choice. The blacks he owned could be said to have made that choice; he was not willing to do so.[33]

—

Thus, we should not see planters as necessarily anxious and fearful, even if they were understandably wary of their enslaved population. Indeed, their actions were more designed to instill fear into enslaved people than to assuage their own anxieties. The key to dealing with Africans, Daniel Defoe argued in 1722, was to make sure they were "ruled with a Rod of iron, beaten with *Scorpions*, as the Scripture calls it." Moreover, slaves, Defoe thought, "must be used as they do use them, or they would rise and murder their Masters which their Numbers consider'd, would not be hard for them to do, if they had Arms and Ammunition sustainable to the Rage and Cruelty of their Nature." This sort of approach was more akin to contempt than to fear and is a strategy intended to break slaves and to keep them terrified.[34]

Let's approach fear in plantation societies from a different angle. Fear was not so much of a problem for planters as a solution. One reason why they so frequently used the language of fear, just as they used the language of slavery, as a way of describing the world they lived in was that the concept of fear was one that enabled them to make sense both of their own feelings and also of the feelings their oppressive behavior engendered in their enslaved populations.[35] In other words, planters used the emotion of fear as a weapon to create white solidarity in environments where white people were outnumbered, often heavily, by enslaved people of African descent. They used the emotion of fear and the very real application of terror against enslaved people to create societies that provided white people with security, comfort, and a degree of release from whatever fears and worries they had. Moreover, they used fear as a weapon to keep enslaved people in check. In short, planters used the insights about how fear could be used to sustain the sovereignty of absolute rulers that Thomas Hobbes explained in his great 1651 treatise on the political economy of the state, *Leviathan*. They used fear as they understood Hobbes used it to develop coercive regimes in which enslaved people were kept in positions of extreme anxiety and insecurity while white people were kept safe. Here it is important to point out that planters' reading of Hobbes was in significant ways the common misreading made at the time. This reading missed much of the nuance that Hobbes placed on denying the people's right to resistance.[36]

Hobbes, of course, was the theorist of politics from the dreadful first half of the seventeenth century, a century in which Europe was convulsed by religious disputes, almost constant war, climate-induced scarcity and famine, and a palpable sense of a world spiraling into disarray.[37] It is not surprising that he is the political theorist who is most illuminating on how fear operated to sustain the power of arbitrary rule. He wrote his work in the middle of a calamitous civil war in his native England and following one of the worst examples of internecine strife in European history, the horrific Thirty Years' War, which engulfed the states of Germany and eastern Europe between 1618 and 1648.[38] Consequently, he had a wary respect for authority and a wry suspicion about the capacity for humans to hurt each other. His Leviathan, a reference to the biblical sea monster supposedly guarding the gates of Hell, was the modern state, a morally neutral sovereign (famously depicted in the frontispiece to his work as a composite portrait of all the sovereign's subjects) who ruled over humans, each of whom was governed by selfish appetites and self-interests.[39]

Individually, people were lost, as their efforts to enact their selfish desires in worlds without law ("the state of nature") led to the life of the jungle, a dog-eat-dog world in which "the life of man was solitary, poor, nasty, brutish and short."[40] That was the world that slaves lived in—a world without protection except that which came from arbitrary decisions made by tyrants without much concern for enslaved people's welfare. Hobbes argued that people could only escape this horrible predicament by pooling their self-interests and in the process creating an all-powerful sovereign whose power rested on the willingness of men to surrender some of their competitive instincts (that unchecked led to the state of nature) in order for the rule of law to frame a civil society.[41] Of course, the idea that the world of slaves was akin to Hobbes's state of nature was not completely accurate. Slaves, Hobbes argued, did not "consent" to their condition. "Consent" was crucial to Hobbes's theory of politics. For Hobbes, a "body politic" or a "civil society" was a body of men agreeing to serve under one person to preserve their common peace, defense, and mutual benefit. Subjects, however, had to agree to their subjugation: the union under a common power occurred by a covenant in which subjects laid down their right to resist the sovereign. Even if proslavery theorists tried to argue that slaves "consented" to their enslavement, it was impossible to say that enslaved people consented to being ruled by a sovereign power because they had no choice in accepting their status.[42] Planters, however, tended to ignore what

Hobbes wrote on consent—unsurprisingly given their commitment to a political system where consent was denied to most people.

When Hobbes envisioned a state of nature, he had especially America, and probably Africa, in mind. He argued that in present-day American "savages," Europeans could see their ancestors living in a state of nature.[43] Hobbes says less about Africans than about Native Americans, but it is probable that he associated the savagery of America with that which contemporary commentators also believed existed in Africa.[44] Important, however, is that Hobbes insisted the state of nature transferred to transatlantic slavery. Slaves were in a state of nature while they were enslaved on American plantations, notwithstanding justifications of slavery based on the idea that people captured in warfare could choose to be slaves rather than be killed.[45]

A primary reason for arguing this point is that Hobbes denied that slaves (unlike European servants) could be said to have entered a covenant where they submitted to the rule of a sovereign. Slaves, he stated, "have no obligation at all" to obey a ruler because while enslaved they remain at war with their conqueror, or master.[46] Because, in Hobbes's conception of politics, slaves can never agree to their condition, they cannot then enjoy anything other than a temporary respite from the tyranny of masters. As slaves are in a state of nature and at constant war with their masters it is only people who are not enslaved who are always free to exercise the power of life and death, an opportunity denied to slaves, who gave up their liberty for survival. That power of life and death comes from slaves being in the war zone that is Hobbes's state of nature.[47] Hobbes even denied that in slavery a master has any obligation to obey laws of natural justice, as a legitimate sovereign is assumed to do. He argued that the master is under no necessity to spare the life of the vanquished.

Masters thus have a right to punish and kill slaves as they see fit, but the corollary is that slaves are under no obligation to obey what masters demand. The vanquished remain enemies and at war even after submitting to becoming slaves rather than being killed because of being captured in war. Slaves, in Hobbes's view, always have the right of resistance, even including the ability to kill their masters or mistresses. It may be the uncompromising manner by which Hobbes accepts both the idea that masters could treat their slaves as they pleased and as inhumanely as they wished, especially if this made their person secure, and also the idea that slaves were entitled to resist their treatment, that stopped slaveholders in the antebellum American South from using Hobbes to bolster their proslavery beliefs.[48] Such rights of resistance in a status-oriented, hierarchical society, where subordinates were meant to be complaisant to established authority, were alarming because they denied that planter authority was rightfully established. It overwhelmed what slave owners might have thought of as positive in Hobbes's dicta, namely that they had a right to absolute dominion over their slaves, no matter how unreasonably they exercised that dominion.[49] As Mary Nyquist comments, in outlining the implications of Hobbes's arguments for American (and West Indian) slave owners, "despotical dominion as Hobbes conceives it—that is, servitude originating in warfare—legitimates the extralegal power held by the slave master at the same time that it contributes ideologically to the militarization and bureaucratization of sovereignty claimed by the state."[50]

Hobbes is one of those thinkers to whom we customarily misattribute ideas that have become part of everyday political discourse. We often call a dystopian, anarchic, violent world where there are no rules "Hobbesian," when what Hobbes wanted to illustrate in his works was how to create states that were viable and that provided a degree of stability and physical safety for its inhabitants.[51] The state of nature—what we often think of as a Hobbesian world—is the opposite of the world Hobbes advocates. That world is one in which the state is the protector of ordinary people, albeit a demanding protector to whom unthinking obedience is owed.

What distinguishes Hobbes from other theorists is his clear-sightedness about the role of fear in establishing respect for authority. He believed that a state founded on recognition of the centrality of fear to all human struggles could rescue individuals from the strictures of living in a brutal state of nature. In the state of nature, people faced not just the real or potential threat of violence but the experience of endless anxiety. Hobbes described people in this state as susceptible to false philosophers who scared already frightened people with empty tales of ghosts and demons. He compared these scaremongers to scarecrows in fields that "fright Birds from the Corn with an empty doublet, a hat and a crooked Stick." A well-founded state freed people from falling victim to these peddlers of anxiety. One might have to give up a few individual rights, but one gained a release from needless anxieties when one was protected by a sovereign power that can provide security against violence. That sovereign power based its authority on demonstrating that there were things that any reasonable person should be afraid of and that the sovereign alone had the power to prevent being realized. People needed to make sense of reasonable fears and rely on the state to stop those fears becoming real.[52]

The political situation that Hobbes describes characterized slave societies in plantation America, especially in the British West Indies, where the ratio of blacks to white was often more than 9:1 and where slavery was especially brutal and arbitrary. A plantation society was a Hobbesian society in the correct definition of the word, insofar as it contained a state that protected individuals (in this case, white people) from reasonable fears (the possibility of slave revolt). The West Indian writer who most recognized the Hobbesian nature of the world in which West Indian planters and slaves lived was Bryan Edwards, a planter-historian who inherited great wealth from two merchant uncles, Zachary and Nathaniel Bayly. Zachary Bayly, coincidentally, was the proprietor of one of the estates in St.

Mary's on which the great slave revolt led by Tacky, Kingston, and Wager erupted in April 1760.[53]

Edwards arrived in Jamaica as a young man in 1759, and thus Tacky's Revolt was one of the first events in Jamaican life that he witnessed. It affected him deeply. The savagery of the treatment of rebelling slaves in Tacky's Revolt shocked him. Edwards noted that "it was thought necessary to make a few terrible examples of some of the guiltiest." He described the tortures inflicted on these "wretches" as follows:

Of those who were clearly proved to have been concerned in the murders committed at Ballard's Valley, one was condemned to be burnt, and the other two to be hung alive in irons, and left to perish in that Dreadful situation. The wretch that was burnt was made to sit on the ground, and his body being chained to an iron stake, the fire was applied to his feet. He uttered not a groan, and saw his legs reduced to ashes with the utmost firmness and composure; after which, one of his armes by some means getting loose, he snatched a brand from the fire that was consuming him and flung it in the face of the executioner. The two that were hung up alive were indulged, at their own request, with a hearty meal immediately before they were suspended on the gibbet, which was erected in the parade of the town of Kingston. From that time, until they expired, they never uttered the least complaint, except only of cold in the night, but diverted themselves all night long in discourse with their countrymen, who were permitted, very improperly, to surround the gibbet.

One of these slaves died after eight days; the other took nine days to die.[54] Contemporary accounts confirm Edwards's account. A Kingston merchant noted in 1760 that "there's scarsely a day passes but some of the Negroes are Executed. We had Two that were hung up in Chains alive & Starv'd to Death. They hung a great while [and] one of them kept his Speech and Senses 9 days & a few Hours." He added, gratuitously, that "I don't Think this Death as Cruel as people imagine as they were not at all impatient for Meat or Drink & scarsely Complain'd."[55]

Edwards wrote in 1792 that seeing the punishments given out to rebels in 1760 alarmed him: "I felt a shock at a scene which presented itself to me on my arrival, that has not yet lost its impression. If it had, a paper which I wrote on occasion of a miserable wretch that was burnt, and which has since appeared in a great number of different publications, would stand in judgment against me." But although he "felt the utmost indignation and horror at such extraordinary punishments," he came to realize, as an older man in 1792, that such punishments were justified, given the crimes those "wretches" had committed as part of their strategy to "take over the country." He noted, for example, that these "fierce and warlike savages" killed whites "in a savage manner and literally Drank their blood mixed with rum."[56]

Less conspicuously racist than his historian predecessor Edward Long, but still a writer highly derogatory about African capacities, Edwards had special insights about the political importance that needed to be attached in slave societies like Jamaica to the principle of white supremacy. He had a well-developed philosophy about how fear could be used as a political strategy that sustained order and harmony. His tenets were expressed most clearly in his account of the slave rebellion in Saint-Domingue, written in the early 1790s. He acknowledged that white freedom was preserved by adopting a very strict color line, in which color distinguished slavery from freedom. He argued that "so long therefore as freedom shall be enjoyed exclusively by one race of people, and slavery be the condition of another, contempt and degradation will attach to the colour by which that condition is generally recognised, and follow it in some degree, through its varieties and affinities."[57]

Having established that whites benefited from prejudices based on color and that political and social entitlements based on color enabled whites of all condition to be politically united as a single master class, he outlined the basis for authority in the sovereign power that governed black and white relations. Slaves were the lowest class of people in West Indian colonies and needed to be controlled rigorously because of their "habitual barbarity." That control came from all-powerful but merciful sovereigns, with each slave owner being a sovereign power in his, or occasionally her, own right. It was a notable feature of political order, he argued, "in all countries in which slavery is established," that "the leading principle on which government is supported is *fear*, or a sense of that absolute coercive necessity, which leaving no choice of action, supersedes all sense of *right*." Foreshadowing later arguments made by American slaveholders that only they could make laws regarding slaves as only they had the specialist knowledge of African behavior that others lacked,[58] Edwards denounced concerns that such assumptions led to tyranny as being naive, arguing that "every endeavour, therefore, to extend positive rights to men in this state, as between one class of people and the other, is an attempt to reconcile inherent contradictions, and to blend principles together which admit not of combination."[59]

In short, slaves had no rights that masters need recognize. The master's need for self-protection justified any sort of behavior, no matter how outrageous or objectionable. Such acceptance of arbitrary rule worked in societies where there was "a marked and predominant character to all white residents" that included "an independent spirit and a display of conscious equality throughout all ranks and conditions." All whites were equal insofar as they had "the pre-eminence and distinction which are necessarily attached even to the complexion of a white Man, in a country where the complexion, generally speaking, distinguishes freedom from slavery."[60] A political order based on fear united all those wielding power: whites needed to be equal because they were equally dependent on each other for protection from their slaves. The underlying principles were expressed best by French officials in neighboring Saint-Domingue, a colony that watched closely Jamaica's experience with slave revolts. In 1771, the French Crown in Saint-Domingue had declared that "it is only by leaving to the masters a power that is nearly absolute, that it will be possible to keep so large a number of men in that state of submission which is made necessary by their numerical superiority over the whites. If some masters abuse their power, they must be reproved in secret, so that the slaves may always be kept in the belief that the master can do no wrong in his dealings with them."[61] Here Hobbesian principles of rule through fear were made transparent.

Slaves had no such protection, at least before the advent of ameliorationist policies toward slaves in the early nineteenth century.[62] Edwards, coming from Jamaica, where legal protections for slaves against tyrannical masters were minimal, gave the faintest of praise to the French Code Noir. In a passage immediately before he made his Hobbesian statements that slave societies needed a political authority based on fear, he allowed that the Code Noir "breathed a spirit of tenderness and philanthropy" but insisted that "there is misfortune attending this." European legislators did not understand the nature of American slavery and had to understand that their "tender" regulations "are inapplicable to the conditions and situation of the colonies in America." Despite the provisions in the Code Noir, French planters did not "treat their negroes with greater humanity and tenderness than the British" and probably fed them less and clothed them worse. In short, making laws about the treatment of slaves was pointless. Slaves had to depend on the goodwill of their masters. "The great, and I am afraid, the only certain and permanent security of the enslaved negroes," he declared, "is the strong circumstance that the interest of the master is blended with, and in truth, altogether depends on the preservation, and even on the health, strength, and activity of the slave." It was the fact that slaves were property, rather than that they were human, that allowed them to survive and, Edwards thought, to thrive.

Because masters had an interest in seeing that slaves were healthy, so they could be worked hard, it meant that slaves were better treated than they would have been if planters had to obey strictly laws that specified the minimum amounts of what slaves were due as food and clothing. If their situation was contrasted with "that of the peasantry in many parts of Europe," whose condition was "by any means, the most wretched of mankind," Edwards argued, in an argument that became a standard trope within proslavery literature, it was clear that slaves' position as property under self-interested and thus reasonably benevolent all-powerful masters was their principal form of protection against an indifferent state. One could lament "the licentiousness of power, the corruption of manners, and the system of slavery," and still think that "the scale preponderated on the favourable side." As Hobbes had intuited a century and a bit before, Edwards believed that "human life, in its best state, is a combination of happiness and misery, and we are to consider that condition of political society as relatively good, in which, notwithstanding many disadvantages, the lower classes are easily supplied with the means of healthy subsistence; and a general air of cheerful contentedness animates all ranks of people."[63]

Edwards was not just channeling Hobbes in making these political assumptions. His political theory also nodded to a political theorist who predated Hobbes, Nicola Machiavelli.[64] Machiavelli was also a theorist of fear, arguing that imposing terror was a legitimate and indeed necessary political strategy to be adopted by rulers seeking to establish new regimes. Indeed, reminding people that the state can adopt policies of terror was, Machiavelli contended, an essential instrument of everyday rule. Thus, a regime predicated on fear, determined on an absolute color line in which whites were considered superior to nonwhites, and on strategies that used terrorizing violence against outsiders (such as slaves) who contested the exercise of such absolute power, made sense in Machiavellian terms. Jamaica thus lived a Machiavellian moment right up until the abolition of slavery in 1834, even if this was less positive than the moment famously described by John Pocock in his account of the development and survival of principles of republican liberty in the seventeenth- and eighteenth-century Anglo-American world.[65]

The problem for Edwards and the Jamaican planters he represented is that they adopted the assumptions of Machiavelli about the appropriateness of fear and terror as political stratagems useful for a political regime under stress, as modified by Hobbes, in a period when a different conception of the political utility of terror was being advanced by Montesquieu and was increasingly being accepted by Europeans as the proper way in which to think about fear, terror, and political violence. Montesquieu changed the terms of the debate about the political meaning of terror established a century earlier by Hobbes. Terror, which for Montesquieu was a synonym for fear, had little positive to recommend it as a political tactic. He thought that terror was "the defining characteristic of the governing principle of despotism." He accepted that in aristocratic and republican forms of government, informed by concepts like honor and virtue, it might be necessary to employ a limited and exemplary violence or force against threatening domestic foes. But rulers needed to be careful that such violence did not devolve into terror, as this degradation was the means whereby "oriental despotism" entered the political realm.[66]

What was crucial was moderation and adherence to the rule of law. Montesquieu rejected Hobbes's view that the alternative to lapsing into a state of nature was submission to an absolute sovereign. The problem with this formulation, Montesquieu thought, was that it refused to acknowledge the potential in humans invested with unchallengeable power to turn into tyrants, as "every man invested with power is apt to abuse it, and carry his authority as far as it will go." If men submitted blindly to the will of the sovereign, then tyranny was likely to occur, as a ruler was just as bent on self-preservation as was his subjects. Such rulers, if places like Turkey were anything to go by, were led by caprice and thrived on uncertainty. A despotic government thus shared many of the features that Hobbes thought characteristic of the state of nature. Montesquieu contended, in an argument that became increasingly accepted by Europeans, that a moderate government with a carefully worked out separation of powers and a balanced constitution, based on adherence to rules of law, was preferable to despotic government because it meant that "one man was not afraid of another."[67]

One way of seeing whether a country was despotic or not was in its manner of punishments. Montesquieu argued that a moderate government might, without danger to itself, relax its punishment of evildoers, especially those accused of treason or riot against the state, supporting itself by laws and by its own internal strength. A despotic prince by contrast, ruled through fear and terror and restrained subjects through examples of severe punishment. Horrific punishment was the province of despotic governments. Civilized countries were marked by punishment regimes that eschewed torture, were not excessively violent, and were swift, certain, and not arbitrary. In Europe, the extended *amende honourable* and then grisly and prolonged torture and slow execution in 1757 of Robert-François Damiens, the half-demented servant and failed regicide of Louis XV, marked a turning point in Enlightenment thinking. Michel Foucault made this clear when he began his account of changing punishment regimes in Europe with a prolonged description of Damien's torments as a gruesome display of the theater of hell, intended to show the awesome power of monarchical rule, before outlining how a new technology of punishment in the shape of the prison emerged in western Europe in the late eighteenth century.[68]

Increasingly, followers of Montesquieu, Rousseau, and other Enlightenment thinkers came to believe that punishment should be corrective rather than vindictive. Societies that punished people in the way that Damiens was publicly butchered were regressive and barbaric. Edwards, by contrast, supported notions drawn from Hobbes that considered the normative response in slave societies, in which fear was the leading principle to challenges to authority, to be state displays of "spectacular terror." He applauded and justified, at least by 1792 if not in 1760, the cruel measures that white Jamaicans took to kill rebels caught after Tacky's Revolt was crushed. Like Edward Long writing in 1774, Edwards believed that such harsh punishments against traitorous slaves were essential to keeping order in slave societies where masters were vastly outnumbered.[69]

British observers, however, were appalled at what they saw as Jamaicans' unrelenting cruelty and barbarous actions toward their slaves.[70] It suggested to them
two things: that white Jamaicans were as savage as the Africans whom they
despised; and that their willingness to use the tools of terror demonstrated that
Jamaicans had moved decisively away from normal standards of British behavior.
As Sarah Yeh notes, "the harsh slave master of the plantation came to be seen as a
hideous parody of the benevolent English lord of the manor" that white Jamaicans
professed to idolize. The violence endemic to the plantation, and the violence of
state-sanctioned punishment meted out as retribution to slave rebels, played
badly in a world where the dictates of Hobbes had lost favor and where the
precepts of Montesquieu were in the ascendant.

Ironically, West Indian planters lost out in two diametrically different ways, as
Yeh intriguingly suggests. Their inability to form and sustain families, their
degraded and debauched lifestyles (including their propensity to breed with black
women), and their ready recourse to violence made them seem either atavistic
throwbacks to some barbaric past or else "oriental despots" corrupted by climate
and luxury. On the other hand, their highly modern methods of plantation management made them anathema to the many Britons who were disturbed by the
values and visions of a rapidly developing entrepreneurial industrial modernity.
West Indian planters seemed, Yeh suggests, "a disturbingly extreme version of the
new 'improving' landlords in rural Britain who callously hired and fired tenants
and laborers according to their needs without any sense of obligation or responsibility to those beneath them."[71]

What is interesting, therefore, about the white exercise of power against slaves
in Jamaica is not so much its excessive brutality—eighteenth-century Europe and
Africa were also violent places—but is how white Jamaicans created a police
state where whites enjoyed a form of absolutism based on their monopoly of the
coercive powers of the state and also on the ideological advantages of having
white skin. There was no pretense that whites governed through any form of
consent. Indeed, blacks were imagined to be so far outside the social system that
it was impossible to conceive of ways whereby any fiction that slaves consented to
their treatment was possible. How could "property" be part of civil society? White
Jamaicans were absolutist tyrants with a torturer's charter that allowed them to
do whatever they wanted to enslaved property. The ownership of enslaved bodies
was placed ahead of any recognition of slaves as people. As the historian Bryan
Edwards put it, the occasional planter kindness "affords but a feeble restraint
against the corrupt passions and infirmities of our nature, the hardness of avarice,
the pride of power, the sallies of anger, and the thirst for revenge."[72]

Here Edwards admits that Jamaica was not, at bottom, a civil society, at least
by the standards that Hobbes had laid down in *Leviathan*. It was less a place of
multiple absolute monarchies than an anarchic society at war, held together by

tyranny. Orlando Patterson, a major modern scholar of freedom and slavery in Jamaica, takes a similar view.[73] For Hobbes, it was a theoretical impossibility that a sovereign could act like a tyrant because a sovereign could not govern contrary to the principles of natural law. Hobbes's adoption of an authorization doctrine, whereby subjects consented to being ruled by a sovereign, and his belief in natural law allowed for a measure of rightful rebellion. If a sovereign ignored natural duties, as Edwards implied a planter might do owing to the "hardness of avarice" and "the thirst for revenge," then even though a subject could not justify being disobedient, it was easy to predict that rebellion might occur. Hobbes warned sovereigns that when they acted unjustly, "the commonwealth is itself dissolved, and each man recovers his right to protect himself at his own discretion."[74] Translated into Jamaican contexts, Hobbes's argument and Edwards's admissions suggested that rebellion was both likely and justifiable.

One reason why Edwards was so clear-sighted and pragmatic about the political sources of authority in a slave society was because he had experienced a real and dangerous slave revolt—Tacky's Revolt. Political thinkers in colonies that had not experienced a slave revolt were able to fool themselves that the basis for white authority lay not in fear but in a love of liberty. Jamaica was different, not just because it had experienced a slave revolt organized by enslaved rebels operating perhaps separately, perhaps in tandem, in multiple locations throughout the island, but also because that revolt had been successfully put down, albeit with considerable difficulty and with maximum bloodshed. Jamaicans knew that just as they could survive hurricanes and famines and just as it was likely that climatic conditions and the terrain of the island made them probably able to fend off foreign invasion, so too they had the physical means and the mental willpower necessary to defeat any attempts by slave rebels to try and overturn white rule.[75]

The actions taken by white Jamaicans in the immediate aftermath of the grisly executions that signaled the failure of Tacky's Revolt suggest a polity emboldened by their ability to defeat enslaved rebels rather than one paralyzed by fear. In 1762, for example, the Jamaican Assembly engaged in a bitter dispute with Britain over a law Jamaica had passed that forbade smuggling and that mandated that convicted smugglers would be put to death. The Board of Trade balked at Jamaica passing laws that were contrary to imperial legislation and even more so at the vicious retribution proposed for wrongdoers. It was outraged that "the legislature of Jamaica could so far depart from the known and established principles of justice, equity and reason, and the laws of the mother country, as to have framed so sanguinary a clause as this appears to be." The Jamaican Assembly was in no mood to worry about the sensitivities of British politicians. It said that Jamaican politicians "do not incline to . . . admit the objection . . . to that act carrying any weight" and that "they are by no means disposed to submit their sentiments to the determination of their lordships nor ever will, at any time, suffer them in any respect to direct or influence their proceedings by any proposition or decision whatsoever."[76] This intransigence continued during the 1760s, as the Assembly took on and defeated an unpopular governor over a matter of parliamentary privilege and joined other colonists in protesting the Stamp Act.[77]

Colonists elsewhere took notice. Something that is little appreciated about Tacky's Revolt, because the emblematic slave rebellion in the Caribbean has become the Haitian Revolution, is that for people living in slave societies in the late eighteenth century the most important example of a slave rebellion in the Americas was what happened in Jamaica in 1760. It had quite different resonances than did the later Haitian Revolution because it was a failed rebellion. What contemporaries noticed was less how slaves nearly brought Jamaica to its knees, though this was acknowledged, than how whites were able to prevent slave rebels from succeeding in their objectives. They noticed also that the Jamaican state employed maximum weapons of terror in the aftermath of the rebellion.[78]

Moreover, reactions to what happened in Jamaica varied according to whether people lived in slave societies or not. We can see this instructively in examining the response to Tacky in France and in its most important American colony, Saint-Domingue. In France, Tacky's Revolt was placed within a growing debate around the morality and efficacy of slavery. The birth of a humanitarian antislavery literature in France owed much to news of events in Jamaica. Jean-François de Saint-Lambert's story "Ziméo," republished seventeen times between 1769 and 1797, for example, was "the first colonial fiction to seize the imagination of French readers."[79] Set on a Jamaican plantation during a revolt led by a charismatic West African, it was clearly inspired by accounts received in France about Tacky's Revolt and Jamaica's Maroon Wars. In the story, Saint-Lambert suggested that it was in planters' self-interest to treat Africans with humanity and even to free them.[80]

Planters in Saint-Domingue were less inclined than Frenchmen in Paris to draw the lessons that Saint-Lambert drew from Tacky. Their view was that Jamaican slave owners' ability to do as they pleased to slaves, including torturing them for information about possible poisonings or conspiracies to rebel, was an essential bulwark protecting them from the depredations of their enslaved property. The planters of Saint-Domingue wanted to emulate how white Jamaicans had put down Tacky's Revolt with a regime of terror after the event. Thus, when imperial officials in 1784 decided to revise the Code Noir so that slave owners might be prosecuted for maltreating slaves, they faced a wave of protests from colonists who believed that this would undermine planter authority and lead, eventually, to disaster. An army officer in Cap Français in May 1785 objected to the law because it "attacks the sacred rights of property and places a dagger in the hands of slaves by submitting their discipline and their regimen to hands other than those of their masters."[81] Saint-Domingue planters agitated for no restrictions over their authority over their enslaved property, as they imagined was the case in Jamaica.[82]

The most notorious example of an insistence on untrammeled planter privi-leges over slaves was the case of Nicolas Lejeune, arrested for egregious violence against enslaved women. To the delight of his supporters and to the consternation of imperial authority, he was acquitted of the charges against him, despite clear evidence both of his guilt and that he was a psychopathic torturer and murderer. Lejeune gloried in his release, declaring that his "cause is become that of every *colon*." He noted that each slave hated his master and that "it was only force and violence that restrains him. . . . It is not the fear and equity of the law that forbids the slave from stabbing his master, it is the consciousness of absolute power that he has over his person."[83] Colonial officials dissented, lamenting, "How many barbarities have the planters worked together to hide from the administration, whose vigilance on this matter is so hateful to them? . . . For one hundred years these cruelties have been exercised with impunity; they are committed right before the slaves because it is known that their testimony will be rejected."[84] Lejeune's victory was thus a triumph for planters who insisted that it was only the exercise of absolute authority and the application of weapons of terror, as in Jamaica during Tacky's Revolt, that kept white people safe. As Gordon K. Lewis noted, this is "the voice of naked power, stripped of any pretense at apology or moralizing justification. It is the language of Hobbes, not Rousseau. Consent yields to fear. There is no room for the social contract; everything is in the state of nature."[85]

—

Whites in Jamaica thus had a complex relationship with the political discourse of enslavement. Fear of insurrection certainly nagged at white Jamaican psyches. Indeed, they used their fear of slave revolt as an ostensible reason not to join with North Americans in rebellion. The Jamaican Assembly sent a formal petition to the British Crown in 1774 protesting against British actions in British North America, declaring that they could not express their discontent as violently as colonists were doing on the mainland because their fear of slave rebellion and their dependence on British defense had reduced them to such a "weak and feeble" state that they could not offer physical resistance.[86]

Nevertheless, nothing suggests that white Jamaicans were so paralyzed by fear of their slaves that this fear compromised their political beliefs. It was not fear of a slave uprising that stopped them from joining the rebellion of their North Ameri-can compatriots. Instead, much of their behavior was designed to terrify enslaved people. If a slave revolt happened, Jamaicans thought they could deal with the consequences. Indeed, when Jamaicans reflected on their behavior during the most dreadful days of Tacky's Revolt, when it looked as if the island might be lost to slaves, they congratulated themselves on their bravery, their resoluteness in face of attack, and their effective use of all the resources they had at their disposal. Tacky's Revolt had shaken Jamaica. But, in retrospect, it was a singular event with limited long-term consequences. Indeed, recent Jamaican prosperity could almost be dated from the end of its suppression. Repression had worked once. It could certainly work again. Planters were less incapacitated by fear of what slaves might do to them if they rebelled than is commonly supposed. That they were wary of slave violence is clear. So too it is clear that enslaved people took advantage of the fog of war to pursue their own agendas at the expense of planters. In the end, however, planters had every reason to think that a slave rebellion would be overcome.[87]

Talk about slave rebellion was part of the discourse in plantation societies around slavery and liberty, and how African Americans were suited to the one but not the other. Nevertheless, how colonists dealt with slave rebellion, especially the example of what Jamaicans had done because of the only serious slave rebellion in eighteenth-century British America, had long-term consequences. Jamaica's repression of Tacky's Revolt confirmed British suspicions about the inherent barbarism of life in the New World. The more Britons thought about it (and they thought of it in terms dictated by Montesquieu, not Hobbes), the more disturbing Jamaican actions seemed. Treating slaves so badly seemed to guarantee further revolts and the potential loss of Jamaica to France or Spain. Britons were aware of both how frequent slave revolts were in the New World and how viciously they were put down. Between 1737 and 1773, fifty-two articles about forty-three slave revolts appeared in the *Gentleman's Magazine*, the most widely read periodical of its time. It suggested to well-read men, like Samuel Johnson, that America was irredeemably violent.

Famously, Johnson quipped about American protestations against British tyranny in the 1760s that it was surprising that Britons heard "the largest yelps for liberty among the drivers of negroes." It followed on from his remark in 1758 that "slavery is now nowhere more patiently endured, than in countries once inhabited by the zealots of liberty." Johnson took the common assertion made by American colonists and made the link that historians now more commonly make, which is to claim that slaveholders were hypocrites—people who were especially sensitive to their claims to liberty being contested because they were themselves very keen on taking liberty away from Africans "patiently enduring" a particularly brutal form of enslavement. Indeed, metropolitan responses to reports from Jamaica in 1760–61 show that Europeans were more concerned about white brutality than about the inequity and immorality of slavery.[88]

It cemented in British minds the idea that Jamaican planters were cruel tyrants. It also suggested to them that people who inflicted these acts of terror on enslaved people had lost essential parts of the British character. It initiated a view of Jamaican planters that was developed and elaborated during the early years of

British abolitionism, which was confirmed during the long war of Britons against American planters during the American Revolution. It matured in the 1780s as Britons contemplated a new kind of empire in which white settlers were far less numerically or ideologically dominant than before. What Tacky showed Britons was that white Jamaicans were natural tyrants who were fundamentally un-British. In short, these men were "oriental despots" on the model outlined by Montesquieu rather than decent scions of Britain abroad.[89]

The irony was that white Jamaicans' adoption of fierce repression toward slaves and their constant complaining about imperial inattention to their needs meant that they created a more formidable enemy—one properly to be feared—than their imagined enemy of frightened and traumatized slaves.[90] That enemy was British public opinion, which turned increasingly against them after the American Revolution.[91] White West Indians' unfettered power over slaves was successful in keeping slaves checked, cowed down by the relentless assaults made on them by masters and mistresses. But planter brutality did not impress a metropolitan audience increasingly attuned to humanitarianism. By the mid-1780s, Britons were convinced that the West Indies was a place of barbarity. By 1788 abolitionists had convinced many Britons that it was a moral necessity to stop the Atlantic slave trade, the bulwark of West Indians' wasteful management practices.[92]

Abolitionists persuaded the British public that it was planters, not their slaves, who were the true barbarians. Jamaicans suddenly grasped that it was their enemies "at home" in Britain who were not only destroying their prosperity through abolishing the slave trade but possibly promoting slave unrest and making slave rebellion likely.[93] For absentee planter Nathaniel Phillips, the events of 1788 meant that slaves could plausibly argue that "the whites in Great Britain are their *friends* & wish to make them *free*," making it hard for slaves "to be kept long in subjection." The London-based Jamaican merchant Robert Hibbert reiterated such premonitions of doom. Enslaved people, he thought, on hearing of abolitionist agitation would think that if they engaged in rebellion then "*they* and not *we* should receive assistance from England."[94] Jamaican planters and merchants and their metropolitan supporters now faced the situation that their policies had been intended to forestall: a hostile and powerful outside force that gave encouragement to enslaved people to take up arms against white authority. Slaves knew that such rebellions would provoke a ferocious white response, further convincing metropolitan Britons that planters were irredeemable, and that slavery was a moral stain on the imperial conscience.[95]

Fear was indeed a potent emotion in the thinking of planters in places like eighteenth-century Jamaica. It could be used successfully against one set of

enemies—enslaved people—who lived in a world defined by the political philosophy of Hobbes. That some enslaved people, in the period before abolitionist
sentiment transformed the politics of slavery in the Atlantic world, were willing to
engage in slave rebellion shows how desperate a few brave men (relatively few
rebels were women) were to escape from the tyranny that slave owners exercised
against them. Their chances of success were miniscule, and the likelihood of
gruesome retaliation when a slave rebellion was discovered or foiled was
extremely high. Their reward was usually a martyr's death, and a death implemented with the sort of cruelty that resembled more the martyrdom of early
Christians than that exercised against other sorts of rebels in the eighteenth-
century world.[96]

Martyrdom, however, is not always pointless. Slave rebels who became martyrs may have done little to improve the chances for freedom of the mass of
African-descended slaves in the Americas in the eighteenth century. But their
martyrdom was noticed in a transatlantic universe where slavery was increasingly
seen as immoral. In a world where people were as much concerned about the
problem of tyranny from above as they were about the risk of revolt from below,
the politics of fear, as employed in late eighteenth-century Jamaica, backfired,
with considerable consequences for planter authority. Bryan Edwards was convinced that planters' utilization of the politics of fear kept them safe in places
where the possibility of slave rebellion was very real. Planters in Saint-Domingue
thought the same.

But by the late eighteenth century metropolitan thinkers had moved away
from believing that societies founded on Hobbesian principles could be "just"
societies. That happened even as the Reign of Terror raged in France in 1793–94,
making it clear that it was not just in the colonies but in the very heart of Enlightenment Europe that fear, horror, and frenzied apprehensions of imminent violence could lead to dramatic political change. It coincided with the warnings of
Whig polemicists like Edmund Burke that giving in to the politics of fear could
lead only to the subversion of government. If the seventeenth century was the
century of Hobbes, in the eighteenth century it was Montesquieu, and his insistence that social organization had to be based on principles of liberty rather than
on principles of fear was pivotal.[97] In this world, as opponents of slavery in the
Americas recognized, one could not defend, as Edwards defended, propositions
that in slave societies governments had to be supported by fear and "absolute
coercive necessity." If the politics of fear worked to keep enslaved people terrified
in places like Jamaica and operated so that Europeans became cruel despots, then
it confirmed to the growing number of Europeans and Americans who believed
that slavery was a sin that there was something desperately wrong with British
American plantation societies.

Chapter 2

Edward Long's Vision of Jamaica and the Virtues of a Planned Society

Near the end of the third volume of Edward Long's impressive three-volume *History of Jamaica*, published in London in 1774, the author expiated on the progress of the French settlement at Cape Nicola Mole. Over ten pages, he discussed this small port on the far northwest tip of Saint-Domingue (present-day Haiti) in detail. Cape Nicola Mole (or Môle-Saint-Nicolas) was in an arid and barren part of northern Saint-Domingue but had great strategic importance given its location on the Windward Passage very close to Cuba and directly in the shipping lines connecting the Greater Antilles with North America. Long's ostensible purpose in this lengthy digression was to disclose to the British and Jamaican public some new materials about the "enormous expenditure" that the French had made on fortifying this port. He wanted to show how Cape Nicola Mole had proved effective "as a restraint upon our navigation and trade" and how it might "in future time" serve as "some very capital machination against our commerce in this part of *America*." But his larger point was to bring home a theme that had preoccupied him in the two previous volumes of this history, especially in volume 1. That point was that the French colonial "machine" was advancing so rapidly in the West Indies, especially in Saint-Domingue, that it was eclipsing the British West Indies in wealth. Even more important, the French, in Long's opinion, had adopted policies and practices in their colonization of the French West Indies that put Britain to shame. As Long argued, the French government's "principles and genius" were so "conspicuous" in the various "departments of their colony-system" that they "manifest a degree of forecast, prudence, and vigour, that are not so observable in any movement of our own torpid machine."[1]

What the French had achieved at Cape Nicola Mole, he believed, was close to "incredible," turning in ten years a place that was "rocky, barren, and incapable of producing any sustenance for man or beast" into a place "crowded with shipping" and "a well-established, secure and opulent *emporium*." Soon, he thought, it would advance "by hasty strides to a superiority and grandeur beyond the oldest and most boasted seats of trade in any of the British islands." The principal reason was the steadfastness and purpose of the French state: "there is a spirit in the French monarchy, which pervades every part of their empire; it has select objects perpetually in view, which are steadily and consistently pursued; in their system the state is at once the sentient and the executive principle." Long considered that a determined state, devoted to careful planning for future benefits, made sure that its aims were closely coordinated to the execution of its well-established plans. The French imperial state, he argued, was "*all soul*; motion corresponds with will; action treads on the heels of contrivance; and sovereign power, usefully handled and directed, hurries on, in full career, to attain its end."[2]

Long was fond of digressions, and many of these were intended to suggest larger points. One point that Long also wanted to make was that the history of Jamaica was littered with ambitious schemes that ended, unlike the example of the French at Cape Nicola Mole, in dismal failure. Jamaica had a different, and inferior, kind of state apparatus to the French colonial "machine."[3] One emblematic failure was the establishment of Bath Spa and resort in St. Thomas in the East during the farsighted governorship of Edward Trelawney in the late 1740s. The creation of a town and hospital around rejuvenating mineral springs was an instant success, although an instant success achieved at vast expense. In a short time "it became the fashion every year for a crowd of company to assemble here from all quarters of the island." They engaged in fashionable pleasures: "The powers of music were exerted; the card-tables were not idle; and, in short, it grew into a scene of polite and social amusements." Alas, what fashion decreed, it soon decried. The political passions of the 1750s made it hard for people belonging to different factions to mix together. People stopped going. By 1768, most of the houses were in "decay; the half-finished frames of some, which were just beginning to rear their heads, have mouldered into dust."[4]

Long provided a poignant scene of an abandoned billiard room with an expensive "superfine green cloth, which covered the table, all besmeared with the ordure of goats, and other animals, who took their nightly repose upon it." Its fate bewildered the author: not only how "after such solemnity and parade in establishing the town; after so much apparent happiness derived from the institution; such munificent provisions for the sick poor; after advancing the plan so far towards maturity" could the scheme have stopped, but also how some short-sighted men could be so opposed to the supporters of the town and hospital that they "devoted their whole fabric to its subversion?" The strong implication is that what happened at Bath, Jamaica, would not have happened in the more enlightened realm of Saint-Domingue.[5] An even stronger implication is that the failure of careful and determined future planning by the British imperial state to make sure that Bath would last as a resort, combined with the inability of inconstant Jamaicans to show any willingness to transform short-lived enthusiasms into schemes with a long-term future, meant that France would not just compete with Britain as a power in the Greater Antilles. It would surpass Britain with ease. Possibly France would do to British possessions in the West Indies what England had done in the mid-seventeenth century to the Portuguese sugar trade. The French were "formidable competitors" who, to the mortification of Jamaicans, "excel us in two of our oldest West India staples, sugar and indigo." Just as superior production a century earlier had allowed the English to "crush the Portuguese," these very same means had allowed the French to "gain the advantage from us."[6]

It is worth investigating why Long felt the French development of Cape Nicola Mole was so successful and why he despaired about the failure of the spa at Bath in St. Thomas-in-the-East. These twin episodes symbolized graphically one of the principal themes of Long's vision for Jamaica—that Jamaicans could profit from emulating the French example of colonization in Saint-Domingue. He was especially keen to stress the ways in which the French used the resources of the state in a patient and assured way to plan effectively the development of Saint-Domingue. These plans were in marked contrast to the make-shift ways in which an indifferent British government and an admirable but sadly impatient white Jamaican society held back improvement. Long's admiration for French colonization in the Greater Antilles has been little appreciated in accounts of his work. He was certain that Jamaica did not meet French standards. This judgment demonstrates that Long was a much more sensitive, and critical, observer of colonial West Indian society than is usually acknowledged.[7]

What differentiated French activities at Cape Nicola Mole from what the British attempted to do at Bath was the extent to which the French had a plan of action that they implemented quickly and carefully, without concern about expense. Even more important was the perseverance that the French showed in making sure that their plans were realized. Long asked his readers to ponder on "the activity and solicitude with which France has persevered in strengthening her colonies, by those instruments wherein her ability and power have most consisted." France realized that Britain's naval resources gave it command of the sea, so it adopted a long-term and carefully planned strategy to make Cape Nicola Mole "a second *Dunkirk*," made "as impregnable, as forts, batteries and cannon, can render it, in order that it might shelter their men of war and privateers." But France wanted more than just to make this port an "asylum to their own fleets." They wanted to make the port "a source of everlasting annoyance to us [the British]." To do this, they imported five thousand Acadians (most of whom succumbed to fever) and three hundred Germans and gave them economic opportunities at Cape Nicola Mole, such as an excellent road and help in providing means for subsistence farming. They made Mole a free port, which allowed North American ships to flock there, "in consequence of which, the towns-people have derived a subsistence, that the land adjacent could not afford." The result was a handsome town of four hundred "good houses," and multiple well-built "public edifices," such as a church, a hospital, and commercial and government buildings. The town was well supplied with water, and the government took care to employ 450 slaves "under the direction of an able engineer, in carrying on the fortifications, and other public works." It was not just a port but a flourishing center of trade and commerce. Taxes were low, and prices for produce provided by North American traders were strong. It was a town designed "to expedite mercantile transactions."[8]

Long was convinced that Mole was a prime example of how advanced French colonization was in the West Indies compared with British colonization. Britons, he thought, could not act collectively in the same way as the French did because British liberty encouraged people toward "independent thinking and acting." The effect of such commitment to liberty might be ideologically beneficial, but it led to disappointments, such as the rise and fall of Bath. The spa town lapsed into decay not just because it went out of fashion but also because "unfortunate political divisions" during the administration of a "hot-headed governor," made people who used to meet at Bath as friends avoid the place when politics "destroyed all the harmony between families, which . . . had been the principal cause of making this place an occasional retreat."[9] Long was not sure whether the "shameful neglect" of Bath was due "to indolence, to caprice, or inconsistence." The effect was the same—the dropping of a great and useful public scheme when "a small fund, set apart for the purpose, would have supported all the buildings erected here for the general use or amusement."[10] It was this inability of Jamaicans to set aside differences for the greater whole, as the French were able to do, that made him think that Jamaicans might "envy, but I fear we shall never equal, this wonderful pattern of French policy, in founding; industry and ability in accomplishing; so truly noble a fabric."[11] France was winning the colonial battle; if it came again to war with Britain it might prevail.

Edward Long compared Cape Nicola Mole with Bath Spa to emphasize that France was succeeding where Britain was not. An examination in detail of how Long treated French colonization in his history shows us his ideas about Jamaica's future at a time when many writers thought the prospects for plantation colonies boundless.[12] This new perspective on Long's *History of Jamaica* shows that Long deserves attention as an imperial thinker, and not just as a racist ideologue. His books were a pioneering, indeed heroic, study in political economy and a polemical brief in favor of large-scale planning and colonial investment in infrastructure in Jamaica. One aim Long had was to convince planters and government officials that a radical plan to reform Jamaica was necessary and desirable, if only to keep Jamaica competitive with Saint-Domingue. This perspective shows how carefully thought-out Long's many policy proposals were. But Long's radical reform ideas were, in the end, impracticable because Long significantly misread the true nature of Jamaican society. There were reasons beyond those that Long himself advanced that explained why Jamaicans were addicted to short-term solutions rather than to long-term expensive planning.[13]

—

Edward Long was born in Cornwall in 1734 and died in Sussex in 1813, but his genealogical roots and the source of the income that allowed him to live as a landed English gentleman lay in Jamaica, where he lived for twelve years between 1757 and 1769 before ill health forced him back to England.[14] Long belonged to the upper reaches of Jamaican society. He was the descendant of Samuel Long, a famous speaker of the House of Assembly who had been cast as a victim of gubernatorial tyranny in the many battles between the Assembly and the executive in the late 1670s.[15] He had inherited a large estate in Clarendon from his father, also Samuel Long (1700–1757). His sister married the Jamaican-born governor, Sir Henry Moore, and Long himself married Mary Palmer, the widow of a wealthy planter, John Palmer, and the daughter of Thomas Beckford, a member of Jamaica's richest and most powerful family. Long served as a judge of the vice-admiralty court and used his time in Jamaica to furnish materials for his history. His *History* was based on considerable research, derived from data largely collected in the late 1760s. What it lacked in originality it compensated for in detail and in its careful exposition of a Whiggish pro-planter position.[16] Long was a proud patriot who believed that Jamaica had progressed wonderfully in the 120 years since English settlement in 1655. He believed that it had the potential, if various obstacles were overcome, to develop even further.

The *History of Jamaica* was the most impressive work in Long's corpus, which included a satire on English game laws written when he was a young man in 1757; a bitter denunciation of Lord Mansfield's decision in *Somerset* to restrict the rights of slaveholders over slaves in Britain in 1772; a curious and slightly strange exposition in 1778 attempting to defend the English from the jibes of Voltaire and Rousseau that the English were a savage people; and a polemical pamphlet on 1784 on the ill effects of British policy toward North America after the American Revolution.[17] Elsa Goveia, who wrote an especially penetrating evaluation of Long in 1956, concentrating on his strongly Lockean-influenced Whig views and distrust of executive power in government, considered Long's *History of Jamaica* "one of the truly great achievements in the writing of West Indian history."[18]

Long's depiction of his work as a "history" was misleading as it was less a narrative history in the traditional sense than an extended commentary on the social, economic, geographical, and political contexts of contemporary Jamaica.[19] Long lost interest in the history of the island as a coherent narrative after he had discussed early settlement in the first volume. He tended to insert historical narrative at points in the text to illuminate contemporary debates rather than to settle historical controversies. It is noticeable, for example, that his discussion of the native Amerindian inhabitants of Jamaica came almost as an afterthought following his exposition on Cape Nicolas Mole, which in itself followed a lengthy annotated translation of France's Code Noir, or the laws governing the treatment of slaves.[20] Readers looking for a survey of Jamaican history in any sort of chronological order would be disappointed.[21] The *History of Jamaica* was a strikingly modern work. Indeed, it is more a work of historical sociology, skillfully interweaving material from a whole variety of disciplines in the social and natural sciences, than traditional history. It was strongly polemical in tone, as seems to be the case for all of Long's work, and was masterly in exposition as well as being based on an extraordinary range of sources over which Long exercised remarkable control. He did not follow contemporary practice by "entering into detail of the characters and speeches of our governors; or reciting the various exploits of generals and admirals" or dwell long on "the gloomy thickets of politics." Instead, he obsessed about the social and economic characteristics of Jamaica in ways intended to answer the "slanders" of Britons while urging reform. As Goveia notes, "he not only narrates and describes" but "constantly argues and criticizes."[22]

Long wrote his *History of Jamaica* at a time of growing interest by British metropolitan writers in the expanding British Atlantic empire.[23] It is within this context that we need to understand his fixation with how Jamaica was failing to emulate the success of French colonization in Saint-Domingue. His obsession with how Jamaica was being prevented by a variety of means from achieving its full potential, while Saint-Domingue was striding forward aided by a benevolent French imperial administration, sets his work apart from other books on imperial topics, most of which adopted a more congratulatory tone. Thus, Long's work was very much a critique of empire, much of it unfavorable to Britain.[24]

Britons were very pleased with themselves and the great victories they had won all around the world between 1759 and 1763.[25] Their success changed the ways in which they talked about empire. One notable feature was the celebratory tone of imperial writings. Adam Anderson exulted in 1764, for example, that Britain had surpassed other European nations so that it would "by the additional wealth, power, territory and influence . . . now thrown into our scale" always be independent from European machinations.[26] The language through which empire was discussed also changed. It was expanded from a discourse based on commerce to one that included what Jack P. Greene calls the language of "imperial grandeur."[27] Britain saw itself as the new Rome and like its famous predecessor imagined its rule extending throughout the world. If Britain was to be a world as well as a European power, then it needed to pay attention to its new and old possessions in the Americas, Asia, and Africa. This task was the more urgent as British subjects in the Atlantic World would help Britain develop what Thomas Pownall declared in 1768 would be "a grand marine dominion" that would "build up their country to an extent of power, to a degree of glory and prosperity, beyond the example of any age that has yet passed."[28]

Celebration was not the only mode of discourse about empire. Some writers were sensitive to the wider context of empire and argued that the celebration of it needed to be accompanied by reflection on where the American and West Indian colonies sat in relation to Britain.[29] This sensitivity, of course, increased as constitutional issues between British North America and Britain exploded into antagonism and eventually into war between 1765 and 1776.[30] Several writers, especially colonial writers, urged Britain to pay more attention to what was happening in America. The famed agricultural improver Arthur Young explained that now that the colonies were so obviously economically valuable to Britain, the old pre–Seven Years' War tendency to ignore the Americas, while it grew "silently but prosperously . . . would no longer do." William Knox, who had property in the southern colonies, was more forceful: "the rapid increase of the people there, [and] the great value of their trade all unite in giving them such a degree of importance in the empire, as requires that more attention should be paid to their concerns."[31]

Long's first contribution to this imperial debate was to confirm Jamaica's economic importance. Jamaica, he asserted, was "a constant Mine whence *Britain* draws prodigious Riches" and was "the most advantageous and profitable colony to *Great Britain*, of any his Majesty's Dominions." His first volume contained a detailed summary of Jamaica's trade with Britain, Ireland, and North America in which he showed there was a large balance in favor of Britain of at least £700,000. Long was convinced that if some of his proposals for improvement were adopted, the gains Britain could make would be substantial. Indeed, after a lengthy tabulation of production, exports and imports in a statistical section of eighteen pages, he concluded that Britain gained as much as £1,249,164 by its commerce with the island, including £200,000 per annum remitted to absentees in Britain and £125,142 made in the Atlantic slave trade. Jamaica was valuable for other things besides its extensive commerce. It was a "nursery for seamen" and a source of income for thousands of "transient traders . . . [who] return full laden to their native hive." "What a field is here opened to display," he proclaimed, "to display the comforts and blessings of life, which this commerce distributes among so many thousands of industrious subjects in the mother country!" Further development, he thought, could only increase the monetary value of Jamaica to Britain as well as "finding employment for artificers and manufacturers of almost every denomination, and for numberless indigent and idle persons, who would otherwise prove a nuisance to their countries."[32]

It was not, however, the idle and indigent who Long thought would most gain from the wealth of Jamaica. He concluded his second volume with a summary of who he thought would most benefit, and for what reasons. Jamaica was a wonderful place, "where many a great fortune" was made, for "those, whose slender patrimony, or indigent circumstances, render them unable to gain a competent provision in their native country." Furthermore, unlike the northern colonies of British North America, which he pointedly argued competed with Britain in their economic activities, Jamaica was a fresh field of opportunity, in which British investment had already paid off handsomely, in making ordinary people rich and the British Empire prosperous. Jamaica was no longer an undeveloped country, and while it could be developed further, it was an "established society" with "every accommodation and convenience that can be desired." What was needed to increase the "daily improvement" of the island was what France was doing in Saint-Domingue: "wise laws and fit regulations, calculated for the protection of property, the encouragement of industry, the abolition of tyranny, the discountenancing of selfish monopolies, and the *conservation* of *health*."[33]

—

Of course, not all the commentary on Britain's Atlantic possessions was positive. As more attention was paid to Britain's colonies, Britons started to take more notice of some disturbing tendencies in the colonial situation and especially in the colonial character. For many writers, the colonies were a place of political and moral laxity. The political solution was for the metropolis to devise stricter and better regulations—an approach, of course, that led to massive resistance in British North America and that precipitated a revolution.[34] The main concern for Britons was that India and the Caribbean were sites of "Asiatic Plunder" and "Creolean Despotism."[35] India was more the center of imperial disdain than the West Indies at the time that Long was writing, but signs of a later and very important discourse over the morality of slavery in the West Indies was starting around the time that Long was completing his researches.[36] Random voices had always been raised against slavery. William Burke, for example, in an important work written in 1757, thought slavery a moral indictment on colonial character: "the negroes in our colonies endure a slavery more compleat, and attended with far worse circumstances, than what any people in their condition suffer in any other part of the world, or have suffered in any other period."[37]

Jamaica came in for attention as a site of colonial brutality. As noted in Chapter 1, the cruelty of Jamaican slave owners toward their slaves was picked up with enthusiasm by French writers, who often focused on Jamaican brutality and on the inevitability of slave revolt in that island, especially after news of Tacky's Revolt in 1760 and its suppression became widely known.[38] Adam Smith joined in these sentiments, acidly commenting in his *Theory of Moral Sentiments* in 1759 that "Fortune never exerted more cruelly her empire over mankind" when she entrusted Africans, whom he thought "a humane and polished people" to "the refuse of the jails of Europe . . . whose levity, brutality and baseness, so justly expose them to the contempt of the vanquished."[39]

Such comments about slavery in the West Indies, however, were relatively rare before the mid-1760s. The book that particularly initiated a significant shift in the direction of intellectual argument was Granville Sharp's remorseless excoriation of slaveholders in his 1769 *A Representation of the Injustice and Dangerous Tendency of Tolerating Slavery*. Sharp took slavery as his main subject, rather than a minor side issue in a larger discussion of colonial delinquencies. He not only made clear his outrage that Britain allowed such an iniquitous system to flourish but also demonstrated deep study into the intricacies of colonial slave laws. He took direct aim against American and West Indian slaveholders, accusing them of "hardness of heart" and "shameful depravity." Slave owners' participation in such "arbitrary, cruel and inhuman" practices made them hypocrites whose commitment to liberty was just a sham.[40]

Long was very concerned in his text to rebut such negative perceptions of the West Indies. He was especially incensed against Sharp, whom he took pains to denigrate on several occasions. One wonders whether he would have devoted quite so much attention to defending slavery and to developing a theory of African inferiority if he had not been determined to counter what he saw as scurrilous attacks from writers like Sharp describing Jamaican slave owners as hypocrites in their professed devotion to liberty while acting as tyrants to their slaves.[41] Long's principal fault as a polemicist was his inability to leave the arguments of opponents to one side. His *History of Jamaica* was diminished by a continual urge to score points against opponents, be they antislavers like Sharp, anti-American scientists like Comte du Buffon, or proponents of ideas he thought to be in error. The most egregious example of his need to show the superiority of his ideas over those of others comes in his notoriously over-the-top denunciation of the Afro-Caribbean poet Francis Williams. Long devoted a whole essay, or ten pages, to proving that Williams—a free black man who was the "subject of an experiment by the Duke of Montagu" about "whether, by proper cultivation, and a regular course of tuition at school and the university, a Negroe might not be found as capable of literature as a white person"—was a poet of limited skills.[42] Long went to near ridiculous lengths to parse Williams's poetry in order to show that it was derivative and showing a "strain of superlative panegyric, which is scarcely allowable even to a poet."[43] The overkill in Long's work is palpable and was intended not just to prove his point but also to answer the "many in England" who thought the duke's experiment a success. It is one of the least edifying sections of the whole three volumes.[44]

Nevertheless, Long did more than just rebut negative perceptions of the West Indies and score points against other writers. Long added a new element to the discourse over comparative imperialism developing in Britain in the aftermath of the Seven Years' War. He reminded British readers that the competition with France as to who was the best colonial power in the Americas was far from over, contrary to the fond hopes of writers who advanced claims of British imperial greatness in the aftermath of the 1763 Peace of Paris. These men enjoyed "the jig, confidence and exultation [that] sat on every countenance; [while] pleasure, luxury and dissipation reigned with an unbounded sway throughout all [of Britain's] triumphant dominance."[45] France, however, had learned from its defeat, Long argued. Saint-Domingue's growth since the Seven Years' War had been even more spectacular than that of Jamaica, with exports advancing from £1,200,000 in 1749 to £2,250,000 by 1773, surpassing by the latter year the £2,063,287 that were Jamaica's total exports.[46] "The prodigious increase of their West India traffic, within a few years," he asserted, "affords the most incontestable proof, that the encouragements and regulations given to their colonies are admirably well continued to render them populous and flourishing."[47]

Their "astonishing increase" in wealth and population showed that the French were "formidable competitors." Their success, Long argued, reflected "the work of a wise policy, and a right turn their government had taken." Unlike Britain, which took Jamaica and its planters for granted, or, worse, treated them with contempt, France "considers a planter . . . as a Frenchman venturing his life, enduring a species of banishment, and undergoing great hardships for the benefit of his country." That view made them offer planters "great indulgences," such as alleviating hardships due to natural disasters, protecting them from avaricious creditors, and subjecting them to few taxes while providing generous loans to "set them forward." France was moving ahead at such a stride that if it joined with Spain (as Long expected it to do), "how fatal the neighbourhood if so potent an alliance may grow, in time, to the interest of our settlements, which do not thrive in a proportionate degree, may justly be apprehended?"[48] Long sent a pointed warning to British imperialists that they could not take their command of the seas and excellence of their colonial subjects in the West Indies for granted as "the growing and united power of France and Spain in these seas should persuade us into the expediency of strengthening ourselves against them." He urged Britain to do so through fostering "colonizing and trade," which he declared to be "the only solid foundations on which we can build a successful opposition in this part of the world."[49]

Usually, Long's work attracts attention because of his attack on the character and intelligence of Africans. He made the first extensive racially grounded argument in defense of slavery through claims that suggested that Africans might belong to a different species than other humans. Alternatively, scholars cite Long as an authority on the character of white Creole Jamaicans. He defended enthusiastically West Indian planters and their political privileges, framing his arguments within a well-thought-out climatic theory that attributed planter behavior to tropical weather patterns.[50] He is thus normally discussed within the history of slavery and racial discrimination.[51] Long's racial schema, however, is the weakest section of his work. It may be the "most extensive and racially grounded argument in defense of slavery," but even by the standards of the time the argument is undone by his "vitriolic linkage of Negroes to an animal world" in which Long's bias against Africans blinds him to any scientific evidence—of which there was a considerable amount in the late eighteenth century—that contradicted his firm prejudices. Unlike the rest of the work, which was based on carefully assessed firsthand evidence and lengthy immersion in vital data, Long was happy to rely on spurious accounts of Africa that emphasized Africans' natural barbarism in ways that undermined the reliability of his polemics. His work was widely read for its racial propaganda, and his themes found a receptive readership among proslavery advocates who wanted empirical evidence to bolster their predetermined arguments. Nevertheless, even when he was writing his work, Long's racial theories were outmoded.[52]

—

Long believed above all that Jamaica could be transformed into an Anglicized society. He supported ambitious and expensive plans for improvement, most of which would have shown benefits only after many years in operation. In compiling this list of planned developments—which would have greatly expanded the role of government in Jamaica and placed a large tax burden on its wealthiest residents—Long was attracted by what he understood about Saint-Domingue. Somewhat surprisingly for an author about to write a book defending British freedom against French slurs, Long saw French colonization in very positive terms. To him, Saint-Domingue's rulers were practicing the benevolent interventionism that Jamaica needed. France, he argued, was "like a skillful gardener" who was "careful in the choice of plants, and treated her colonies as a favourite nursery, in which none should be fixed that were not vigorous, healthy, with all the promising appearances of thriving luxuriantly, and producing good fruit." By contrast, Britain, he lamented, "treats her plantations as a distant spot, upon which she may most conveniently discharge all her nuisance, weeds, and filth, leaving it entirely to chance, whether any valuable production shall ever spring from it."[53]

This indifference, Long thought, was because colonial governors were more interested in self-advancement than in the island's public welfare. If only such "hirelings, fools, and sycophants" treated the worthy citizens of Jamaica as "fellow-subjects" rather than people to be exploited for immediate gain, then Jamaica would reach its full potential.[54] What especially disturbed white Jamaicans, he believed, was living in "an unsettled mode of governing," where there was the "apparition of freedom without the substance." It made "honest and industrious" white Jamaicans consider removing to the French colonies as they preferred "a settled, absolute form of establishment" rather than the "fleeting, painted shadow" of liberty white Jamaicans got from an uncaring British imperial government.[55] Indeed, the absolutist government of France was not as bad as it seemed, Long thought, given France's commitment to a British-style balance of power. According to Long, who was not well informed in this matter, the French colonies had a well-established system of checks and balances that provided means whereby "the people are not oppressed" by any one of the governor, the intendant, or Saint-Domingue's royal council.[56] In the British colonies, inhabitants had the blessings of British liberty, but these benefits were compromised by the lack of "systematic order" in administration and by the sad tendency of officials to leave things to "chance."[57]

Long believed that Jamaica's potential was remarkable, despite its poor government, because Jamaica was a productive place and because the general character of the inhabitants was admirable. In general, Jamaica was "getting forward rather than declining in its most valuable settlements." Its history over the hundred years since first establishment, he concluded at the end of volume 2, made it an ideal place to restart one's life and gain a fortune, being "the asylum of the distressed and unfortunate, where all may enjoy sustenance." Recent developments showed how much it had "improved" as a place of residence, meaning that recent arrivals had the "advantage, unknown to our ancestors, of coming to an established society, which, from the number of towns and settlements, has every accommodation and convenience that can be desired."[58]

Nevertheless, Long was not blind to faults in the people and the place. He waxed lyrical in his praise of Creole men, seeing them as "sensible, of quick apprehension, brave, good-natured, affable, generous, temperate and sober; unsuspicious, lovers of freedom, fond of social enjoyments, tender fathers, humane and indulgent masters; [and] firm and sincere friends." And he extolled Creole women as "lively, of good natural genius, frank, affable, polite, generous, humane, and charitable; cleanly in their persons even to excess . . . [and] faithful in their attachments [and] hearty in their friendships." But he deplored the tendency of white Jamaicans to avoid schemes of colonial improvement. They were lukewarm to plans to develop public roads, bridges, and irrigation works because they were overly addicted, he believed, to short-term expediency and to the narrow pursuit of self-interest over the longer-term interests of the public.[59]

Jamaica's biggest problem, in Long's view, was its inattention to white settlement. Jamaica should concentrate on attracting and retaining "the sort of men, best qualified for increasing the number of whites." The colony set the deficiency law (which fined planters who did not retain enough white employees in relation to their slaves) so low that planters did not bother about employing the "sober, frugal and industrious artificers" that Jamaica needed. These men, along with "the poorer farmers and graziers" were "a hardy useful people, and most fit for occupying the unsettled deserts" that Jamaica still contained. One might see Long's *History of Jamaica* as intended for the consumption of men like Thomas Thistlewood, a penniless but highly intelligent and hardworking migrant from Lincolnshire who arrived in Westmoreland in 1750 and who managed to achieve an enviable independence and competency as a small proprietor, moderately large slaveholder, and local worthy by the time of his death in 1786.[60]

Long advanced several schemes for retaining the island's existing white population, such as government-sponsored schools, and the promotion of immigration from Protestant Europe. His most ambitious plan was to develop Jamaica's unsettled interior regions by building a new town every year, a scheme he estimated would cost £8,000 per annum. He aimed to improve the health of residents by taking them away from the poisonous air of pestilential Kingston, as settlers living in central regions "are as healthy as . . . in any part of Great Britain."[61] His plan, most importantly, was a scheme that developed uncultivated land, and that forced land monopolists to give up their land-aggrandizing schemes. He thought that quickly growing St. James Parish in the far Northwest of the island would be much better as a site of development if it had more than its current 132 white settlers. He proposed plans, expounded at length, about how the woods of the interior might be opened to industrious settlers who could grow coffee or indigo. The plans for such towns were very well considered. Each town would have 640 acres for twenty-eight planters with twenty-eight mulattos sent to serve as soldiers while each year twenty-eight "new negroes" would be offered up for distribution to families. His scheme was intended to appeal both to potential settlers and to the more industrious of the British soldiers stationed on the island, many of whom were "ready trained in arms for the defense of the country."[62]

Long may have had in mind here the system in Saint-Domingue in which each region had to have at least a hamlet and preferably a village to run local government.[63] If the Jamaican government built extensive roads and other improvements to support these new centers, each would serve as a hub for the surrounding countryside. These actions could greatly increase Jamaica's European population and, Long thought, nearly double the cultivation of arable land from its 1771 extent. He predicted that Jamaica could triple its population, making a population of eight hundred thousand, of which a greater percentage than hitherto would be European. Even an extra four hundred thousand people, he predicted, would be highly beneficial to Britain. It could send the colony "artificers and manufacturers of almost every denomination, and . . . numberless indigent and idle persons, who would otherwise prove a nuisance to their country."[64]

All that was needed was a firm government hand in forcing men owning undeveloped land to either develop that land or else lose it. A strong government was also needed to encourage industrious settlers. Long commented with approval on how the French government "oblige[d] all ships to carry several indented servants who are directed to be sound strong bodies, between the ages of eighteen and forty."[65]

These new settlers needed to be supported by slaves. Although Long accepted that not all countries should have a system of servitude, African slaves were "*inevitably* necessary" in Jamaica. They had to be kept affordable, therefore, as "the dearer the implements of labour . . . the less will be the inducement for men to begin up on new settlements." The high taxes currently placed on new slaves were, he thought, indefensible for a country that needed cheap labor.[66] The possibility of owning slaves attracted immigrants, but the conditions of slavery needed to be improved, to avoid another devastating revolt like that of 1760. One way to do this was to prevent slaves from being sold off the estate, especially for the settlement of debts. Because slaves were attached to the place in which they were born and raised, selling them as if they were items of money "is by far the highest degree of cruelty annexed to their condition." He also made a firm distinction between newly arrived Africans, whom he thought savages almost totally without merit, and established or Creole slaves, whom he believed could be made very useful members of society, when treated, as he thought they invariably were already, by a kind master.

One way to increase the number of Creole slaves and to remind people to treat slaves well was to place an onerous tax on slaves acquired in the slave trade for four or five years, "except for re-exportation." Planters would learn to take more care of "present stock" and they would have less reason to run into debt through hiring or buying slaves. And the slave trade would be diminished, with a consequent decline in the amounts of money leached out of the country to British merchants. The payoff, Long insisted, would be a rise in prices of all Jamaican slaves.[67] He dreamed of using the deficiency tax to fund an ambitious settlement scheme in which industrious Britons would be sponsored to come to Jamaica. These people, he thought, would replace the current unsatisfactory migrants to the island, whom he designated as "idlers, or vagabonds" obtained from the crimp's office in London and "contaminated with every vice and disease."[68]

He also wanted Jamaica to be economically self-sufficient. He hoped that improvements in slave conditions would cut slave mortality rates and lessen Jamaican dependence on the Atlantic slave trade.[69] Similarly, he advocated that Jamaicans turn away from trade with North America (much of which was diverted into illegal commerce with Saint-Domingue, to Jamaican disadvantage) for "necessaries" and start to grow their own produce, especially livestock. Long was obsessed with the ludicrousness, as he saw it, of allowing livestock, especially mules, to be imported at great cost from other places rather than cultivated in Jamaica. He thought that a tax on such imports would produce more money and be more efficacious than keeping the forty shillings per head capitation on new African slaves.

Another form of self-sufficiency was to provide education to white Creole children. His encouragement of local schools was designed to keep white Jamaican boys from becoming absentees obsessed with living in Britain and was intended to train girls into women whom Creole Jamaicans might think interesting enough to marry.[70] His section on education is especially revealing. The absence of a system of education on the island, he thought, was a matter of national shame, and "one of the principal impediments to [Jamaica's] effectual settlement." That Jamaica "remain[ed] unprovided with a proper seminary for the young inhabitants to whom it gives birth" was an unfortunate reality that "at once excites our pity and regret."[71] Long outlined at length the evils of sending children abroad to be expensively but uselessly educated in activities that were likely to lead them to despise their home country and to "return with a thorough aversion to, or incapacity for, . . . laudable employments." Educating sons of planters in England brought some financial benefits to Britain as children "spent their fortunes in Britain, and learned to renounce their native place, their parents, and friends." But, as this acerbic comments suggest, the bargain was a bad one: "Their industry is, in general, for ever lost to the place where it might have been usefully exerted; and they waste their patrimony in a manner that redounds not in the least to the national profit, having acquired a taste for pleasure and extravagance of every kind, far superior to the ability of their fortunes." In short, he concluded, "the education of the youths *remitted* from this island is, in general, so mismanaged that, was it not for their innate good qualities, not one in ten would ever arrive at the age of discretion, or return to his native country with any other acquisition than the art of swearing, drinking, dressing, gaming, and wenching."[72]

He compared the practice of sending children overseas, in ways that would immediately mean something to Jamaican readers who had absorbed the rest of his recommendations, as a form of remittance from Jamaica to Britain that wasted one of the primary assets of the island—in this case the human capital inherent in young boys and girls. Opponents of his admittedly expensive schemes to restore the seven charitable foundations for establishing schools on the island to their proper levels of funding were "selfish and illiberal" with "bosoms so steeled with avarice, as to have lost all feeling for their fellow-subjects in these remote parts." In both cases, Jamaica was buying its labor needs from overseas. That was fine for Africans, whom Long felt ideally suited for enslavement and whom could not be procured any other way than through the slave trade. But it was both immoral and uneconomic to adopt the same policy for mules or cattle or indeed for native inhabitants. If it was the height of folly to spend thousands of pounds each year on buying cattle and mules from the Spanish when such animals could be produced in Jamaica, it was even more foolish, and wasteful of the most precious resource Jamaica had—its Creole white inhabitants. To send them to be uselessly educated in Britain deprived Jamaica in the long term from the fruits of an investment in schooling but also diminished the number of useful native inhabitants who were committed to the country and well trained in what the country needed. In a telling phrase, he compared sending native young Jamaicans to Britain to be educated to be similar to having them "go like a bale of Dry goods, consigned to some factor, who places them at the school where he himself was bred, or any other that his inclination leads him to prefer . . . where the poor wretch undergoes as much neglect and ill usage as if he was a charity-boy."[73]

One of Long's fondest plans was to make Jamaica self-sufficient both in livestock and in people. Regarding livestock, he proposed spending £10,000 per annum on "breeding camps" for cultivating cattle, horses, and mules. Jamaicans wasted precious currency on buying inferior mules and cattle from the Spanish when, with a little foresight and investment, they could cultivate their own livestock farms. He drew attention to the benefits Jamaica had gained from the establishment of cattle- and mule-breeding farms in a part of central Jamaica, between St. Ann and Clarendon Parish, called Pedro's Cockpit. In this rich region, he noted, industrious settlers of moderate means had established a wonderful district in which "the cattle and mules bred here are larger and finer" than elsewhere on the island and certainly better than those from Spanish America while even the butter made in the district was "so excellent in flavor and firmness, that I have never met with any in England superior to it."[74]

What could be done in Pedro's Cockpit with livestock could be done with schools in other parts of central Jamaica. Long had an elaborate plan to establish a seminary in the center of the island, with land, slaves to work the land, a well-paid schoolmaster, an effective board of governors, and white servants to keep pupils from "a too early familiarity and intercourse with the Negroes, [so as] to adopt their vices and broken English." He advocated teaching subjects of use to colonization on the island, so that students would become "surveyors, book-keepers, mechanics, useful members of this community." Educating children at home would keep Jamaicans on the island, all for around £1,090 per annum. The cost seemed large, until one considered, as Long argued, that vast sums—he estimated £70,000—have been spent in the previous nine years fortifying the island.[75] Long was making two points. First, if Jamaica could spend so much on forts, then they should be able to spend a smaller sum on their most precious resources—their white children. Second, if the white population was encouraged to settle and breed on the island, through having decent schools to send their children to, then the need to spend more on defense would be less necessary, as white population growth augmented the size and significance of local militias.

Despite his laments that Jamaican planters were prepared to spend vast sums on defense while skimping on necessary funds for education, Long also wanted to reform Jamaican defense so that in this area too it was self-sufficient. He made elaborate proposals for a strengthened militia and argued that Jamaica should refocus its strategic priorities away from coastal defense and toward defending a greatly expanded realm of settlement in the interior from slave revolt.[76] His plans for a strengthened militia led to his most radical and costly proposal, which involved a transformation of the position of free people of color. Long was generally not in favor of giving advantages to free people of color, in part because he disliked the way that most people of African descent became freed—because of sexual liaisons between white men and their enslaved concubines. He wanted to maintain a clear distinction between whites and free people of color, with the latter serving as an intermediate class of skilled workers and artisans, prevented by law from gaining overmuch property and stopped from any political participation.[77]

But he recognized, as did his contemporaries in Saint-Domingue, that if Jamaica was to preserve itself in the face of danger from foreign or internal attack, it needed a larger and more effective militia than it currently had. The only way to raise the necessary number of troops was to emancipate up to ten thousand of the most reliable slaves—"tradesmen, drivers, and other head men." These men, "properly armed," could supplement the militia in times of war and could return to their former "trades and occupations, and support themselves by their skill and industry, instead of growing burthensome to the public," once war was finished. The cost of freeing so many slaves would be considerable, amounting to £37,714. Long optimistically felt that this cost "would be most chearfully be paid by Great-Britain, if the inhabitants, after the desolation of war, should find themselves disinclined to raise it by taxes." He also advocated forcing free people of color into the militia, although he recognized that principles of equity and liberty might be considered violated if freemen were forced into service.[78]

—

Long was greatly influenced by what he thought was occurring in nearby Saint-Domingue. He feared that the formidable economic, political, and social growth of this neighbor would soon eclipse Jamaica's productive energies. Jamaica's main advantage was its more efficient slave trade system, but he feared that even that advantage was quickly eroding, as France developed better contacts in Africa.[79] He listed several ways in which Saint-Domingue was superior to Jamaica. It had more fertile lands, cheaper cost of goods, lower wages given to servants, extensive and illegal trade with North America, and lower rates of taxation. But he attributed the success of Saint-Domingue principally to two factors: "their ability to furnish double the number of European hands [as Jamaica], and [their] wider regulations." It was the government of Saint-Domingue that especially impressed him, not only because of its "judicious precautions taken to secure the good government of their colonies" but also because of its practical measures to improve the colony. Long believed that Saint-Domingue's government was interventionist but also benevolent and focused on the colony's long-term interests. The French, he argued, were attentive to planter interests in ways that the British government was not. France was more determined than Britain to get their policy mixes right to advance the economic interests of their colonial possessions.

He argued that the French developed expensive and successful schemes of settlement in order that they could get "middling and petty" settlers into good positions in Saint-Domingue. The result was undying gratitude on the part of the fortunate beneficiaries of government largesse. For Long, the French authorities in Saint-Domingue could do virtually no wrong, acting toward its more indigent citizens as "a merciful creditor, that will never distress them" rather than, as in Jamaica, a hostile force that led to "many of our settlers" being harassed and ruined from deceitful and litigious con artists masquerading as merchants. Long was no fan of the laissez-faire private-citizen approach to colonial development that Britain favored. Instead, he wanted London to emulate Versailles, arguing that "the encouragement and regulations given to their colonies are admirably well contrived to render them populous and flourishing."[80]

Long believed that this government intervention gave Saint-Domingue a more balanced society than in Jamaica. Saint-Domingue had, he argued, a considerable body of ordinary settlers and was a society devoted to "improvement," as seen in their numerous towns, their commitment to ambitious irrigation schemes, and their cultivation of scientific societies devoted to agricultural reform.[81] He even considered whether the French were better slave owners than the British, admitting that many people thought so as a result of the Code Noir, the rules and regulations instituted from 1685 to control slavery in the French Empire. Long praised the Code Noir, providing the full details of its provisions in an appendix to the last volume and describing it as "a very just and sensible mixture of humanity and steadiness." He noted that for many observers these regulations gave the French "a reputation for good discipline and clemency" as opposed to cruel British planters. He was not prepared to concede that point, citing French authorities for claims that the French were often guilty of "wanton, diabolical cruelty" and arguing that the Code Noir was honored in the breach as much as it was followed. Such an argument, however, was hard to uphold in the light of repeated comments about Jamaican ill usage to slaves and to the awkward fact that Saint-Domingue had never experienced a slave revolt while Jamaica suffered them often.[82] In addition, Long believed that Jamaicans would do well to construct their own set of regulations along French lines, arguing that "there can be no dishonor in borrowing, and intermingling with our own system, such of its institutes as the difference of our constitutional principles has not excluded."[83]

By contrast, Long saw Jamaican society as deplorably committed to short-term considerations. Long admitted that its chief feature was the overwhelming desire to get rich as quickly and with as little trouble as possible. Jamaicans were characterized by a "narrow selfishness and total Unconcern for everything that doth not regard their Immediate interest," as an earlier commentator lamented.[84] Such self-centeredness was unfortunate, in Long's view. Their inconstancy was evident in their inattention to developing fences or proper systems of manuring, their willingness to let gardens go "ruinate" at a moment's notice, their failure to plan for additional towns, their inability to think ahead about how the militia might be best organized, their reluctance to systematically explore Jamaica's possibilities for mineral wealth, their dereliction in their failure to preserve public records, their failure to maintain and improve roads and other forms of transportation, and most of all in how they started and then ended schemes of white settlement without giving them time to come to fruition. They preferred immediate self-gratification over delayed benefit. Jamaica's past, he lamented, was replete with "various schemes, both in business and pleasure that had been eagerly started, and then suddenly Dropped, and forgotten as if they had never existed." Jamaicans spent large sums on starting projects. But they would not finish them, owing to a want of "sufficient steadiness and energy."[85]

One problem with Long's adulation of France and French colonization methods was that it was difficult to reconcile his enthusiasm for the French system of government with his patriotism as a proud British Jamaican. Long recognized the contradictions in his argument and tried to resolve the contradictions, only partially successfully, through making a distinction between the natural genius of the British and the superior government of the French. The strengths of Jamaica, he thought, also should number among their weaknesses. Jamaica had a "well-regulated spirit of industry" and an even greater attachment to liberty. That love of liberty made Jamaicans susceptible to conspiracy theories and to the lure of factional politics and political dispute. It also contributed, however, to that "persevering spirit of industry so peculiar to the English." That "genius for industry" was aided, he thought, "by the spirit of national freedom!" Long speculated that if Jamaica were fortunate enough to have farsighted government, in the way that Saint-Domingue did, the natural qualities of the British would have them overtaking in "vigor and opulence" their Caribbean neighbors.[86]

The current advantages that the French had over the British were ironic, he believed, given that the French had achieved what they had done by envy of Britain and "the true source of our greatness," which was the talent and dedication of the ordinary Briton. It was time, Long argued, that instead of France emulating Britain in colonization schemes, Britain should copy them, "mix[ing] a little of

French policy in our system of colony government." The means to unleash Britons' great capacities was to have a sympathetic administration, as in Saint-Domingue, rather than an oppressive one, as in Jamaica. He lamented that in Jamaica "people practice knavery . . . which would be thought criminal in England." One solution was to copy what he considered a French masterstroke, which was to have colonials (in this case "twelve principal merchants on the French council of commerce") serve in imperial bodies. He argued that "the surprising increase of French trade, shipping, and colonies, has been dated from the erection of that council." Local patriots knew best about local matters and could assist imperial officials in providing what was best in the way of long-term planning for a colony. It was astonishing, he exclaimed, that a country "whose subjects are bold and enterprising, and exceed most other people in the spirit and success of their colonization" was not allowing such subjects a say in how they were to be governed. Britain needed to trust white Jamaicans more. Otherwise the "most valuable men and the best supporters [of government] (who are the most honest and industrious) will leave, even to go to Saint-Domingue, where "men bringing their families and effects, would be well received."[87]

The sheer range of improvements that Long suggested for Jamaica and Britain to consider was so expensive that Long's analysis was never likely to do more than inspire spirited discussion.[88] The list of improvements he advocated was staggering. In addition to the schemes outlined above, he wanted to establish societies of agricultural improvement at government expense, advocated increasing the rate of interest from 5 percent to 6 percent, so that British merchants would be more attracted to dealing with Jamaican planters and Jamaican planters could be weaned away from avaricious local merchants (notably Jews—Long was a fervent anti-Semite); large sums to be spent on roads and irrigation schemes; and the development of a commercial center in Kingston. Even those items that he priced (£8,000 per annum for new towns; £10,000 for making Jamaica self-sufficient in livestock through developing livestock pens in central Jamaica; the establishment of a new light coin to make up the shortfall of £85,000 that he thought was the shortage in internal currency) were so expensive as to be highly unlikely to be agreed to by the Jamaican Assembly. The unpriced items, such as a massive new road scheme, a duty equal to a prohibition on all slaves imported from Africa for a space of four or five years, and taxes on imported animals and ruinate lands, would have been more expensive still. Long argued that planters "should accept higher taxes," as would certainly have been the result of his reforms being accepted, even partially, as this would mean that their "property was well guarded" and secure from attack by unscrupulous moneylenders and encouragers of speculative schemes. It is hard to see that Jamaican planters, who were notorious for their lack of this sort of public spiritedness, and who were more willing than other British American planters to tax themselves reasonably heavily, would have taken such massive decreases in their income because of Long's changes with equanimity.[89]

There were other problems with Long's analysis besides its cost. Long was an acute observer, but there were three empirical facts that he would not acknowledge and one policy prescription—an insistence that Britain could not interfere in the absolute control that masters and mistresses exercised over their enslaved people—that cut across his overall analysis and weakened his overall argument. That white Jamaicans should have untrammeled authority in all areas of slave management is covered in Chapters 3 and 6. First, he refused to accept that Jamaica was naturally unhealthy for white people, even distorting population figures to fit his argument. Consequently, the white population was much higher in the *History* than it was in fact; white deaths were systematically underreported; the number of absentees was exaggerated, mainly to blame them for various evils to which Long thought Jamaica was subject; and the numbers of free coloreds and mixed-race slaves were heavily understated. Long convinced himself that Jamaica was a healthy place for whites who did not drink to excess or fornicate with too much enthusiasm. He believed that natural white population growth was possible and that the ratios of whites to nonwhites could change so that the proportions of whites more closely mirrored those in other settler societies based on slavery.[90]

Second, he denied the violence inherent in the slave system. He asserted, against all evidence, that masters were humane and even indulgent to enslaved people. He argued that whatever brutalities occurred were the responsibility of the worst kinds of overseer and did not characterize most Creole relations with slaves.[91] Yet the Jamaican slave system was notorious for its violence. There was violence committed in hot temper and violence played out as part of an increasingly efficient and ever more ruthless system of slave management, as noted in Chapter 3. Planters worked slaves as hard as they could, and if physical violence was indeed less than it had been before (which is hard to quantify), as Long suggested, the demands on enslaved bodies were just as onerous as ever.[92]

Finally, Long underestimated the importance of the Atlantic slave trade.[93] Long dealt relatively infrequently with this crucial business. What aggrieved him most about the slave trade was neither its inhumanity nor the ways in which it inflated slave prices and encouraged planters to take on too much debt. He insisted that the slave trade was efficient and thought it probably the only area in which Jamaica excelled Saint-Domingue. But what bothered him about the slave trade was that it maintained and increased the number of "imported Africans." His plan of improvement included measures designed to discourage planters from their tendency to buy rather than breed through making slaves obtained from the slave trade so expensive as to be not worth having.[94] But what Long did not appreciate sufficiently was how a vibrant slave trade underpinned the whole Jamaican economic machine. Without it, slaveowners would have had to change their whole method of slave management—as occurred to some extent after 1807.

—

Besides these empirical blind spots, Long failed to understand the logic to the short-term perspectives that white Jamaicans adopted. There was an alternative vision of Jamaica's future. Planters, merchants, and their paid employees were perfectly happy with short-term thinking and all it implied. Their main ambition was to increase productivity every year, and they did not care overmuch how those gains were achieved. They sought these gains less through long-term investments than through slave management techniques that emphasized immediate returns and the relentless exploitation of enslaved people. Planters increased their crop by working enslaved people to the limit. They reduced expenses by training slaves to do work previously reserved for white tradesmen. They practiced a form of self-sufficiency, but it was not that as envisioned by Long. They made slave men into artisans and slave women into field hands. They did so to maximize their investment in their human capital, as explained in Chapter 3. This, in turn, meant accepting low birth rates and high mortality rates.[95] A dynamic slave trade was central to that process. It allowed planters to easily replace their slaves and made a hard-driving slave system possible. They welcomed the way this system inflated slave prices, for this effortlessly increased their net worth. Long did not appreciate that in the casino capitalism that characterized Saint-Domingue and Jamaica short-term decision making made a great deal of sense. The plantation system in these places was dynamic and risky but flourishing.[96]

Long's failure to understand the true dynamics animating Jamaican life can be seen in his long discussion in book 3, essay 6 of his second volume on how to "preserve health in Jamaica." Long's understanding of medicine in the tropics was based on the humoral system, in which bad air "admitted into the lungs and circulation, may induce a disposition to putrecency, and render those disorders of the frame malignant, which otherwise, perhaps, the efforts of nature alone, but slightly assisted, might have thrown off."[97] He advanced a number of practical measures Europeans could take in the tropics to ward off the effects of bad air, including living in healthy inland districts, sleeping in the highest parts of houses, eating less red meat and imbibing less strong liquor, dressing in loose clothing, becoming an early riser and avoiding the heat of midday, and making sure that one perspired freely and abundantly. Where he had trouble, however, was in assessing the effect of "the *passions* upon health." He admitted, and celebrated, that white Jamaicans were passionate people: "men are more *feelingly alive* to joy or inquietude; where the system is far more irritable than in a Northern climate." The passionate nature of Jamaicans contributed to their hospitable character and

to making the country an agreeable residence. It also had some negative consequences. He noted that "men of lively imaginations and great vivacity (and such
are the natives of this island) are more liable than others, to sudden and violent
emotions of the mind, and their effects; such strong and sudden transports may
actually throw men into acute diseases." The principal disease he thought afflicted
Jamaicans was anxiety, which he thought "once it has taken a firm hold, it is
generally productive of mortal consequences."[98]

But what Long attributed to bad air and the working of the passions on health
could easily be diagnosed as a reaction to the risk-taking, entrepreneurial, capitalist nature of life in a society devoted to the main chance. Jamaicans were anxious
because they lived so close to the edge. "The life of an industrious planter," he
noted, "is one continued sense of activity, both of body and mind." Men were
anxious because they were "hurried by levity of disposition or want of thought,
into an expensive way of living, or imprudent schemes and pursuits; distress has
poured in upon them at once like a deluge." Harassed by creditors, they sank into
an early grave. The reason for "multitudes" expiring "under the pressure of this
fatal cause" was not, however, humoral peculiarities that could be alleviated by
more attention to "diet and general regimen of life." It was the result of the "chaos
of men, negros and things" that was the principal characteristic of life in the
tropics.[99] Planters had to juggle numerous things together while working to make
a profit from often overencumbered estates. The "industrious planter" was very
busy, and his "slumbers are often disturbed with corroding cares, the failure of
seasons, the casualties to which his property may be liable, and the importunity of
creditors." It was hard to take a long-term view and plan when such immediate
concerns were so pressing and "the day is often insufficient for the multiplicity of
business which he finds himself obliged to allot to it." Long recommended a
"moderate oeconomy" and an avoidance of "temporary expedients" but acknowledged that there was a narrow gap between "hope and despair." "Misfortunes
here, in planting and trade," he thought, "are necessarily very frequent, where
men adventure without limits; give, and take credit; are subject to be hurt by
misplaced good-nature and confidence; and liable to various calamities and
losses."[100]

What Long did not want to consider was that "temporary expedients" and a
culture of risk in which "misfortunes . . . are frequent" was normal rather than
exceptional. Here, he might have profited from some of the more hardheaded
attitudes of writers on contemporary Saint-Domingue, where the true nature of a
society that had been extensively commodified and monetized were more obvious than in Jamaica. In 1776, for example, Michel-René Hilliard d'Auberteuil
wrote a two-volume work, *Considérations sur l'état présent de . . . Saint-Domingue*.
He extolled the transformative power of free trade and investment and how the
cycle of capital and reinvestment allowed some colonial profits to return to the

metropolis with "the rest . . . put back into the system. . . . Therefore there is a seed of fortune in the colony, which will grow as long as its lands are fertile."[101]

Like Long, Hilliard emphasized that changing Saint-Domingue required focusing on the capabilities of the Creole population. His vision of Saint-Domingue, however, celebrated the capitalistic modernity of Saint-Domingue's economy and proposed another modern idea to reinforce it—a racial caste system that would unite "whites" across inequalities of wealth and class. His *Considérations* captured the spirit of Saint-Domingue's emerging modernity, insisting that macroeconomic considerations should override every other constraint in fostering colonial growth. Another writer at the same time put the issue even more succinctly. Lieutenant Colonel Desdorides, stationed in Saint-Domingue in 1779, was among the many visitors struck by colonists' willingness to allow the pursuit of profit and pleasure to overwhelm their sense of proper decorum. He lamented that "it seems that in Saint-Domingue violent agitations of the heart take the place of principles, with the exception of illusions of love, Dreams of pleasure, extravagances of luxury and greed, the heart knows no other adorations." Desdorides believed that an obsession with money had fundamentally changed the character of French colonists: "In Saint-Domingue, men and women behave in a manner totally opposed to what I have described [of the French character]. Men there reduce everything to financial gain."[102]

These kinds of insights were not ones that Long was willing to make. He was an advocate of large-scale planning that utilized the characteristics, as they existed, of Jamaicans and their society to make Jamaica more valuable to Britain. He thoroughly disapproved of notions that in a plantation society like Saint-Domingue or Jamaica everything had to be reduced to money and that money was the transformative agent that shaped the dynamic of colonial growth. In short, he did not see how the logic of capitalism and the commodification of humans in an early modern plantation economy was problematic.

—

Despite his blind spots, Long's impressive summary of Jamaican history and culture and his keen analysis of the comparative colonization efforts of France and Britain point to larger conclusions about conceptions of empire in Britain in the decade after the end of the Seven Years' War and before the start of the American War of Independence. The rapid increase of territory, wealth, and population by the British after 1763 encouraged many polemicists and statesmen to ponder what the proper state of empire might be. It led to some adventurous and politically implausible schemes, such as imagining the British Empire as a place without slavery or as one that recognized the sovereignty and claims to British justice of

Native Americans.[103] As a number of scholars have argued in recent years, moreover, the end of the Seven Years' War started a number of crucial struggles in several parts of the British Empire on what it meant to be a British subject when many individuals subject to the British Crown had different social, racial, and religious identities to white Protestant men. Subjecthood, it has been argued, was a protean and porous concept that fluctuated per circumstances and local understandings of what loyalty under the Crown meant.[104] In North America, Benjamin Franklin, for example, argued that in a "negotiated" empire the relationship between Britain and its colonies should be one where Britain allowed the colonies to develop with little interference (and by implication with minimal imperial financial support). Franklin believed that the British Empire might have lasted indefinitely if Parliament had just left colonists alone, allowing "the King's subjects in those remote countries the pleasure of shewing their zeal and industry."[105]

But Franklin's were not the only views of the role of the state in British America held by colonists. Long, unlike Franklin, believed in a strong and activist state, which would take action to "improve" Jamaica and provide the material support for white Jamaicans' absolute control of their enslaved population.[106] The West Indies was more tied into British cultural and economic circles and more willing to support an active and interventionist state than was British North America. Jamaica was a case study of how colonial prosperity was tied into imperial military and fiscal strategies for defense and economic growth. Wealth from sugar was considerable, but the plantations required considerable local and international investment, both to buy machinery and supplies and to keep the slave trade functioning. As John Darwin notes, West Indian "commercial success and the wealth it brought depended on empire—the assertion of rule."[107]

Unlike in British North America, the presence of the imperial state was impossible to ignore in Jamaica or elsewhere in the British West Indies. Planters depended on the British mercantilist system to compete with the more efficient French sugar producers and on British commercial regulations to keep them involved with Britain's highly sophisticated commercial facilities. West Indian assemblies paid large sums to assist in the cost of imperial garrisons and not only got accustomed to having lots of soldiers and sailors in the islands but clamored to have more. Their barracks, fortification, and governmental buildings, such as the magnificent parade in Spanishtown or the impressive military barracks in Antigua, cost West Indian legislatures a huge amount of money. Assemblies paid the imperial government large sums in duties and compensated colonial governors handsomely. These were not places with minimal government but were places where it was expected that government would be intrusive and expensive. West Indian planters maintained their support for giving generously to government throughout the American Revolution. The Rodney memorial erected in

Spanishtown to commemorate the victor of the Battle of the Saintes in 1782, for example, was easily the costliest memorial ever erected in British America in the eighteenth century.[108]

Long's advocacy for an assertive, centralized, and interventionist imperial state in Jamaica that was like what he imagined existed in Saint-Domingue was therefore more believable in a West Indian context than in a North American context. Long was a Whig, but he was not the sort of radical antigovernment Whig that one found in Massachusetts.[109] He was contemptuous of the "hirelings, tools and sycophants" whom Britain sent out to be governors, but it seems that his main complaint was that these were men of undistinguished background out for the main chance. What Jamaica needed (and got in the nineteenth century) were the sorts of high-born aristocrats, cousins of the king, and senior generals and admirals that the French foreign minister, the Duc du Choiseul, sent out to Saint-Domingue and whom Britain customarily put in charge of Ireland.[110] In the eighteenth century, however, Long believed that the poor quality of the men sent to govern Jamaica was typical of the penny-pinching attitude Britain had to its colonies. His view of empire was that it should be the locus of British investment and a place where the British state (one, moreover, committed to the expansion of slavery) played an active role in fostering improvements and colonial development.

That view was not what imperial officials wanted to hear in the 1760s and 1770s, when they wanted to cut costs as much as they wanted colonies to conform to imperial dictates. But it was a view that was adopted by Henry Dundas, the influential right-hand man to Prime Minister William Pitt, in the difficult days of war against France in the 1790s and 1800s. It was also, according to the powerful analysis of the economics of British imperialism made by Peter Cain and Tony Hopkins, the policy adopted for nineteenth-century Britain and its empire. Dundas, who thought the defense of Jamaica crucial to British credit and reputation, was determined to attack France in the Caribbean and to destroy its colonial trade. He also wanted to increase British possessions in the eastern Caribbean and to harness the forces of British finance to fund the sorts of improvements that Long had suggested twenty years previously.[111] Like Long, Dundas was intensely concerned about the French threat to British trade in the Caribbean. The colonies played an important part in the end stages of the Second Hundred Years' War with France—an intermittent conflict that shaped British politics and British identity throughout the eighteenth century.[112]

But, as this analysis of Long's attempt at comparative colonization shows, competition with France in imperial matters was not just confined to the battlefield. Long was intensely proud of what Jamaica had achieved since England had conquered it from the Spanish in 1655. Nevertheless, he felt anxious about what he saw as the threat of an increasingly successful French colonization policy, centered on Saint-Domingue, Jamaica's flourishing neighbor. As historians are increasingly demonstrating, the French colonial system in the eighteenth century was not just growingly important to ancien regime France; it was also rapidly advancing to challenge other European nations, notably Britain, for economic and political dominance.[113] What Long argued was that Britain could not just continue its colonial practices as before when it faced such a formidable competitor in an area that it thought its own and in which it believed it should be superior as a result of humbling France in the Seven Years' War. His specific policy prescriptions were not heeded. Ironically, however, Britain did become a much more interventionist and coercive imperial power and even started to invest more heavily in empire after the American Revolution. The irony was, however, that where Britain most wanted to interfere in how Jamaicans ran things was in slavery. Long got his interventionist imperial power: the problem was that it interfered in the one area—slavery—that Long thought had to be the province of white Jamaicans. The challenge of abolition forced Jamaica planters in the 1780s to adopt some measures of amelioration, which followed some of Long's policy prescriptions.[114] But white Jamaicans did what they did in respect to improving slave conditions unwillingly, incompletely and ineffectively. The profit margin and the prevalence of short-term thinking over long-term planning continued to plague the island—just as Long had predicted in 1774.

Chapter 3

A Brutal System

Managing Enslaved People in Jamaica

Edward Long (1734–1813) was an eloquent proponent of an idealized resident Jamaican plantocracy and the owner of a large, well-established, and highly productive sugar estate in the central parish of Clarendon. As noted in Chapter 2, his 1774 *History of Jamaica* was, as Elsa Goveia attests, "one of the truly great achievements in the writing of West Indian history."[1] Yet, as also outlined in the previous chapter, Long had significant blind spots, especially when it came to what he considered foolhardy attempts to depict Africans as anything other than barbarians incapable of improvement and Jamaican planters as anything other than "humane and indulgent masters."[2]

Against a mass of evidence, Long insisted not only that the Africans sent to Jamaica in the transatlantic slave trade were rescued by British benevolence from the Hobbesian world of African cruelty and superstition, but that they generally experienced a mild slavery under caring and paternalistic masters who, Long asserted, had a strong pecuniary interest in making the work and lives of African enslaved people in Jamaica relatively gentle. It was in the planters' best interest to be "humane," Long insisted, not just because they were naturally generous people—"there are no men, nor orders of men, in Great Britain possessed of more disinterested charity, philanthropy and clemency, than the Creole gentlemen of this island." They looked after their enslaved property also out of self-interest, "so as to preserve their vigour and existence," knowing that "since Negroes are the sinews of West Indian property, too much care cannot be taken of them."[3]

Long went further than just defending planters against how "they have been unjustly stigmatized with breaking their Negroes with barbarity." He thought it typical of the British to paint their valuable colonial cousins "in the most miserable colours, and represent their enslaved people as the most ill-treated and miserable of mankind."[4] He argued that "a planter smiles with disdain to hear himself calumniated for tyrannical behaviour to his Negroes" when his enslaved property saw him "as their common friend and father." Indeed, a planter's authority over his enslaved people was "like that of an ancient patriarch: conciliating affection by the mildness of its exertion and claiming respect by the justice and propriety of its decisions and discipline." Such wise management, he thought, "attracts the love of the honest and good, while it cures the worthless into reformation."[5]

Long was convinced that Africans were much better off in Jamaica than in Africa. He noted that he "once interrogated a Negro, who had lived several years in Jamaica, on this subject." The man replied to Long that in Jamaica he had "food and clothing as much as he wanted, a good house and his family about him, but in Africa he would be destitute and helpless."[6] He repeated such a story in evidence to the House of Commons in 1791 when he told a story of the late Speaker of the House of Assembly, Sir Charles Price, who had a slave who "had been a man of consequence in Africa." Price, conscious of social hierarchy, even among Africans, told the enslaved African that he could return home. The slave supposedly refused the offer, saying that he wanted to stay in Jamaica "with a young negress" he was attached to. The slave, Long noted, was freed later in Price's will mainly because in Tacky's Revolt in 1760 he persuaded Coromantees owned by Price "not to join with his mutinous Countrymen."[7]

Enslaved people, Long argued, followed their masters in their disposition, and when masters were "humane and indulgent" enslaved people were happy; when masters were tyrants they were unhappy. For Long, it made little sense for masters to be tyrants as "it being so opposite to the interest of any planter thus barbarously to treat, or inhumanely to work his enslaved people to death; if ever such instances of cruelty happens, the owner is, without doubt, either a fool or a mad-man."[8] Indeed, "many Negroes in this island, the tradesmen and such as are casually called House Negroes live as well, or perhaps much better, in point of meat and drink, than the poorer class of people do in England and not one of them, even to the plantation labourer, goes through half the work; for even those who cultivate the lands, are not without indulgence, and frequent intervals of recreation."[9] Such good treatment, Long insisted, resulted in a grateful and well-off slave population. In evidence presented on his behalf to the House of Commons committee investigating the slave trade in 1791, he told two stories about enslaved people supporting planters down on their luck through monetary and other gifts. One planter, he had been told by a member of Jamaica's Council in 1779, had felt so sorry for their master being unjustly sued that they assembled in front of the planter's house and "offered him a large sum of money in bags made of old stockings."[10]

I argue in this chapter that Long was wrong. The critics of Jamaican slavery were right, and indeed Jamaican enslaved people were "the most ill-treated and miserable of mankind."[11] If this refutation of Long is all that were attempted here, however, the result would be disappointing. Long's attempt to describe planters in the best possible light and his assertions that he had "never known and rarely heard of any cruelty either practised or tolerated by them over their Negroes" while arguing that "the just subordination within the line of which our Negroes must be kept, does by no means dispense with our loving and treating them humanely" can be quickly dismissed as self-evidently self-serving and incorrect.[12] His belief that Jamaican slavery was mild and Jamaican planters "humane and indulgent" are betrayed by constant evidence to the contrary from contemporaries and modern historians. Charles Leslie's condemnation of Jamaican planters in 1740—"No country excels them in a barbarous Treatment of Slaves, or in the cruel Methods they put them to death"—and Governor Edward Trelawney's anonymous comments in 1746 that Jamaican planters were so cruel that they should not be allowed to colonize other islands and "that nowhere else were slaves so completely at the mercy and caprice of their of their masters"[13] are more accurate than Long's whitewash of Jamaican slaveholders. Modern studies of eighteenth-century slavery suggest that Jamaican slavery tended to the "Hobbesian" rather than the "Panglossian" extreme of slave treatment. Richard Dunn's claim that "Caribbean slavery was one of the most brutally dehumanizing systems ever devised" is apt.[14] Edward Long was wrong but influential. His views formed the basis of proslavery belief and a persistent argument that Jamaican slavery was mild rather than one of the most extreme forms of exploitation. In one way, outlining his views is useful in understanding how white Jamaicans were able to mount a counter-defense to abolitionist arguments that Jamaican planters were naturally cruel.[15]

One part of Long's analysis, however, makes apparent sense in a modern climate, which is the assertion that it made no sense for white Jamaicans to mistreat their enslaved "stock" to an extent that this "stock" would decrease, necessitating frequent and expensive annual purchases of new laborers from Africa. But what Long thought was logically foolish behavior actually made sense in the dark logic of Jamaican slavery before the start of abolitionist pressure in the late 1780s. I explore here the logic of a slave management system, perfected in Jamaica in the economically booming years between the Seven Years' War and the start of the American Revolution, that was highly destructive of the lives and health of enslaved laborers, especially on sugar plantations.

Michael Craton estimates that around 1770 perhaps as much as 75 percent of Jamaica's enslaved population worked in activities associated with sugar.[16] The slave management practices adopted by Jamaican planters and their plantation operatives were distinctive and debilitating. The significant increases in wealth enjoyed by fortunate owners of Jamaican property after 1750 depended on an increasingly hard-driving slave management system in which enslaved people in Jamaica experienced some of the worst standards of living in the early modern world.[17] The poor treatment of enslaved people was neither accidental nor, as Long suggested, the result of unfortunate but unrepresentative sadism carried out by British-born overseers with a "savage disposition"—"barbarians" who were "imported from the liberty-loving inhabitants of Britain and Ireland."[18] Cruelty was perhaps less common than it had been by the 1760s and 1770s, but it was everywhere evident. Mistreatment was systematic, deliberate, and intended to maximize the productivity of the plantations at the expense of enslaved laborers.

In short, Jamaican planters and their employees developed a system of "just-in-time" slave management that placed short-term imperatives designed to maximize profits through a ferociously hard-driving slave system over the long-term interests of maintaining slave populations by moderating the amount of work enslaved people had to do.[19] Planters and their employees were encouraged in this approach to slave management by being able to augment labor forces through buying enslaved people in the Atlantic slave trade. Moreover, planters consigned to enslaved people the greatest responsibility for feeding and housing themselves and their dependents. This system of maximal crop production and limited attention to self-sufficiency depended on supplies coming from North America and Europe to sustain slave populations, meaning that Jamaica was highly vulnerable to any supply-side disruption, either in provisions coming from abroad or in how enslaved people were able to maintain themselves. What was paramount was the production of tropical crops for market, an ambition that was pursued even at the expense of rendering the gap between the system working and the system failing perilously slight.

That Jamaican wealth increased considerably during this period was remarkable, given that the second half of the eighteenth century saw a long-run fall in sugar prices accompanied by a sharp rise in slave prices and a probably great increase in planter indebtedness resulting from planters having to fund these expensive new additions of labor.[20] This rise in slave prices relative to the price of sugar meant a steep increase in labor costs. Growth in profits implies that the slave system became increasingly efficient. The average productivity of enslaved people working in sugar cultivation, as measured by the pounds of sugar produced per worker, doubled between 1750 and 1810, with productivity increases especially apparent between the end of the Seven Years' War and the start of the American Revolution. Eltis, Lewis, and Richardson estimate that if we assume a production function for productivity as 1.00 in 1700, then the production figure for the western Caribbean increased by 600 percent between 1700 and 1790, despite the region having to absorb extraordinary numbers of new enslaved people. Between 1750 and 1790, the production function increased from 4.88 to 7.[21] It rose again after the end of the American Revolution, with a peak in the early 1790s as Jamaica took advantage of the implosion of the plantation system in Saint-Domingue.[22]

The American Revolution, as well as a series of devastating hurricanes from 1780 to 1785, proved a major challenge to the viability of this hard-driving system, as also did the advent of abolitionism from 1787–88. Historians are divided about whether slave owners responded to these difficulties by voluntarily adopting policies of amelioration designed to preserve slave health through providing them with better conditions of work and maintenance.[23] What is very clear, however, from testimony presented to the House of Commons in their investigation of the Atlantic slave trade and West Indian slavery in 1791, is that slavery in Jamaica before the beginnings of any amelioration in the 1790s was particularly harsh.[24] That harshness was also exemplified in the little empirical evidence we have about the physical welfare of enslaved people. Jamaican enslaved people had stunted heights compared to other working populations, enslaved and free, in British America.[25] Jamaican slavery was always hard-driving, but the evidence we have suggests that the slave management strategies adopted between the Seven Years' War and the end of the American Revolution were especially deleterious to slave health and welfare. The only worse period of slavery in Jamaica was in the late seventeenth century when plantations were being formed out of the tropical forests that covered the island in the early settlement period.[26]

—

Evidence presented, by people with long experience of Jamaican slavery, to the House of Commons in their exhaustive investigation into the Atlantic slave trade and into conditions of slavery in the British West Indies indicates that not every Jamaican resident shared Long's low opinion of the moral and physical qualities of enslaved Africans. Moreover, the evidence presented in volume 82 of the *House of Commons Sessional Papers*—a volume heavily devoted to accounts of the conditions of enslavement in the British Caribbean—points to a different interpretation of what life was like for enslaved Jamaicans than that promulgated continuously to a skeptical British public by proslavery representatives like Edward Long. This evidence suggests that the assertions made by abolitionists such as James Ramsay were correct insofar that the system of slave management in the British Caribbean was designed explicitly to keep enslaved people overworked, overpunished, malnourished, morally deprived, and desperately unhappy.[27]

Of course, we need to remember in analyzing these testimonies that they are selective evidence insofar as the House of Commons committee chose witnesses who were inclined to answer along lines of inquiry carefully chosen beforehand by the committee.[28] The witnesses called before this committee were thus far from impartial. Their testimony reinforced conclusions already arrived at by parliamentarians opposed to the slave trade. Nevertheless, the evidence they presented was powerful, compelling, very detailed, and specific, based as it was on lengthy personal experience in Jamaica as military officers, planters, doctors, and overseers. Some witnesses were longtime residents with at least as much experience of enslaved Jamaicans' lives as Long. Indeed, their testimony is usually more convincing than Long's, not only because it was less obviously based on racist prejudice (the witnesses when asked to give an opinion on enslaved persons' capacity tended to say that it was equal to that of unlettered white people) but because many of the witnesses, notably William Fitzmaurice and Henry Coor, who gave the most convincing and lengthy answers to the parliamentary committee, had extensive firsthand experience of managing enslaved people. The evidence presented by these witnesses to the committee was damning about the effects on enslaved people of what was described repeatedly as an especially hard-driving and pernicious slave management system in the years before and during the American Revolution.[29]

A constant theme in the testimonies was that Jamaica was a place of shocking violence, of extreme treatment of field enslaved people as workers, and of pervasive white immorality. The nature of the island took some getting used to, as witnesses stressed when asked what their first impressions of Jamaica were. Henry Coor, a millwright who lived in Westmoreland Parish between 1759 and 1774, made explicit how migrants became accustomed over time to the peculiar nature of Jamaican slavery. "At my first coming to the island," he stated, "a common flogging of a Negro would have put me in a tremble, and disordered me so, that I did not feel myself right again generally the remaining part of the day." "But by degrees and customs," he admitted, "it became so habitual that I thought no more of seeing a Black man's head cut off than I should now think of a butcher cutting off the head of a calf."

That enslaved people were like cattle was a common first impression. Captain Thomas Lloyd of the Royal Navy, who was in Jamaica in 1779, declared that "the impression made upon my mind was, that they [Jamaican enslaved people] were very generally considered as black cattle and very often treated as post-horses." Similarly, Mark Cook, who lived in Jamaica between 1774 and 1790 as an overseer, clerk, and schoolmaster, was "shocked very much" at the violence meted out to field enslaved people whom he thought "are hardly better looked upon than beasts." Indeed, "the cattle are often treated better than enslaved people" because at least cattle were fed and did not have to cultivate provision grounds for their food. Dr. Jackson, from Savanna La Mar in Westmoreland, opined that "negroes were generally esteemed as a species of being inferior to the whites [and] as a race of people whom the right of purchase gave the owner the power of using according to his will."

Several witnesses commented on how they were accosted on first arrival by signs of extreme cruelty by slave owner to slave. Captain Alexander Scott of the Royal Navy witnessed a shocking scene on the second day after his arrival when a white man chased an enslaved man into the water and having brought him out "hung him by the hands by a crane" and attached large weights to his feet before having him whipped "with a kind of prickly bush, I believe ebony." The extreme punishment came about merely because the man had taken too long on an errand. His fellow naval captain, Thomas Lloyd, related a similar incident on his first day where two plantation enslaved people were executed "on the wharf in the sight of the ship's company," the man for running away and the woman for harboring him.

One poignant vignette was the first impression of Thomas Claxton, who was full "of horror on viewing the Field Slaves, some of whom worked in irons, under the lash of an inhuman Negro driver, and their backs, in general, much lacerated from the blows they had received." Field enslaved people, it was agreed, had the worst lives in Jamaica. Robert Ross, who lived in Jamaica between 1762 and 1786, noted that most field enslaved people were marked with a whip, a statement confirmed by Henry Coor, who thought, moreover, that "it was impossible for a Field Slave to be possessed of any property." The soldier, Captain Giles, resident in Jamaica between 1782 to 1790, thought field enslaved people' treatment "generally severe" and worse than in the army. Mark Cook believed that field hands were punished "worse than domestics" and noted that he had seen "a field Negro receive 200 lashes by order of the overseer" while another field hand who had stolen part of a turkey was held down and his master knocked out four of his teeth "with a hammer and punch." The slave had stolen the turkey, Cook thought, only to manage his hunger—"he was nothing but skin and bones." And the master who knocked out his teeth was not considered a monster. Cook thought that he was "not reckoned cruel."

Field hands suffered most because they were extremely overworked. Thomas Lloyd was told by Captain Cornwallis that "at a dinner with some of the principal planters, the conversation turned upon the profit and loss of sugar plantations and that a Planter of consequence, then present, said that during the crop season he worked his Negroes twenty hours out of 24" and while some of them "must have died of fatigue" that "upon the whole it answered." Mark Cook believed that during crop, "negroes work 18 hours and have injuries through fatigue." They were forced to work in moonlight and had to pick grass after a day's work. They were sometimes so tired that they fell asleep at work—to dreadful consequences on occasion, as happened to one girl who lost her hand when she fell asleep while feeding cane into the mill. It was during this period that it seems that planters started making the work of enslaved people during harvest creep into Sunday, keeping sugar mills running constantly, so that enslaved people did not get a full day of rest. Keeping Sunday for rest became a major abolitionist objective, but evidence from early nineteenth-century Berbice suggests that enslaved people were often worked on Sundays well into the ameliorationist period.[30]

James Towne of Liverpool, who had been in Jamaica in 1772, stressed that field enslaved people in Jamaica were worked very hard, "turned out by four o'clock in the morning and worked to a very late hour in the evening, besides having other work to do at home, after they cannot work any longer by daylight." William Fitzmaurice, who lived in Jamaica between 1771 and 1786, gave a comprehensive summary of the work that field enslaved people commonly did.[31] Fitzmaurice described the common hours of working, starting with shell blows for turning out at either four or five o'clock and work until 10 A.M., before fifteen minutes for breakfast, then more work until 1 P.M., when the shell blew for dinner. After a one-and-a-half-hour break to make and eat their dinners, and then a further half hour to return from their provision grounds to the field, enslaved people worked until dusk at around 6 P.M. They often had to run to their grounds at dinner, as these grounds might be far away from where they were working. After working in the field, they had to tend to cattle pens or pick grass. They returned home at eight P.M., constrained by having to wait "until the slothful Negroes are brought up," which caused, Fitzmaurice believed, "uneasiness to the Negroes in general." He noted, moreover, that enslaved people had many other tasks to do besides sugar cultivation, such as helping tradesmen in preparing their work or in carrying moulds to cattle pens. These tasks were called "before-day-jobs which must be done so far as to not break in upon the general work of the plantation." Fitzmaurice noted also that on "estates that are weakly handed, which is the case by far of the greatest part of the island," at crop time enslaved people also had to work long shifts in the boiling house. The boiling house operated all night and day during crop time, without intermission, stopping only on Sundays.

One problem in the way that enslaved people worked was gang labor, as the custom was to work to the speed of the most skilled and able bodied. Fitzmaurice believed this led to many enslaved people being "hurried to their graves" as the able enslaved people left the weak behind, leading to the weak slave being severely flogged, considered worthless and kept all afternoon "to bring up his row." James Towne expanded on this theme, noting how enslaved people were punished if late to the field and that drivers were focused on preventing delays in work in the cane fields. Thus, he contended, enslaved people who are lame are "yet obliged to work." They were thought lazy when they got behind and were "immediately flogged." Henry Coor expounded at length on the exhaustive discipline exercised in the fields. He noted how enslaved people arriving late into the fields received a "slight whipping" of ten to twenty lashes, usually standing up or occasionally stretched on the ground if a slave proved recalcitrant. "Flogging time" was eight o'clock in the morning when the overseer ordered a driver to flog "criminals" kept in the stocks overnight. He thought that sometimes "these poor creatures" received 100 to 150 lashes and occasionally "two cold hundreds, as they are generally called by the overseers."[32] After being whipped, many enslaved people were sent to the field to work all day "without meat or Drink, save water." Coor believed that "this cruel treatment of whipping, hardworking and starving, has, to my knowledge, made many of them commit the dreadful act of suicide." Dr. Harrison, resident in Jamaica between 1755 and 1765, added that "I have frequently seen Negroes in chains, emaciated and half famished, scarce able to drag one leg after the other, and yet compelled to go into the field."

The heavy demands of field work influenced how enslaved people were able to work their provision grounds, usually the main source of their food. Dr. Harrison argued that enslaved people had only holidays and Sundays to cultivate provision grounds. He argued that this time release from sugar cultivation was not sufficient, as it meant enslaved people having to work in the dark. The result was that "enslaved people looked very indifferent and often plunder estates." Fitzmaurice believed that overseers paid "particular attention to the Negro grounds," punishing enslaved people who "were slothful in cultivating their grounds or who are troublesome in coming up to slaves to ask for victuals."

To the normal problems of overwork and tiredness could be added three special problems specific to Jamaica. The pressure to produce ever greater quantities of sugar and rum led owners to convert provision grounds into cane fields, a trend that intensified in the 1790s when the introduction of new varieties of cane (Bourbon cane) meant that sugar could be grown in less fertile soils. They also made provision grounds ever more distant from the places where enslaved people worked during the day.[33] Planters also used the least fertile grounds on plantations as provision grounds. Mark Cook noted that enslaved people often had their grounds taken away, which he thought was "very hard upon them." Fitzmaurice argued similarly, stating that provision grounds near cane pieces were always converted into sugar land, which meant that enslaved people then needed to cultivate new land. He believed that this practice "is attended with the greatest destruction to the Negroes" as "they go about new grounds with great reluctance." He added that he himself had "changed Negroes to live in a healthier situation, and I have lost many of them from the effect of the change upon their spirits."

Africans who found it hard to work their grounds for whatever reason often caused dissension within slave communities, according to white witnesses.[34] Henry Coor commented that "the poorer sort, who have never had spirits or ability to cultivate their own allotment, depended upon some of the plantation Negroes, for whom they worked all the little time that they were excused from their master's business." But if a quarrel arose between them and "the Master Slave" then "they are turned out of doors, and obliged to steal or beg, or get their food any way that they can come to it." In Coor's opinion, "this is the greatest reason why there are so many bad enslaved people." The principal problem, he felt, was not that enslaved people did not have sufficient ground to work but "they have not time to work it." Modern historians would emphasize that it was not just time but energy that was a major problem for enslaved people. Enslaved people, Roberts argues, were regularly worked to exhaustion. What planters saw as laziness or inherent bad qualities was exhaustion compounded by malnourishment.[35]

As Roberts also notes, power hierarchies and wealth inequalities operated powerfully on sugar plantations and disrupted slave communities. These intra-slave contestations put some enslaved people, notably the poorer, unhealthier, and older enslaved, and especially African-born enslaved people, at considerable material disadvantage compared to usually native-born enslaved people with extensive kin or communal networks. The provision ground system in Jamaica accentuated social strife between enslaved people and made slave communities often tense places where enslaved people strived to obtain advantages, especially when resources were strained. Privileged or "confidential" enslaved people acted as shop stewards, regulating slave behavior. Watchmen, in particular, were notorious for their brutality, reputedly being cruel and willing to kill and beat hungry enslaved people caught stealing food.[36] The privations of Jamaica between 1779 and 1786 were extreme. Hurricanes destroyed provision grounds, which had to be recreated from scratch, often when imported provisions from North America were in very short supply. The effect on enslaved people was to greatly increase their workload while they were famished. It made competition for scarce resources intense.[37]

Simon Taylor recalled in 1789 how after the hurricane of 1781 he met enslaved people from a western sugar estate "hunting about in the Savannas for Sweet sops who told me they were starving." Hector McNeill recalled in 1788 how in 1780 he had been horrified by "the misery of beholding hundreds of wretched beings wasting around you, clamouring for food, and imploring that assistance which you cannot bestow." The Jamaican House of Assembly summed up the effects of the recurring hurricanes, storms, and droughts: "the plantain walks, which furnish the chief article of support to the negroes, were generally rooted up, and the intense Droughts which followed, destroyed those different species of ground-provisions which the hurricane had not reached. . . . The sufferings of the poor negroes, in consequence thereof, were extreme."[38]

In the Blue Mountain estate, for example, a hurricane in 1786 destroyed much of the plantain grove, meaning that enslaved people had not enough food. The "better" sort of enslaved people—tradesmen, drivers, and their families—led the way in establishing new provision grounds. They made the "poorer sort" work these grounds in return for produce. But the effort failed as poorer enslaved people did not receive enough provisions to sustain themselves and as the better-off enslaved people hoarded provisions for themselves.[39] William Beckford, a wealthy proslavery sugar planter who wrote a handbook for fellow planters, argued that such arrangements were common. "Old Negroes" often forced "new

Negroes" to work for them in systems of coercion or tribute. New enslaved people, Beckford thought, were overworked in the fields and then had to learn how to
cultivate their grounds and establish a home.[40] African-born enslaved people were
often abused by established slave families, seldom became tradesmen or drivers,
and, as Roberts notes, "were destined for decades of service in the field where it
would take time to build the strategies, skills and stamina necessary to work in
slave gangs." He concludes that "inequities in the material conditions of life[,]
competition for privileges and limited resources could produce discord and violence within slave communities."[41]

Henry Coor also felt that the increase in jobbing—the hiring of additional
enslaved people on estates, usually owned by overseers or men without estates—
was pernicious. Enslaved people in jobbing grounds were put in very tough jobs
so that their owners could maximize profits, which Coor estimated as being at
least 14 percent on capital and usually much more. Captain Giles, a soldier in
Jamaica between 1782 and 1790, thought that the labor assigned to enslaved
people in jobbing gangs "was more than human nature could support for any
considerable time, because their allowance of provisions, which I had daily an
opportunity of seeing was not equal to support them under it." He estimated that
"a jobbing gang would last seven years, so as to bring a profit to the owner." Such
an estimate suggests excessive mortality in these gangs over even mortality in
sugar estates.

Jobbing gangs did the most arduous activities in plantation labor, notably the
back-breaking work of "holing," or digging deep holes for the planting of sugar
cane. Sugar planters did not like using their own plantation enslaved people for
such activities as it harmed their health. To an extent, the prevalence of jobbing
gangs showed that planters had some concern for the welfare of enslaved people,
or at least some concern about not having excessive mortality among their
enslaved labor forces in a period of rising slave prices. Planters tended to think of
enslaved people as additions to capital stock and not as an operating expense.[42]
But what was of most concern was to maximize production while not destroying a
labor force. Thus, planters subcontracted out arduous jobs like holing to men
from lower socioeconomic classes who saw owning enslaved people as a pathway
to wealth that was impossible to get through waged labor. Jobbers bought
enslaved people at high prices on short credit. The only way that they could
justify their investment was to work them incessantly, getting high profits from
their gangs (they charged premium rates, costing twice as much as an equivalent
team of day laborers hired as individuals), and relying on those profits to be able
to afford to replace worn-out enslaved people with fresh inputs of labor. The
health of the one in ten enslaved people in jobbing gangs was generally poor and
their demographic composition unusual—they were almost all prime working
hands with virtually no children or elderly enslaved people and were invariably

African rather than Creole. The jobber was thought to be especially "cruel," though that cruelty is better described as ruthless pragmatism, with jobbers making careful decisions about how hard to push or preserve enslaved people depending on how much profits they needed to make to serve the large debts incurred in buying a jobbing gang. What they did was to make working in sugar as a jobbing slave even more onerous than it was for enslaved people in general, given relentless short-termism. One planter-attorney commented that the "jobber thinks only of immediate profits; he never thinks of the slow mode of increasing the value of his gang by natural increase." It often worked. Lewis Cuthbert told Parliament he knew "many proprietors of sugar estates, now in opulent circumstances" who had started out as jobbers. The effect on enslaved people, however, was disastrous. It was not uncommon for a jobber to lose half his enslaved people to an early death in just a few years' hard labor. William Beckford marveled that "in the hands of jobbers, it is amazing what numbers of negroes die."[43]

It was not just that the work was onerous beyond the endurance of many enslaved people. The food that they received was inadequate and the housing substandard, and enslaved women were preyed on by whites relentlessly. Coor noted that enslaved people were fed with putrid herrings and otherwise found it hard to feed themselves through their provision grounds. Mark Cook noted that enslaved people ran away frequently mainly "due to hunger." Thomas Clappeson, resident in Jamaica for fourteen years between 1762 and 1789, admitted that he had sold provisions "of very bad quality for the use of Negroes, such as damaged corn and flour, and the like." Enslaved people were not fed enough, and slave owners bought provisions "for cheapness, when there is better to be got." He cited enslaved people telling him when provisions were scarce that "hungry da kill me" and argued that "negroes complained of [bad provisions] every time they came to fetch it from the wharf." The provisions were so bad that Clappeson said that "a neighbour of mine told me his hogs would not eat it" and that enslaved people had to "steal what they can eat or Drink and necessaries they want." Bad food, Fitzmaurice argued, led to death from enslaved people getting fluxes from eating rotten vegetables and bad flour.

The system of food maintenance for Jamaican enslaved people was predicated on importations from North America and Europe and on extensive slave cultivation of provision grounds. It was precarious in the extreme. The dangers inherent in this strategy became clear during the American Revolution when a halt to provisions coming from North America, captures at sea of ships from Europe, and privation through hurricanes caused what Richard Sheridan has called a "crisis of slave subsistence" when perhaps thousands of famished and emaciated enslaved people perished.[44] Thomas Irving, inspector general of the exports and imports of Great Britain, testified that during the American Revolution "the planters, impelled by necessity, were obliged to deviate from their previous system, and to turn their attention more to raising provisions upon their own estates." Irving believed that "the good effects of this plan has been so forcibly felt, that the importation of Indian corn, which may emphatically be stiled the bread of life, with respect to the food of the Enslaved people, is reduced from 600,000 bushels, the quantity imported before the war to somewhat under 300,000 bushels." Irving thought that the lesson of the war was to illustrate the foolishness of Jamaica's lack of concern with self-sufficiency: "in a political sense, I conceive that no country capable of producing corn to feed itself, ought to be dependent upon any other, for any article which it cannot do without, even for a day."

The enslaved people who fared worst from this inadequate method of feeding and housing enslaved people were elderly enslaved people ("superannuated" in the parlance of the times), pregnant women, and those enslaved people who were owned by planters who were in excessive debt. Several witnesses commented on the sad fate of superannuated enslaved people who, "if they have no immediate relations among the other Enslaved people," must live "at the corner of a cane field" where they received only a few plantains "to preserve them from dying of hunger." They were "generally dirty and emaciated to the last degree." Mark Cook lamented that superannuated enslaved people were given no allowance "except what they can get among their children or relations." He had "seen them wandering about on the beach, left to take care of themselves as well as they could." On occasion, this lack of care turned into sadism. Both Henry Coor and Thomas Clappeson told stories of slave owners persecuting "an old decrepit woman slave." Clappeson remonstrated with the owner "without effect" that it was unfair he "would not allow anything" to this old woman.

Coor thought that pregnant women suffered badly from overseers never giving "any proper preparation for the reception of infants." The houses that enslaved people lived in were so bad that neither mother nor child had protection "against the cold damps of the night." He thought "it would be proper to leave it to the gentleman of the faculty to judge what effect the milk from a woman who has been exposed to hard work and poor living and drinking nothing but water under the perpendicular rays of the sun would have, whether that could be suitable for a tender infant, just come into the world." Most infants, he thought, "die convulsed, and generally about the eighth day." The House of Commons committee was very interested in whether planters who had taken on large debts, as was alleged to be often the case from the 1760s onward, were prime movers in making Jamaican slavery overly hard driving.[45] Witnesses were unanimous that the enslaved people owned by planters who were financially embarrassed faced bad treatment because, as Dr. Harrison noted, "their distressed circumstances obliged them to work their enslaved people beyond their strength, to make sugars to pay their debts." Lieutenant Baker Davidson, of the Seventy-Ninth Regiment, resident in Jamaica between 1771 and 1783, noted that enslaved people in such situations "used to come in the night time and rob of us of provisions of any kind they could lay their hands upon."

Their terrible treatment made enslaved people very unhappy. In addition to a multitude of cruel punishments meted out to enslaved people regularly, as almost every witness testified to at length and with plenty of specific examples, and in an exploitative sexual culture where black women were fair game for lusty overseers, overwork weighed on enslaved people' minds. Dr. Jackson stated that enslaved people "complained frequently that they were an oppressed people; that they suffer severely in this world; but expect happiness in the next." Baker Davidson argued that "Negroes express a great deal of pleasure when they think they are going to die, and say, that they are going to leave this Buccra country."

Suicides, it was alleged, were common. Most witnesses were able to give anecdotal examples of enslaved people killing themselves. Dr. Harrison told of "a negro who had been a man of consequence in his own country" and who "refused to work for any white man." He was punished and was being sent to another plantation when "going over a bridge, he jumped headlong into the water and appeared no more." Robert Ross, who lived in Jamaica between 1762 and 1786, believed that enslaved people killed themselves on first arrival, "when they understood that they were in a state of slavery." Some made away with their lives when "they saw their fellow creatures punished and that they thought it might be the case with themselves in a short time." Witnesses believed that only Africans, not Creoles, committed suicide. Henry Coor believed that enslaved people from the Gold Coast cut their own throats while enslaved people from inland Africa hanged themselves. William Fitzmaurice thought that the principal means of suicide was hanging and dirt eating. He noted that he lost in one year a dozen new enslaved people through dirt eating. "When I remonstrated with them," he commented, "they constantly told me, that they preferred dying to living."

Jamaican slavery, therefore, was particularly dreadful and much worse, witnesses believed, than slavery elsewhere in the British American world. Dr. Harrison had lived in South Carolina after 1765 and thought that Jamaican enslaved people "were in general treated very ill," while in South Carolina they "were in general treated very well." He summarized his views as follows: "in South Carolina they were well fed, well clothed, less worked and never severely whipped; in Jamaica, they were badly fed, ill clothed, hard worked and severely whipped." Henry Coor added that he thought that enslaved people in the northern provinces of British North America were treated so well compared to Jamaican enslaved people that they might be compared to farmers' servants in England.

—

It was a mystery to the interlocutors in the House of Commons that slave owners would tolerate such a hard-driving system that was so destructive of their very valuable slave assets and that meant that planters were always having to replenish their human capital through purchasing expensive new Africans in the slave trade. The prices of enslaved people bought from the Atlantic slave trade soared in the 1760s and 1770s, rising from £31 on the eve of the Seven Years' War to £43 on the eve of the American Revolution and to £60 between 1785 and 1789. Sugar prices hardly rose between the Seven Years' War and the start of the American Revolution, making enslaved people relatively more expensive over time. It was only in the 1790s and the advent of the Haitian Revolution that sugar prices rose enough to justify ever more expensive prices for enslaved people.[46] Parliamentarians questioned witnesses vigorously about whether enslaved people who were well treated were likely to have longer life expectancies and thus obviate the need for planters to make constant purchases of new enslaved people. The witnesses, without exception, affirmed the prejudices of the committee, giving detailed examples of plantations presided over by "humane and indulgent masters" in which favorable treatment of enslaved people led to populations increasing without the help of the Atlantic slave trade. They also argued that "the use of the plough would save a great deal of labour of the Enslaved people" and "that if more encouragement was given to the Enslaved people, by treating them with more humanity, they would do considerably more work." Much of the blame for why planters never thought about treating enslaved people humanely and indulgently was attributed to widespread owner absenteeism. There was widespread agreement that "the residence of the Planter on his own estate was of the greatest advantage to his Enslaved people, as they were always used better for it and were shewn more attention in every respect."[47]

Witnesses believed that one major problem with Jamaican slave management was what they considered perverse incentives given to overseers and attorneys. Attorneys were particularly to blame, as they were paid by commission based on the size of the sugar and rum crops produced and thus had a strong incentive to prioritize the making of large crops over preserving the health and welfare of the enslaved labor force. Coor argued that if an overseer produced an outstandingly large crop, "the attorney winked at his pressure to perform more work than human nature was able to do." But owners could be equally blamed for this set of priorities, depending as they did for their luxurious lives in England on the large remittances they received from their plantations in Jamaica.

Little was done to make work easier so that enslaved women could have children. Captain Hall, who had lived in many places throughout the West Indies, including Jamaica and Saint-Domingue, argued that "breeding was by no means thought desirable." Overseers, he believed, thought that "the charge of rearing a slave to the state of manhood more troublesome and greater than the buying of a slave fit for work." He continued, associating Africans with animals, that it was "no uncommon thing for them to give away a child of two years old, as you would a puppy from a litter." "It was the overseer's business," he contended, "to make as much sugar as possible, and to effect that purpose, it was his business to work the Enslaved people to the utmost, it being of no concern to him whether they died under their work or not, as he was sure to be supplied with others."

Overseers, it was universally agreed, were encouraged by the structures of how they were rewarded and judged to push enslaved people as hard as they could if that resulted in maximizing the size of an annual sugar and rum crop. If this process meant that along the way overseers indulged in sadism and cruelty toward enslaved people and were liberal in how they punished enslaved people, this was of no matter, first, because enslaved people had no redress to a magistrate but could complain only to attorneys and overseers, and, second, because few people condemned people who were deemed cruel. Robert Ross was the only person who claimed that "overseers were frequently turned out of their places for over-whipping, when there is a complaint to the master or magistrate, and on that account the overseers are more lenient about punishment than they were in for-mer time." He contradicted himself almost immediately, however, in claiming that he "never understood that that the evidence of a Negro would be good against a white Man" and that "often overseers gave over 200 lashes" to enslaved people and "were never called to account for it." John Shickle, a wealthy planter and attorney to the extensive Pennant estates in Clarendon Parish, was reputed by Ross to have "flogged a Negro at three different times in one day," and "no public notice was taken of it." He accused a man whose name was redacted in the records by order of the committee of having "hanged a Negro Man on a post close to his house, and in the course of three years he had destroyed 40 out of 60 by severity and was not called to any account."

Henry Coor laid out the reasoning behind overseers' lack of attention to the rearing of children with crystal clarity. He stated that he had "heard many of the overseers say, that they would far rather the children should die than live." "It was more the object of the overseers," he thought, "to work the Enslaved people out, and trust for supplies from Africa." He had "heard many of the overseers say, "I have made my employer 20, 30, or 40 more hogsheads per year than any of my predecessors ever did, and though I have killed 30 or 40 Negroes per year more, the produce has been more than adequate." Baker Davidson added that owners of estates say "that a Creole slave when fit to work costs them more money than a new Negro." For Mark Cook, the argument was essentially circular: the overseer's job was "making larger crops in general, and working the Negroes hard; if he does not work them hard, he cannot make large crops." Overseers sometimes worked for a stated annual salary but also worked "for so much per hogshead of sugar and so much per puncheon of rum."

William Fitzmaurice repeated an "old story in St. Thomas-in-the Vale that if a Negro lived seven years he paid for himself." This story refers to imported enslaved people rather than native-born enslaved people, who became profitable to planters only when in their teens. He repeated statements "from several propri- etors" that "they considered every child born and reared upon the estate to be a dear Negro" and that overseers disliked women becoming pregnant "on account of its interfering with the work expected from the Negro women" although he admitted that "other proprietors I have known take a pleasure in seeing the women breed." Fitzmaurice was forceful about what needed to change in order that "breeding" could be encouraged. "Breeding won't work," he concluded, "without changing the present system of management and adopting the plough and putting in the canes after the plough."[48] At present, he thought, overseers were "encouraged to push Negroes." He related a story of "a gentleman in the parish of St. John" who employed him and with whom he later quarreled because of an excessive punishment to a slave whom Fitzmaurice had thought had done nothing wrong. His employer had "a shorthanded stock of Negroes and cattle." He ordered Fitzmaurice "to drive them without mercy, as the loss of a few Negroes and stock was no object in comparison to sending home his crop in time." This "gentleman" seems to have been a resident proprietor: it was not just absentee owners who pushed enslaved people hard to maximize income.

———

The results of such a hard-driving system were highly positive for the small per- centage of the Jamaican population who were involved in plantation agriculture and highly deleterious for most of the population who were enslaved. The free population prospered in the second half of the eighteenth century. The period between the Seven Years' War and the start of the American Revolution and once again after the 1780s until the end of the slave trade in 1807 saw Jamaican wealth at its height.[49] Some Jamaicans were seriously wealthy by the 1730s, but wealth was much more widespread from the 1750s. The Spring estate in St. Andrew Parish is indicative of increase of wealth for medium-sized estates. In the mid- 1740s, the estate made £414 per annum, but investment in additional cane fields and more enslaved people catapulted profits in the 1750s to £1,518 per annum and, after more investment after 1788, to annual profits of over £4,000, peaking at a magnificent £5,819 in 1792—one year after the House of Commons investi- gated slave management practices on the island.[50] Great estates were even more profitable. The huge estates of Richard Pennant made net profits of £6,400 in 1765, over £10,000 per annum in the 1770s, and £23,382 in the super-profitable year of 1792. In that last year the profit per laborer from 1,228 enslaved people was £19.04, meaning that even with slave prices rapidly rising and spiking at £60.23 between 1785 and 1789, an African paid for himself or herself in a little over three years.[51] Jamaica's wealthiest man, Simon Taylor, made the enormous sum of £56,000 in that auspicious year of 1792.[52]

The average personal wealth of inventoried Jamaicans increased by 45 percent between the period 1725–49 and the period 1750–84. As land prices also rose steeply in the period, increasing from £1.59 per acre in the period 1745–54 to £5.84 per acre in the period 1765–80 in St. Andrew Parish, the wealth of landowning Jamaicans increased even more than these figures suggest.[53] Sugar estates became seriously expensive to purchase. The vast sums of money and great risks required to buy an established sugar plantation in these economically booming years are illustrated in Simon Taylor's purchase of John Kennion's St. Thomas-in-the East plantation of Holland in 1771 for the enormous sum of £71,459. It was prime sugar land, probably the best estate on the island. Yet Taylor took an enormous risk in buying Holland, despite being the heir to a large fortune and probably the most skillful and hard-working planter of his generation. He needed to pay £10,000 Jamaican currency per annum for six years without interest and then eight further payments of £5,000 per annum with 5 percent interest. The debt was underwritten by Jamaica's largest and most sound commercial firm, Hibbert, Purrier, and Horton. Taylor agreed that he had paid "an Amazing Sum of money" for Holland, but he believed that he had bought "the best estate in the West Indies for size."

Results justified the risk: by the time Taylor died, in the same year as Edward Long, whose life he paralleled, he was the richest man in the British Empire, with an estate worth over £1 million. But Taylor could do what he did—Holland was making only around £4,500 per annum when Taylor bought it—only because he was thinking in terms of decades rather than of years. Taylor was one of the few white Jamaicans to think long term rather than living in the moment. He could buy Holland because he owned more than one sugar estate, which allowed him to spread his risk, enabled him to reduce costs, and caused him to be able to sustain his investment through economies of scale and the advantages resulting from some integration of activities. Nevertheless, the risk he took on was enormous. As Christer Petley argues, Taylor was engaged in Britain's riskiest and most expensive branch of capitalism, a planter's arcadia that offered glittering rewards to entrepreneurial risk takers who had good access to credit. It was always an undertaking that could lead to disaster—fortunes were as easily lost as gained in Jamaica. Taylor called the years between 1771 and 1776, when he was paying the large annual installments for Holland, his "years of purgatory" when he was constantly anxious and uneasy. "All my views tend to one object," he told his brother in 1775, "that is, completing the payments of Holland . . . my last thought when I go to bed, the first when I rise."[54]

White Jamaicans became wealthy in this period, sometimes, as with Taylor, very wealthy, but much of this wealth was built on a precipice of debt. It is difficult to get precise figures on this much-debated topic, the amount of debt held by West Indians at the start of the American Revolution, but it seems clear that

planters were highly leveraged, both to local moneylenders and to British merchants. Astute and hardworking planters like Taylor were able to use excess profits
from large crops to reduce their debts to manageable levels. Not all Jamaicans in
debt, however, were as well resourced, capable, skilled, and lucky as Taylor.
Thomas Thistlewood's employer, John Cope, wasted a substantial estate through
incompetence and bad luck so that he had to sell one of two sugar estates by the
1770s and ended up relatively poor at his death.[55]

I estimate that Jamaican moneylenders probably lent out £2.4 million in the
1760s and £3.5 million per annum in the period of the American Revolution.[56]
British moneylenders were just as willing to lend large sums to Jamaicans, based
on the soundness of the Jamaican economy.[57] They often needed to foreclose,
however, on planters who had overestimated their ability to service large mort-
gages. Between 1772 and 1791, 80,021 executions for £22,563,786 were lodged
in the provost marshal's office in St. Jago de la Vega on judgments in the Jamaican
Supreme Court.[58] There were strong reasons why overly leveraged planters
wanted to push the enslaved people on their estates to produce crops that provided a high annual return on their substantial investments. They also tended to
wait until after bumper crops to buy enslaved people and used jobbing gangs in
the interim between slave purchases, when estates were shorthanded, to make up
the difference in labor needs.

Simon Taylor railed against what he considered the short-term foolishness of
delaying necessary slave purchases until planters could afford them. A constant
theme in his letters as an attorney to his employer Chaloner Arcedeckne was the
need to buy more enslaved people. In 1768 he urged him to buy more enslaved
people "as they are excessively wanted for to carry on the estate" as Golden Grove
in St. Thomas East had "a great number of old Superannuated Negroes and young
children there, and the whole Plantain Garden River is deemed unhealthy even in
this country." If enslaved people were not purchased, he argued, it would be "a
pity" that "the estate should fall off for want of a sufficient strength of Negroes,
which if not put on must infallibly be the case." He believed that regularly "adding
to the strength" of an estate was "prudent management." He thought that the
purchase of "new negroes" was a predictable and occasional part of the running
costs of an estate, rather than an unexpected drain on resources, and unlike most
of his contemporaries he paid great attention to the preservation and reproduction of the workforce through buying, as in 1789, "new negroe women, or girls, to
see if they would breed." He bought ten enslaved women for Arcedeckne that
year, exclaiming that "they are Eboes which we think are the best breeding people." What he especially disliked was the common practice of making what he
considered "jumping crops" or sudden increases in crop size that he believed
could be made only "by distressing and harassing both negroes and stock." Such
excessive "pushing" of an estate happened because overseers were willing to have

enslaved people "worked to death" in order "to aggrandize an overseer's name by saying he made such and such a crop for a year or two." Short-term expediency led, he believed, to long-term problems for it is "very easy to destroy a good gang of negroes but very difficult to raise one" and each "jumping crop" meant that "it takes three years to bring matters into their proper channel."[59]

A principal component of increasing Jamaican wealth was human capital improvements. Investment in enslaved people increased from 45 percent of slaveholder wealth in the early 1730s to 57 percent of slaveholder wealth in the 1770s. David Ryden estimates that the average annual gross returns on slave investment between 1752 and 1807 was 17.8 percent, suggesting annual returns on slave investment on a conservative estimate of 8–10 percent, taking credit into account.[60] It made sense for slave owners to try and improve the human capital that produced those high returns. "Breeding" was one way of improving human capital, but, as the evidence of witnesses to the House of Commons suggested, this approach was too slow, depending on children growing to adulthood. It was also too uncertain, given high enslaved infant and adult mortality.[61] The most expeditious means of improving human capital was in focusing on making these already expensive slave investments even more valuable by improving the skills of enslaved people so that they fetched a larger price in the marketplace. Planters adopted this approach systematically by training their most promising young male enslaved people, usually Creoles, to be tradesmen and drivers. It made financial sense to do so as tradesmen and drivers brought a strong premium in the market. In the 1770s, slave tradesmen were valued at 27 percent more than prime male field hands and 65 percent more than the average adult male slave (which included unhealthy and superannuated enslaved people).[62]

Such practices had a significant impact on the composition of slave labor forces. Field hands were increasingly both African and female.[63] These enslaved people were the least acculturated enslaved people to Jamaica, with little knowledge about the rules of Jamaican slavery, and were sometimes taken advantage of by more experienced and healthier enslaved people.[64] Women faced a special pressure because they were worked hardest during their most important reproductive years. That most female enslaved people worked in the harsh working environment of the sugarcane fields even when they were pregnant greatly harmed female reproductive capacity, as Kenneth Morgan has detailed convincingly.[65] Sasha Turner, dealing mostly with the period after 1780, argues that enslaved women valued motherhood and used community resources to create communities in which child-rearing rituals were employed to assist the raising of children, but the constraints that enslaved women faced when their health was so damaged by their conditions of work were substantial and militated against either getting pregnant or giving birth safely.[66]

Britain and Jamaica in the second half of the eighteenth century were places of considerable advances in scientific agricultural management and labor discipline amid growing nutritional stress for the poorest section of the population.[67] In Britain, a rising population outstripped the ability of farmers to increase productivity, creating an embryonic Malthusian crisis. Robert Fogel claimed that by 1790 a large sector of the English population was so malnourished that it was effectively excluded from the labor force.[68] David Meredith and Deborah Oxley's exhaustive survey of calorie consumption in eighteenth-century England shows that after considerable nutritional improvement for ordinary English people in the first three-quarters of the eighteenth century, there was a precipitous fall in calorie consumption per person per day of one thousand calories in the thirty years after 1770. As they conclude: "massive population growth furnished the economy with many young, productive workers, liberated from the land and available for industry and other urban pursuits." But at the same time, "domestic farming was left unable fully to feed the people and imports were slow to fill the gap." The result, they argue, was that "the English population did not succumb to famine, yet they did suffer sickness, stunted growth, hunger and premature growth, and they watched far too many of their babies die, in part for want of adequate nutrition."[69]

Proslavery advocates like Edward Long had a point when they argued that instead of tormenting West Indian planters in parliamentary commissions, British leaders, many of whom were themselves improving landlords, should have concentrated their efforts on bettering the condition of the English poor.[70] Long believed that imports from Jamaica provided the additional nutritional inputs that prevented widespread starvation. Jonathan Hersh and Hans-Joachim Voth suggest that by 1850 the average Englishman would have given up 15 percent or more of his income to maintain access to sugar and tea, with sugar helping to raise average real incomes through both providing a cheap nutritional supplement and also improving working-class welfare (and probably happiness).[71] Kenneth Pomeranz famously argued that the domestic deficit in food production after 1770 was met only by production from the "ghost acres" of empire.[72] Britain, in this view, should have been applauding the ingenuity and inventiveness of West Indians in increasing productivity to such an extent that it helped keep England and Scotland, though not Ireland, from famine.

What Jamaican planters had done in their small island in the second half of the eighteenth century was indeed remarkable. As Fogel observes, the development of a new industrial labor system in the West Indies was a major technological innovation. Not only did the discipline of work consist of combined and consistent labor under a regimen of coercion that ensured high returns on capital; it also extended the duration of labor beyond the traditional daylight hours, contributing to a rethinking of time as less a constraint and more an advantage within systems of production.[73] Such agricultural innovation did not always attract positive commentary. Jamaican planters were certainly "improving" land owners, eager to invent new methods of human capital management to increase crop production, but, as is evident in the tone of testimony in front of Parliament in the slave trade investigation of 1791, they seemed the sort of "improving" landlords whom in Britain were becoming widely despised as money-grubbing moral bankrupts determined to disrupt customary understandings of mutual moral obligations in the countryside through increasing work intensity and by widening the participation of children in the labor market.[74]

It was the cruelty and callousness of Jamaican plantation slavery that most offended British observers. Even writers favorable to white Jamaicans, such as J. B. Moreton, who wrote a racy guide to life in the West Indies intended for British migrants intending to become overseers, in which the opportunities for fornication, drunkenness, and easy money were played up, noted how Jamaican slave management was destructive of slave welfare. He noted that "the weak and sickly" enslaved people were placed with the "stout and healthy" enslaved people and forced to work at the pace of the latter, adding that "when any of these miserable and probably half-starved wretches lagged in their work they were flogged and treated "as if they were mules or oxen."[75]

The results of such treatment were reflected, as Moreton indicated, on the slave body.[76] We can measure this dreadful standard of living in several ways. Using Robert Allen's pioneering methodology, Laura Panza, Jeff Williamson, and I constructed a standard basket of goods and priced it by local Jamaican costs.[77] This basket provided "bare-bones" subsistence. We then computed welfare rations based on relative purchasing power. This methodology allows us to show that the standard of living for enslaved Jamaicans circa 1774 was absolutely and relatively awful. Rapid increases in the cost of essentials made the standard of living in 1779 worse than dreadful: the cost of essential goods for survival was so high as to lead to starvation. The average slave in 1774 was maintained on a diet of 1,995 calories, and with just sixty-nine grams of protein, mostly fish. Such calorie consumption was woefully insufficient for enslaved people working perhaps eighteen hours a day doing hard labor. Enslaved people received precious little of the high and growing national income of the island and had a welfare ratio much lower than any white person—the ratio being in the order of 24:1 between enslaved people and overseers and 612:1 between enslaved people and sugar estate owners. Enslaved people had a welfare ratio of 0.268 where 1.0 signified subsistence, suggesting that even at the best of times, enslaved people able to supplement their provisions with food from provision grounds were on the very edge of starvation. Compounding poor welfare rates was the high cost of living, with the cost of living in 1774 in Jamaica being 74.1 percent higher than in London and 139 percent higher than in Boston. In 1779, these figures jumped to an extraordinary 957.9 percent and 944.8 percent respectively. Rural enslaved people in Jamaica earned a pitiful £4.81 per annum in 1774 compared to overseers' £541 and sugar estate owners' munificent £3,960. Gini coefficients by occupation indicate that Jamaica was a highly unequal society with the gross poverty of enslaved people contrasting with the extreme wealth of planters and merchants.[78]

Their sufferings varied during the year, with the best time being at Christmas, when provisions were abundant and slave owners relatively generous, and worst in the "starving time" from the end of harvest in June to the end of the hurricane season in October.[79] J. R. Ward shows that between July and September slave mortality peaked, with mortality on Mesopotamia in August at 150 percent of the average annual mortality rate and on the estates where Thistlewood was overseer or owner at nearly 200 percent of the average mortality rate. It is possible that planters welcomed the "starving time." Robert Dirks argues that enslaved people who were hungry were "docile, tractable and empathetic, isolated and withdrawn." Food abundance led to "gregariousness and aggressiveness." The Christmas season saw feasting, dancing, but also a strong possibility of rebellion.[80]

The poor health of enslaved people and the effects of hard-driving slave management on slave welfare can be seen in how the considerable annual natural attrition of the slave population in Jamaica required planters to resort frequently to the Atlantic slave trade for additional inputs of labor. It can also be seen in anthropometric evidence on slave heights, heights of individuals being a strong indicator used by historians to measure nutrition and nourishment.[81] The attrition of the Jamaican slave population in the eighteenth century is hard to document properly, as data on annual slave mortality rates is seldom available, except patchily in a few estate records.[82] Yet all the evidence we have suggests that there was a constant excess of deaths over births in the enslaved population, with the population growing as fast as it did only through massive importations of Africans through the transatlantic slave trade.

Philip Morgan estimates that the enslaved population of Jamaica reached 145,100 in 1750; 201,700 in 1770; and 275,600 in 1790. Using a formula devised in 1973 by Richard Sheridan, in which population estimates are derived from slave imports, the annual attrition rate for Jamaican enslaved people who died every year can be established with a degree of certainty. Slave mortality rates were especially poor when plantations were being formed in the seventeenth century. Morgan estimates that nearly 5 percent of the slave population died annually in the last quarter of the seventeenth century. Mortality rates improved in the first quarter century of the eighteenth century, with mortality rates more than halving from the late seventeenth century. But mortality rates increased appreciably after 1725, peaking between 1751 and 1775 at 3.4 percent per annum.[83] The spike can be partly explained by the large numbers of new Africans brought to Jamaica, as death rates in the first three years of "seasoning" were always high, with perhaps

15–20 percent of new Africans arriving on the island dying during this initial period of adjustment to plantation labor.[84] It also, however, was the result of the hard-driving system described above. The harder enslaved people were worked, the more likely that some of them would perish. Rising mortality rates suggest that Jamaican slavery was getting worse, not better, as the American Revolution approached. Despite the travails of the American Revolution and hurricanes in the 1780s, mortality rates improved between 1776 and 1800, suggesting an immediate impact of the adoption of amelioration policies in the 1790s by Jamaican planters.[85]

Estate lists give some indication of how unhealthy Jamaican enslaved people were, even if they survived the passage to Jamaica and three years of seasoning. Richard Dunn's exhaustive examination of the copious records of the Mesopotamia estate in western Jamaica shows that between 1762 and 1782 the estate grew from 269 enslaved people in 1762 to 300 enslaved people in 1789 but did so only through 160 purchases, 107 of which came through the transatlantic slave trade. The 109 births on the estate (less than one birth per woman of reproductive age) were far outnumbered by 230 deaths. Of those enslaved people who reached adulthood who lived on the estate between 1762 and 1833, the average life expectancy was 45 for men and 48.5 for women, with men being considered "prime Hands" (in other words, enslaved people able to work in trades if they were men, in the house if they were women, and in the field for both sexes) for 11.9 years on average. Women were considered "prime hands" for 12.1 years on average. Sugar work wore enslaved people out quite quickly, with the average working life of 256 male field hands being 17.7 years and of 305 female field workers 19.6 years. Drivers, by comparison, almost all of whom were male, worked for 25.4 years on average. Tradesmen had working lives of 19.8 years, 2.1 years more than male field workers, while female domestics worked 21.3 years, which was 1.7 years longer than field workers.

Moreover, male drivers, tradesmen, and female domestics spent longer of their working lives classed as "able" than did field workers—the 101 enslaved people in the former category had average working lives of 14.3 years while 561 field workers were classified as "able" for an average of 9.7 years. Overall, enslaved men at Mesopotamia, including enslaved men unable to work or "marginal workers," spent 9.2 years as "able" enslaved people, 8.5 years as "weak" enslaved men, and 1.2 years as invalids. Enslaved women fared worse than men, being classified as "able" for 8.6 years, 8.7 years as "weak," and a lengthy 3.3 years as "invalids." The harder toll that work took on females than on males can be seen in how twice the percentage of adult women were classified as nonworkers than were men. It also made motherhood tougher. Between 1774 and 1833, 276 women at Mesopotamia gave birth to 376 children or 1.36 children per adult female. But there were 147 women who had no children at all and a further 51 who had only one child.

Policies of amelioration worked a little: 119 women at Mesopotamia in 1802 had 224 children or 1.88 per woman. Adult women working at Mesopotamia before the 1790s, by contrast, had barely one child on average each.[86]

As these figures suggest, Jamaican slavery was especially brutal for women. Its brutality can be seen in David Ryden's careful analysis of twelve large sugar estates belonging to William Beckford II and John Tharp in 1780 and 1805, estates containing 2,985 enslaved people. He shows that while the expected productivity of teenage girls and young women was like that of male counterparts and that their probated value was equally similar, this similarity in work and value declined dramatically as women aged. Women were seldom released from field work and as a result their health collapsed. Male field workers, on the other hand, were often moved away from field work, customarily from their middle thirties, being retrained as tradesmen or promoted to being drivers. Consequently, fewer men than women had to work in sugar cultivation for long periods of their life. This relatively short time spent as field workers meant that male field workers were consistently classed as healthier than female field workers, as they tended to work in the fields only when they were likely to be "able" hands in their twenties. By contrast, women continued doing arduous field labor well after their midthirties. The result was that women's health broke down. Two-thirds of women on the Beckford and Tharp estates who were aged over forty were classified as "weak" or "invalid" compared to under 50 percent of men aged over forty.

The reason for women's bad health was that about 85 percent of women in their twenties and thirties worked in the field, dropping to a still high 70 percent for women in their early forties. By contrast, the percentage of men who were field hands dropped from 80 percent in their twenties, to 65 percent in their thirties, and to 45 percent in their forties. Ryden notes that 91 percent of women on these estates who were not field workers after the age of forty were classified as either ill or lame. He concludes that "promotion" into specialized nonfield work was a matter of choice for men (planters selected them to be tradesmen or drivers) but was a consequence of ill health for women. Women worked in the field until they collapsed and then were put into marginal work until they died. Ryden speculates, intriguingly, that one reason why women were treated so brutally was part of a larger attempt to deny enslaved women any power within plantations. He argues, echoing arguments made by myself, that planters contemplated these actions of favoring men over women not just because they shared with African men a number of patriarchal principles that stressed women's subservience to men but also because they assumed that men were dangerous and thus needed to be assuaged in various ways (being given easier jobs, having access if they were leading men to polygamous relationships, and having power within slave communities) if they were not going to rebel. Ryden notes that planters also assumed that women were "natural servants" who were expected to be subservient, not just to slave owners but also to enslaved men.[87] Just as plausibly, planters moved men out of field work because they anticipated that training them to be tradesmen was an easier way to maximize increases in the value of slave property than any form of human capital improvement.[88]

One way to assess the poor living standards of enslaved people in the period before the end of the slave trade and the start of amelioration is to examine slave heights. The height of individuals is well known to be a proxy for welfare. J. R. Ward, in an article mainly concerned to say that we can measure real improvements in slave welfare because of amelioration from the 1790s onward, shows that slave heights at the start of his analysis, in the period 1788–99, were remarkably low. Using observations drawn from workhouse records, he shows the height of Creole adult male enslaved people was 162.6 cm and the height of Creole enslaved men was 154.9 cm. These heights were very low, even lower than African-born enslaved people, where the average height of adult men was 164 cm and adult women 155.1. They were also much lower than heights elsewhere. The heights of rural Jamaican enslaved men born circa 1745 were a little bit lower than the heights of adult male Mexicans (164 cm) and appreciably lower than that of North American enslaved men (171 cm) and adult men in Great Britain (170 cm).[89] These dismal figures are stark testimony to just how deleterious were the material conditions within Jamaican slavery before the start of amelioration policies in the 1790s. Maintenance standards improved from the 1790s, leading to an immediate improvement in slave heights. That improvement, however, reflects incredibly poor conditions for Jamaican enslaved people before amelioration, with insufficient calorie intake and excessively long hours of work preventing enslaved Creole men from growing as they should have done.[90]

—

Edward Long was thus very wrong about how well Jamaican enslaved people fared under slavery in the 1770s. At the time he wrote and published his encomium to the generosity, humanity, and indulgence of Jamaican planters, between 1768 and 1774, enslaved Jamaicans were living lives of extreme hardship, as can be measured empirically in their nutritional deficiencies, in their stunted growth, and in their dreadfully low welfare ratios. Commentators with more direct experience than Long of how overseers, attorneys, and slave owners treated enslaved men and women made it clear that they endured sadistic punishments; inadequate food and housing; and arduous, relentless, and dangerous conditions of work. The extraordinary wealth of Jamaican whites was built directly on the suffering of enslaved people of African birth or heritage. It was a system that could be sustained only through a dynamic and efficient slave trade, which provided every year new additions of labor to replace enslaved people who had died from mistreatment and overwork.

It was also a system of slave management that before the American Revolution was dangerously dependent on a whole series of external conditions—a flourishing slave trade, a large trade in livestock, and reliance on imports for much of the island's food needs. It was a just-in-time system founded on a reliance on a series of precarious conditions. That precariousness made planting "an anxious pursuit" even when everything went very well.[91] When—as during the American Revolution—the slave trade virtually stopped, provisions from abroad became hard to obtain even at vastly inflated prices, and hurricanes destroyed local provision grounds, then the system nearly ground to a halt. Slave owners were massive risk takers, and Jamaica was built on the willingness of planters and merchants to take enormous risks to gain windfall fortunes. Sometimes, of course, those risks meant that people came undone, with extremely serious consequences for the enslaved people who underpinned white people's risky behavior through their undercompensated labor. Enslaved people also took risks, such as when they resisted slave managers' demands or when they ran away. The consequences of slave risk taking, however, were far more serious than for nonslaves, resulting usually in physical punishment.

Simon Taylor was one such risk taker. He believed that planting was a difficult and time-consuming activity, ill suited to short-term decision making. He was a practical, pragmatic man, determined to make an immense fortune, committed to staying in Jamaica, whatever the slings and arrows of outrageous fortune that British abolitionists flung at the island, and contributing to its improvement.[92] He demonstrated throughout his life no concern about the welfare of enslaved people as people. His interest was in how slavery contributed to the business of planting. Indeed, his most telling comparison was not comparing enslaved people to cattle but enslaved people to machines. He thought it madness to work enslaved people so hard that they could not cope: enslaved people could wear out just as machines did. Arcedeckne's enslaved people were being overworked and were as a result dying, especially when boiling sugar in the mill was in full swing, between December and August: "This is 8 months wherein the poor wretches do not get above 5 or at most 6 hours out of 24." He had to try and convince his owner that while "more cattle more Negroes etc" was the means to increase output and profits, it required a large outlay and that he should think hard about why he was so determined to produce large crops when that killed enslaved people. "I leave it to you," he informed Arcedeckne, "whether you choose to have more sugar and rum than the natural strength of your estate will make."

Taylor was tested in his resolve by the different imperatives about planting that were advanced by his highly capable and very ambitious overseer on Golden Grove, John Kelly. Kelly was as good at his job as Taylor was at his. He was able to produce remarkable crops. In 1775 he produced 740 hogsheads of sugar on Golden Grove, which was more than double the 350 hogsheads of 1769 and "the most extraordinary crop ever made on any one estate in Jamaica."[93] It was worth over £14,000, which was an immense sum for one estate. He achieved this stunning result, however, by driving his enslaved people extremely hard. They suffered heavy mortality, above that which was normal on a Jamaican sugar estate. While the fortunate heir was gallivanting about Europe as a twenty-six-year-old rich man, his Jamaican enslaved people were working eighteen hours a day, or 5,083 hours a year (more than one thousand hours more a year than during the period of late amelioration around 1830) and suffering ill health and dying before their time.

How did Taylor and Kelly double Golden Grove's production of sugar between 1765 and 1775? An investigation of this particularly well-recorded case study of slave management on a large sugar estate helps us pinpoint some of the issues addressed in this chapter about how concerns for slave welfare inevitably took second place to the hunt for profit on Jamaican plantations. Barry Higman argues that the remarkable improvements in productivity and produce at Golden Grove was achieved mostly from the hard-driving practices of slave management exercised by both men, but especially by the overseer, John Kelly. In addition, Kelly and Taylor approved the judicious application of hired labor at times of stress for resident workers, allowing other people's enslaved people to be worked very hard and thus saving some resident enslaved people from extreme overwork. It was also the result of better use of technology, careful plantation management, and the investment in and manipulation of human resources. The one thing that did not contribute to productivity increases was any rethinking of how enslaved people worked. The sole concession Taylor made in this area was to replace white skilled labor through training enslaved people to be tradesmen. This practice probably had more to do, however, with maximizing increases in slave values rather than in assisting with slave welfare, especially as more of the burden of plantation labor was placed on women of reproductive age.[94]

Taylor recognized how important Kelly was to this process implicitly rather than explicitly. He never tried to stop Kelly from overworking enslaved people to present to Arcedeckne a bumper crop, at least before the American Revolution. He may have confided to his employer that Kelly worked the "poor wretches" too hard, and he might have kept on arguing that his employer needed to invest in continual additions of labor through the slave trade. But Kelly's success in increasing the size of the sugar crop kept Taylor quiet. Moreover, Taylor was also obliged to keep Kelly employed because Arcedeckne was more interested in increasing his annual income than in long-term investments in slave property. He was thus happy to continue and increase the short-term exploitation of enslaved people. Kelly gradually increased his areas of authority at Golden Grove and, despite Taylor's objections, was given a power of attorney by Arcedeckne, so he could act (or interfere, as Taylor saw it) in how goods were shipped to Britain.[95]

The privations of the American Revolution changed Taylor's attitude to Kelly, or at least emboldened him to be more critical of him to his employer. He complained that Kelly worked enslaved people so hard that "their hearts have been broke," that he favored his own enslaved people over Arcedeckne's, and that he was very ungenerous to enslaved people so that the enslaved people on Golden Grove "were starving" and a "very feeble set indeed." He concluded that if Kelly continued these practices that Golden Grove would be eventually denuded of enslaved people as they would be "killed by overwork and harassed to Death." Soon this profitable property would be a "Land without Negroes." Taylor persuaded Arcedeckne to get rid of Kelly in 1782. Nevertheless, Kelly's sacking was less due to how he treated enslaved people than to what Taylor thought were continual intrusions on his authority. Kelly himself suffered little from falling out with Taylor, soon being employed as an attorney (a promotion from overseer) on the adjoining property to Golden Grove, which was Duckenfield Hall.

At bottom, Taylor had only a slight and superficial concern for the welfare of enslaved Jamaicans. He was, like most white Jamaicans, more concerned about himself than about his enslaved people. The advent of the American Revolution made him feel very sorry for himself. Richard Sheridan estimated that as many as twenty-four thousand enslaved people (one in twelve enslaved people) may have died in the subsistence crises of war and hurricanes between 1780 and 1787.[96] But Taylor thought it was him who was a victim of the conflict. He lamented the "very distressing circumstances" of "a war with the whole world and not a single friend." He felt so badly treated and betrayed by British actions after the war ended that he grumbled, "If we are the most favoured subjects, God help the rest." Indeed, he moaned in a private letter in 1784 that he might sell his properties "and move to some other government where the laws will allow me to buy bread from foreigners when I am starving."

Taylor never came close to starvation. But his enslaved people did. Moreover, they starved not just during times of austerity but in prosperous times.[97] There was a cruel logic to the hard-driving slave management style practiced by Jamaican planters in the booming years between the end of the Seven Years' War and the start of the American Revolution—a style that came to grief in the more difficult environment of the late 1770s and early 1780s. That logic was predicated on a stunning lack of empathy for enslaved Jamaicans, who not only were treated as less than human but were conceptualized in ways that equated them constantly to animals and thus as people who did not warrant being treated with any respect. The result was that most of the Jamaican population was indeed "in Miserable slavery" with the time of white Jamaicans' greatest wealth and influence coinciding with a nadir in the standard of living of the people who allowed them this wealth and influence.[98]

—

Was Jamaican slavery in the period of the American Revolution the worst kind of slavery experienced in colonial British America? It seems to be at the far edge of exploitation if we consider the material conditions of the enslaved. Morgan's estimates of annual population decline in Jamaica in the seventeenth and eighteenth centuries suggest that if we take the health of a population as the principal indicator of material comfort and standards of living then life for enslaved people in Jamaica before the start of amelioration and pronatalist policies in the 1790s was very bad. A literature on slavery in the nineteenth-century United States suggests that slavery involved a form of "soul murder," in which enslaved people suffered psychological trauma that had long-term historical consequences.[99] Being a slave led to real murder in the Caribbean. Radburn and Roberts note that the job of holing given to enslaved people in jobbing gangs "murdered enslaved people, something that contemporaries knew well," and which skeletal remains confirm. Randy Browne argues from his study of Berbice that "historians have long known that Caribbean slave societies were death traps," arguing that "the rare men and women who lived into their forties or fifties, old age by the standards of plantation slavery, had watched hundreds if not thousands of shipmates, friends and kin perish."[100] Other slave societies emulated but probably did not surpass Jamaica in the brutality of their slave management system. Early nineteenth-century Berbice was especially brutal after overextended British planters took over from the Dutch and pushed enslaved people extremely hard in a colony in which sugar could be made all year round. In Berbice, however, at least enslaved people were protected by the law through the agency of the Fiscal, which placed some small constraint on how slave owners could treat enslaved people. So too did enslaved people in Saint-Domingue have the mild protection of the Code Noir. Jamaican enslaved people were singularly without supporters against what their masters and the representatives of masters did to them.[101]

The demography of slavery in the Caribbean was appalling and was worst on sugar plantations. The plantation economy was an ecological disaster. The architects of this economy built a paradise for pathogens and thus made the Caribbean a hellhole for humans. The owners of sugar plantations cared little about this because, as Adam Smith argued in the 1770s, sugar planting was the most lucrative form of agriculture and the West Indies was ideal for its production. They pushed on with putting more land into production for sugar, making the plantations they owned among the most hazardous disease environments on earth in the early modern period. Jamaica, where almost everyone lived within five kilometers of the coast, was an ideal breeding ground for pathogens, being warm, wet, and well connected to Atlantic shipping. Working on a sugar estate at any time—and three out of four enslaved people imported from Africa were sugar workers in the eighteenth century—was unhealthy for all and especially for newly arrived Africans who left an unhealthy African environment to arrive in Jamaica via the dreadful Middle Passage teeming with pathogens and suffering from poor nutrition. Perhaps 45 percent died within the "seasoning" period of the first couple of years working in plantation agriculture.[102] The result was one of the most catastrophic sustained demographic calamities in world history, with mortality rates on a sugar plantation being comparable to mortality rates in the Soviet Gulag during its worst period between 1934 and 1953.[103]

John McNeill argues that the ecology of the plantation economy should be interpreted as a syndemic event.[104] A syndemic is when two or more diseases form a cluster of epidemics affecting a given population in social, political, and environmental contexts that perpetuate that disease cluster and exacerbate its effects. It combines biological and social variables into a single concept.[105] This chapter does not deal with the disease aspects of a syndemic, although in my previous work I have examined the empirical evidence around the poor demographic performance of the Jamaican population in considerable detail.[106] The concept of a syndemic provides support to the social context through which poor demographic results occur. The bad health of enslaved people in Jamaica was partly due to the rigors of working in sugar, a truism in slave studies that is attested in studies of eighteenth-century Surinam and nineteenth-century Danish West Indies and Martinique.[107] But social oppression, as detailed above in how white Jamaicans treated enslaved people, played a part in aggravating health problems. Malnutrition, for example, weakened the immune system of enslaved people, making them vulnerable to infection. The frequent use of the whip also increased the risk of infection as it sliced open skin, making enslaved people more likely to get yaws and tetanus. Sexual infection through sexual exploitation of women enslaved people by diseased white men lowered conception rates. And sugar plantations had a plenitude of livestock, which increased sanitation problems and accentuated cross-species infection.[108]

Temporality, however, matters. Slavery has a history, and conditions under slavery in some periods were worse than others. Philip Morgan's estimations of enslaved population decline show that the adoption by planters of pronatalism and ameliorationist policies under abolitionist pressure after the revelations of how bad slavery in Jamaica was were presented to the House of Commons in 1792 led to concrete results. Between 1801 and 1825, the rate of enslaved population decline fell precipitously to 0.5 percent per annum, reversing from the 1790s what from 1700 to 1780 had been an increasing rate of annual enslaved population decline.[109]

J. R. Ward argues, convincingly, that in the 1790s and 1800s West Indian planters responded to abolitionism with a change of tack in their slave management policies and with a greater effort to implement pronatalist and ameliorationist policies. He suggests that planter-led amelioration in this period led to some significant successes in countering enslaved demographic decline. Planters adopted such policies when they had resisted them previously for a variety of reasons, including rising slave prices, accumulating management experience, the influence of humanitarian ideas about better treatment of enslaved women, pro-natalist incentives for "breeding women," improved medical practice among more skilled doctors, the threat of abolition forcing action, and the gradual increase of native-born enslaved people who were healthier than African-born enslaved people. Ward concludes that planter policy, determined above all by metropolitan influences, meant that the Jamaican slave regime improved considerably in its later years.[110]

To this interpretation should be added the activities of enslaved people themselves to improve their material conditions. Mary Turner and Sasha Turner accept that material conditions improved but attribute such improvement less to planter benevolence than to slave-led insistence that maternal health be more strongly supported.[111] Sasha Turner notes how ideas of amelioration of slave conditions led to a coalition of interest between Jamaican slaveholders and doctors on one side and enslaved mothers and abolitionists on the other, both arguing that enslaved women's labor, discipline, and health care should be adjusted so as to increase birth rates and to improve infant mortality. The alliance was an uneasy one, as enslaved women did not trust their masters, and slave owners remained wedded to making as much money from plantation agriculture as they could, and most important, to the maintenance of the plantation economy with slavery as its linchpin. Masters wanted to reform rather than change slavery, and even as they adopted ameliorationist policies, they did not get rid of their negative opinion of enslaved person's capacities. Enslaved women, conversely, took advantage of planter concessions—concessions they themselves had won by their own hard efforts—to create rituals and customs around child rearing that allowed them to reduce their workload and dictate to some extent the pace and nature of the work they did. And while much of the amelioration improvements came from more committed white doctors, the efforts of the enslaved to use African pharmacopeia to counter the inadequacy of much European medicine became more accepted, even among whites.[112]

But before the start of ameliorationist policies, first planter led and then adopted by a revitalized abolitionist movement following the abolition of the slave trade through the creation of a slave registration system in the 1810s and a raft of legislation about slave welfare in the 1820s, Jamaican enslaved people faced a perilous position in a slave system designed to let the plantation machine power ahead no matter the cost to slave health and well-being.[113] The testimony presented to the House of Commons in 1792 provides a chilling account of the logic of short-term slave management in Jamaica in the period of the American Revolution. The voices of a few planters, such as Simon Taylor, who thought that destroying the health of enslaved people through overwork and bad treatment was economic madness, were drowned out by the actions of the mass of slave owners seeking quick and regular profits from plantation agriculture and from attorneys and overseers determined to make their name by producing bumper crops of sugar and rum, no matter how much harm this did to the lives and material conditions of enslaved people. Edward Long refused to see what was in front of his eyes and what he recognized was true also for livestock. What he argued about the ill-treatment of mules and the economic foolishness of working mules to death could have been applied with even more force to what he and other planters offensively termed human stock. The conditions of slave management in Jamaica in the period of the American Revolution made the island rich for fortunate owners but a living hell for most of its residents.

Chapter 4

Tacky's Revolt and Its Legacies

The American Revolution is customarily held to have started in 1763 with the Peace of Paris concluding Britain's victories in the Seven Years' War (1756–63) and with the Proclamation of 1763, blocking westward expansion in British North America. The war proper starts in 1775 and ends with the British acknowledgment of the independence of the United States in 1783.[1] Such a chronology is unsurprising given that American historians, as Stephen Conway notes, see the American Revolution as "one fought almost entirely in and for America," with the final act being the British defeat at Yorktown, Virginia, in October 1781.[2] The American Revolution, however, was a global conflict, which spread from Boston to West Africa, the cape, and southern Asia.[3] The Caribbean was an area of major conflict, especially after France and Spain entered the war on the American side in 1778. For Jamaica, the war ended in April 1782, when Admiral George Rodney's victory at the Battle of the Saintes saved the island from Franco-Spanish invasion.[4] Grateful Jamaicans marked the occasion with a toga-clad statue of Rodney, erected at great expense in the Spanishtown square.[5]

When the American Revolution started in Jamaica is less obvious. The Proclamation of 1763 did not affect the West Indies, and the tumult over the Stamp Act was less pronounced in the Caribbean than in British North America. The Sugar Act was of more moment. It was the first overt imperial tax raised by Parliament and an act that caused a considerable rupture in the relations between British North America and the West Indies, though, contrary to North American opinion, the act, while overall beneficial to the British West Indies, did impose some burdens on West Indian planters.[6] The start of a new phase in imperial relations in Jamaica, I argue, which determined patterns of interaction between the imperial and colonial states and white Jamaicans, began in the pivotal year of the Seven Years' War, in 1760. The key event was a massive slave rebellion that goes under the name of Tacky's Revolt, but which was more plausibly a revolt concentrated in Westmoreland Parish, in which a Coromantee enslaved person called Apongo or Wager, with extensive military experience in both West Africa and the eastern Caribbean, was the likely mastermind.[7]

This slave revolt made real concerns previously expressed by observers such as Edward Trelawney about the security of the island. Those concerns derived from the harsh methods of slave management practiced by slave owners; the disproportion between a diminishing white population and a rapidly expanding black population; and strategies for preventing slave uprisings that were shown up to be highly deficient. The operations of the imperial state were managed by Lieutenant Governor Henry Moore, a native-born Jamaican and later governor of New York, who was one of the most impressive governors in British America in the 1760s. In this retelling of the revolt, Wager and Moore were the principal protagonists. It was an uneven contest. Wager was captured and gruesomely executed as his revolt failed. Moore received a baronetcy as a thank-you from a grateful British state for his efforts in quelling the dispute. He was awarded the governorship of New York in 1765, where he was instrumental in skillfully lessening imperial tensions arising from the imposition of the Stamp Tax.

It was, however, an uneven contest only in retrospect. Jamaica was saved from disaster (at least for its white inhabitants) by the fact that the rebellion occurred during the Seven Years' War, meaning that experienced troops and naval forces were present on the island, and by the Maroons holding to their side of the bargain negotiated in the Treaty of 1739.[8] This combination of factors, and Moore's adept response to crisis, allied to the probability that the slave revolt of April 1760 that erupted in St. Mary's under Tacky undermined Wager's plans for an island-wide revolt in that month or the next one, allowed Britain to overcome an existential threat to Jamaica's existence.[9] It is easy to see from a distance that the balance of power between rebellious slaves and a determined imperial state could have been reversed if contingency had worked out differently.

Tacky's Revolt was a crucial event in the history of Jamaica. It made manifest what had only been feared before: that enslaved people hated their masters and were willing to risk the relatively near certainty of a gruesome death to challenge white authority. They nearly succeeded. Thomas Thistlewood, whose diaries are the single best source on the revolt and whose management strategies before and after 1760 will form part of this analysis of the revolt and its legacies, summed up white Jamaican feelings on 24 October 1760. He dined with a fellow overseer, John Stewart, who told him of an old proverb "which frights many people: One thousand seven hundred and sixty-three, Jamaica no more an island will be." Thistlewood added in parentheses, "not for whites."[10] This chapter examines how white Jamaicans came to experience the horror of a major slave revolt, assessing two possible interpretations for how the revolt came about—the first, held by Edward Long, whose account of the revolt has formed the basis of all subsequent historiography, that it was a concerted campaign by Akan or Coromantee slaves (enslaved people from the Gold Coast) to destroy white authority; the second, held by Thistlewood, that it was less an ethnic conspiracy than an event emanating from collective action by enslaved people of various nations who shared residence in particular plantations (Frontier estate in St. Mary's and Masemure in Westmoreland, owned by absentee planters Ballard Beckford and Commodore Arthur Forrest, respectively).

It then looks at one aspect of the response to Tacky's Revolt by Jamaican legislators, as they aimed to prevent revolt from happening again. This response involved two strategies. The first strategy was the immediate and unrelenting application of terror to show enslaved people thinking of rebellion how ferocious the Jamaican state would be in defense of white privilege. The second strategy was a dramatic rethinking of the security needs of the island, so that slave managers were supported in their management of enslaved people by a coercive, intrusive, and expensive imperial and colonial state, combined with the active enforcement of the provisions of the Treaty of 1739 agreed with Maroons. That arrangement lasted until 1795 when the Second Maroon War and Jamaica's increasing involvement in global geopolitics because of British involvement in the Haitian Revolution disturbed a working modus vivendi about securing white Jamaicans' safety. The end of the Second Maroon War ended one phase of Jamaican history, which coincided with the "long" American Revolution in British America and the United States, but which worked out differently in Jamaica than in other parts of the British Atlantic World.

The significance of Tacky's Revolt for discussions of the American Revolution is that it, and its repression, highlighted how the concerns of colonists in Jamaica were not those of colonists in British North America. Andrew O'Shaughnessy has shown that the constitutional issues that provoked West Indians were questions around the prerogative rather than parliamentary sovereignty.[11] The more pressing concerns revolved around security, preserving white rule from internal and external attack and the system of slavery that sustained Jamaican prosperity. As Jack P. Greene has argued in his evaluation of a December 1774 petition by the Jamaican House of Assembly to the British Crown that they might "become a mediator between your European and American subjects" in disputes over parliamentary sovereignty, Jamaicans recognized that their reliance on British defense made it impossible for them to jeopardize their security through joining with the mainland colonies in imperial dissent. In what George Metcalf calls a "sadly misguided" attempt at colonial mediation, Jamaicans urged Britain not to adopt measures that reduced it to "abject slavery" while admitting it could no nothing to stop Britain doing what it did, given that Jamaica was "weak and feeble" because of "its very small number of white inhabitants, and its peculiar situation, from the encumbrance of more than two hundred thousand slaves." Given this situation, the Assembly argued, "it cannot be supposed that we now intended, or ever could have intended, resistance to Britain."[12]

But contrary to Greene, who stresses that this position illustrates how Jamaica's need for British defense reduced it in fact to that state "of abject slavery" that it purported to despise, I argue that the arrangement that Jamaicans forged with the imperial state after the shock of Tacky was one that suited their instinctive loyalism. It shaped their experience of the American Revolution in satisfactory ways, even in the harshest times of high taxation and privation in the 1780s, keeping them safe, both from the threat of slave rebellions (the foiled revolts of 1765, 1766, and 1776) and from external attack (Franco-Spanish invasion in 1782).

—

In making 1760 a major dividing line in Jamaican history, I follow Claudius Fergus.[13] Fergus argues persuasively that the dread of slave insurrection in abolitionist politics can be traced to Edward Long's *History of Jamaica* and his belief that the 1760 insurrection was caused by an influx of unassimilated Africans following the declaration of peace between Maroons and the Jamaican state in 1739. Long thought many of these Africans were "warriors in Afric or criminals."[14] Fergus notes that understanding the "dread of insurrection in abolitionist politics" requires engaging with "the dialectic between resistance by the enslaved and changing management practices of planters or the actual engagement of legislators with the fear of "black power." He skillfully connects Long's racist advocacy of plans to transform the enslaved population from African to Creole (native-born) to widespread concerns in the colony about internal security. Long thought that the slave trade needed reform because it introduced into Jamaica barbarous savages who were inclined to conspiracy and who were prone to "the most bestial vices" that hindered natural reproduction. He argued that slave owners should encourage natural reproduction so that they could stop buying Africans who endangered the island's safety. Increasing taxes on enslaved people acquired in the slave trade might be one way of achieving that natural increase, making African slaves so expensive that planters would adopt voluntarily schemes of slave amelioration.[15]

Such ideas formed the "pragmatic rationale for abolition of the slave trade." Fergus cites the prominent early abolitionist James Ramsay as heavily dependent on Long's arguments about internal security and the virtues of creolization. Ramsay argued that amelioration achieved through the abolition of the slave trade would make slavery more humane, would improve economic returns from plantation agriculture, and, most important, "would add to the strength and security of the colonies." Ramsay agreed with Long that "Creoles or native-born West Indian negroes are universally acknowledged to be more hardy, diligent and trusty than Africans." Fergus's main interest is to show how Long's views, as repeated by Ramsay, demonstrate how abolitionism was linked to fear of rebellion. He shows how from the mid-1790s warfare in the Caribbean, as part of Britain's epic contest with France in Europe and Haiti, focused British minds, including those of William Pitt and George Canning, on how prudent slave management in Jamaica might make the island more secure from foreign attack. Prudent slave management revolved around eliminating the threat to white rule from imported African slaves prone to conspiracy with martial backgrounds, like the Coromantee warriors Long fingered as the leading rebels in 1760. Canning especially favored "creole colonization" as the means of increasing "wealth" and preventing "weakness." Fergus argues, "by 1802 creolization had become the lynchpin of abolitionism." Its association with issues of security helped the abolition of the slave trade become a reality by 1806.[16]

Fergus's formulation of Long-inspired ideas of "creole colonization" as "a panacea for the principal evils of the colonial system—insurrections, high cost of production, ill-treatment of enslaved laborers and dependence on the Atlantic slave trade"—illustrates that Long's discussion of Tacky's Revolt and its suppression was highly political.[17] Long may have been correct that the leaders of the rebellion were Coromantee and that these naturally rebellious enslaved men were also involved in later revolts, notably in St. Mary's in 1765. Other writers also placed responsibility for the revolt on Coromantees. Yet we need to recognize that blaming Coromantees for this breach of security served an important polemical purpose. It was convenient that Coromantees were blamed for the rebellion. It allowed Long to argue that if Jamaica lessened the number of Coromantee slaves imported onto the island, as was proposed but defeated in a motion in the Jamaican Assembly in 1765, it would prevent slave insurrection, keep whites safe, and save the colony hundreds of thousands of pounds. Long's argument, however, as shown below, did not originate with him but was part of long-standing discourses around how slave insurrections could be prevented that were current in Jamaica since at least the end of the First Maroon War and prevalent in the West Indies as a whole, especially after the 1736 slave revolt in Antigua was attributed mostly to the influence of Coromantees.[18]

The evidence is not sufficient to discount Long's theory that the slave rebellion of 1760 was part of a Coromantee, or Akan, conspiracy. It is a theory that remains popular within contemporary scholarship. Enslaved people from this region have been argued to have had an ideology and cultural orientation that made them eager to challenge white authority. As Marjoleine Kars argues, Amina slaves (the Dutch name for what the British called Coromantees and what Iberians termed Mina and what Africans thought of as either Akan or Ga) were behind many New World rebellions, such as St. Jan in 1733, Antigua in 1736, Jamaica in

1760, and Berbice in 1763. Vincent Brown's forthcoming book focused on Tacky's Revolt extends this insight to look at Coromantee leadership in insurrections more generally.[19] Other contemporary Jamaicans besides Long thought Coromantees especially rebellious. James Knight wrote in his so far unpublished 1746 two-volume history of Jamaica that Coromantees were "ingenious" but "fractious" people who were "in their Nature Deceitful, Revengeful and blood thirsty." Wise planters needed to be "cautious" of having too many of them on a plantation because "there never was as I have heard of in this or any other colony any Ploy or Conspiracy, but they were at the bottom of it."[20]

Nevertheless, Long's interpretation of the causes of the rebellion is not the only interpretation that we can form from limited evidence. Maria Alessandra Bollettino argues that historians' adoption of Long's account is inadequate because "it has served to obscure the full range of enslaved and free Blacks' actions in what contemporaries understood to be a battle of the Seven Years' War." She notes that while Long's account rendered the revolt "intelligible and potentially preventable" if the importation of Akan slaves onto the island had been banned or reduced, it had "averted any sustained inquiry into the brutality that undergirded the institution of slavery in the British West Indies."[21] It was not just Coromantees, of course, who had a reason to hate their masters and wish them dead: that was a common feeling among all enslaved people.[22] As Emilia Viotti da Costa notes in her magnificent study of the Demerara Revolt of 1823, "rebellion was the product of many contradictory historical forces," best described as emerging from "voices in the air," from the multiple experiences of being enslaved in a particular place under particular conditions of labor. Richard Sheridan concludes in his masterly evaluation of the 1776 Hanover slave "conspiracy" that "slaves needed no borrowed ideology and motivation but only favorable circumstances to rise against their oppressors."[23] As Bollettino notes, it is probable that whites in Jamaica did not experience a single slave rebellion in 1760 but suffered from waves of revolt that started in the northern and sparsely populated parish of St. Mary's, and spread to the newly developed and populous sugar-producing plantations of Westmoreland, Hanover, and St. James, and then extended into central parishes such as St. Dorothy, Kingston, and the eastern parish of St. Thomas in the East. She argues that slave rebels exploited the anxiety of whites following the April outburst of violence in varying ways and to different ends.[24]

Thistlewood downplayed the importance of Coromantee influence in favor of plantation solidarity. The plantation world was a dangerous one for enslaved people, and one of the few protections slaves had from attack by other slaves or whites was collective unity derived from common residence on a plantation. Over time, within slave communities, "an embryonic social consciousness and awareness of the necessity for slaves to join for mutual safety developed."[25] Thistlewood notes that the rebels were Coromantees, declaring on 31 May 1760 that Mr. Reid had "been warned off his estate as the Coromantee rebels were coming." His main emphasis, however, in describing the progress of the rebellion was to stress how most of the rebels came from the one estate—the absentee naval officer Captain Forrest's Masemure estate.[26] On 2 July, the day before Wager was taken, he noted that "114 Negroes [are] wanting this day from Forrest's estate besides those that are known to be kill'd or took." His argument as to how the revolt occurred centered around a complex series of events where "Addison's Cuffee stole a gentlemen's Trench with a great deal of money" and gave the money to Wager to fund a rebellion by buying a barrel of powder from traitorous Jews (Thistlewood shared the general anti-Semitism of his countrymen) rather than be discovered in his theft by "another Negro who was privy" to the theft. He claimed that "the Treasurer of the Rebells" had been caught by "three of the Retrieve Negroes" when he tried to buy powder in Savanna la Mar with seventy doubloons (or £333). Elsewhere, he claimed that the plan of the rebels had been to rise against whites on Whitsuntide (falling on 26 May 1760, the day after the night that rebellion started in Westmoreland), setting fires in town, and murdering whites who came to extinguish the flames while killing overseers on estates. The rising by Tacky and Jamaica in St. Mary's at Easter had come early when "a Negro carrying the wooden sword adorned with parrot's feathers (being the signal of union some part of Guinea) was discovered by a Capt. of a Guinea man who saw it carrying in a procession at Spring Path, had the fellow seized and discovered the affair."[27]

He mentioned Forrest's estate as a hotbed of rebellion in other diary entries, commenting that Quacoo, who had worked for him, shot a slave, Abraham, from Forrest's estate. He lamented that "Captain Forrest's Fortune" whom he considered "a principal offender" returned unpunished to Masemure. Thinking back about the causes of the revolt, he noted that Jackie had shaved his head following a visit from "two of Capt Forrest's Negroes at our Negro Houses." A shaved head was, he believed, a signal of war. On 25 May, Thistlewood heard "about 9PM a blast off a horn at our Negro house" and the next day news came of an uprising at Masemure that started at 9.45AM with Mr. Smith "being murdered by the Negroes . . . Capt. Hoar sadly chopped, Capt. George Richardson and Thos. Barnes running to the Bay on foot, a narrow escape they had." Two days later, he reflected that his slaves were ready to rise, one sign being that Lewie had visited Forrest's plantation on the 25 May. At least four of the principal ringleaders hanged in gibbets or burnt—Goliath, Davie, Wager, and Cardiff—were enslaved people on Masemure estate. Escaped slaves from the estate continued to be hunted down by troopers, with most captured by mid-August 1760 and some hunted down by 6 October.[28]

Thistlewood was not alone in seeing Masemure as the epicenter of revolt. Francis Treble wrote that three days after the rebellion started troopers had taken

"60 Stand of Arms from Forrest's Estate."[29] A letter written by David Miller to Brigadier General William Lewis in 1765 suggests that Forrest's estate remained a center of enslaved disaffection long after Tacky's Revolt. Miller had been told by Mr. Parr "in this neighbourhood" that "there was a Rebellion Concerted, between the Negroes on Capt Forrests, Mr. Crawfords and Midgeham Estates." Each of these estates had been centers of the 1760 revolt and had had enslaved people executed for their roles in the rebellion. Parr had been told by one of his "wenches" (Nancy) who had heard a rumor from a "wench" on Forrest's estate (Lucetta) that several meetings had been held "to put their Scheme into execution" in late November 1765. Miller "prevailed" on Parr to send his "wench" to find out more from Forrest's slaves, giving her the promise of a reward. She confirmed that three enslaved men (Crawford's Fancy, Forrest's Bacchus, and Forrest's Cuffee) were in a conspiracy to rebel and that the three all went to Midgeham's estate "to consult with their accomplices there." Miller asked Lewis to assure him of his "prudence and vigellance to have such measures taken as I hope will not only frustrate their Intentions but Save the Country."[30]

—

Tacky's Revolt may have been a plan that Long argued "was projected, and conducted with such profound secrecy . . . without any suspicion from the whites," but that Jamaica was bound to have a slave revolt at some stage of its eighteenth-century history was something many whites rightly feared. It had not experienced a slave rebellion since 1690, when three hundred to four hundred slaves in Clarendon rose up and killed a few whites before being quelled. Long argued that Jamaica was kept safe almost by accident: before 1739 Maroons had tried to attract disaffected slaves to their cause, thus preventing active rebellion but encouraging running away.[31] The success of the large integrated plantation in transforming Jamaica's economy; the settlement with the Maroons in 1739; the rapid expansion of the Atlantic slave trade, bringing in as slaves every year almost as many as were in the white population, resulting in an enslaved population of 106,592 and a white population that was no more than 10,000, all meant that white Jamaicans were riding their luck in the second quarter of the eighteenth century when no slave rebellion happened. The island was a powder keg, ready to erupt at any time.[32]

The writers who tried to make sense of Jamaica's situation acknowledged that the disproportion of white and black numbers exposed white Jamaicans to enormous danger, a danger that was made worse by the ill-treatment that slaves faced in a brutal system of slave management. It is likely that the nadir of enslaved experience in Jamaica came during the second quarter of the eighteenth century.[33] Charles Leslie, who wrote a brief history of the island in 1739, argued, "No Country excels [planters] in a barbarous Treatment of Slaves, or in the cruel Methods they put them to death." He tried to explain away such cruelty by stating that brutal treatment was inevitable "given how impossible it were to live amongst such Numbers of Slaves, without observing their Conduct with the greatest Niceness and punishing their Faults with the utmost Severity."[34]

James Knight was a more sophisticated writer than Leslie, especially about the character of Africans. He was aware of the security issues raised by excessive slave importations. He commented, "A Country . . . that is Cultivated with Negroes or Slaves and a small Number of White People may indeed become Rich . . . but it cannot be deemed safe or secure from an intestine or Foreign Enemy." Yet, he also argued, that except for Maroons in the 1720s and 1730s, Jamaica was "never in danger." He thought that enslaved people could be kept under control by scattering, mixing, dividing, intimidating, and monitoring. Enslaved people were impressed by white firepower and were controlled by useful laws that stopped them from communicating with each other, and their most significant members were co-opted into slave management and given privileges that allied them with their masters. Most important, he thought, wise planters mixed slaves from different regions of Africa together, knowing that these varied "Countries or Nations" had "as great and Natural an antipathy to each other, as any 2 Nations in the World." If planters kept a "vigilant Eye and a strict hand" over enslaved people, that would "prevent them doing wrong or mischief."[35]

Knight finished his unpublished history about 1746. In 1746, an anonymous pamphlet likely to have been written by Governor Edward Trelawney was published. It laid bare how Knight's assumptions were wrong, owing to planters' inability to control their impulses toward placing short-term profit making above the security of the island.[36] Planters needed enslaved people to make sugar and got the slaves they needed from an increasingly efficient slave trade. "Planters," he asserted, "have a Rage for buying Negroes and have little Care and Conduct that is used in the Management of Them." Like children "playing with Edge-Tools, which they cannot manage," planters' "Rage to push on their estates" meant that soon "Jamaica would be over-run and ruined by its own slaves."[37] Jamaicans should be "prevented from cutting themselves" by Parliament enacting a ban on the slave trade. Britain should also send as a matter of urgency a regiment and naval squadron, presumably to be paid for by the colonists themselves, so that Jamaica would not be taken over by slaves "even tho' the Enemy should not take the advantage of joining with them." Trelawney did not share Knight's confidence in planters' good sense, had the normal governor's disdain for the capabilities of the local militia, and was more conscious that laws were ignored than the fact that the Jamaican Assembly kept on passing laws to control enslaved people. He noted that slaves were increasingly numerous, whites were diminishing in number, and planters showed a remarkable inattention to laws ensuring their safety "due to a narrow Selfishness and total Unconcern for everything that doth not regard their immediate Interest." As he acidly noted, "what signifies making new Laws until the old ones are obey'd." Given that planters were unable to see wider imperial contexts and given that "the Island of *Jamaica* being of the greatest Importance to its Mother Country, and at the same Time so Insecure, that the Inhabitants are not only alarm'd by every trifling Armament of the Enemy, but under the greatest Apprehensions frequently from their own Slaves," Parliament needed to take action. "I cannot think that the Parliament of *Great-Britain* ought to risqué the Security of so valuable an Island as *Jamaica*, and defer doing what is absolutely necessary for its Safety, out of a Deference to the Humour of a few Planters."[38]

He advanced a story of a "Conspiracy in St. *John's* Parish" recently discovered and "the Ringleaders seized" to show that while planters briefly obeyed slave laws, after a month "the Masters and very Overseers were weary of signing their Names" on tickets slaves had to get to travel. Soon "the Conspiracy [was] itself forgot, and looked upon as a Dream, though several poor Negroes were in reality burnt alive for it."[39] The result was that thirteen slaves took advantage of the laxness of slave management and became a bandit band "till they were first routed by the late rebellious Negroes that submitted in 1739." The same pattern as before happened—alarm, enforcement of laws, then a slackening of controls and forgetfulness about the recent danger. Trelawney exploded about planters' foolishness: "Have they not Sense to know how precarious their Condition must be, when thirteen Rogues only out of so many thousand are able to bid them Defiance and give them such Alarms? Can they be ignorant that their whole Constitution must be quite crazy, their whole Mass of Blood vitiated, when such small Eruptions can endanger so much the whole Body, and when they break out so fast one after another?" The cause was less stupidity or even indolence but Jamaicans' chronic addiction to the short term: "Many see the Symptom of a Country approaching to its Ruin, but they fancy it may last their Time, they may sell out and get Home first, and what comes afterwards they care not." The solution, he thought, was abolishing the slave trade, which "will effectually hinder People from throwing away their Money in such Sales and people would have "to make the most of the Negroes they have, when they find they cannot have more." It would also mean that "lazy House-Slaves" (who can be assumed to have been native-born, although he does not say so) "would be sent into the Field, their proper Place" as "the Slave is for the field only.[40]

Trelawney hammered home his point in an unusual comparison, one that suggests that his discourse about the rightness of slavery was less important to him than that "the Sugar Manufactory" that kept the "Almighty Ship of Trade" afloat was maintained. He related from Montesquieu's *Persian Tales* a story of a nation that killed every foreigner they caught and turned them into grease useful for boiling their soap, "human Tallow giving it a Beauty and Fineness which that of no other Animal could do." Jamaicans, he argued, were doing something similar, in their hard-driving practices because "to kill a Man for the sake of his Grease, or to make him melt it away in hard Labour for another's Profit, is not so very unlike." "An ingenuous Man," he added, "might perhaps say as much for those Savages who *kill* to better their Soap, as for us who *inslave* to make our sugar, and defend their Practice of dispatching a Man at once, and clapping him in their Cauldrons, as well as we can ours of forcing a Man to consume his Life in Attendance upon our Coppers." He concluded: "And both no doubt in Policy are right, for what signifies the Life or Liberty of an Outlandish Man, if you can but send better Goods to market?" This whimsy fantasy, read in conjunction with the rest of the pamphlet, indicates that an acceptance of the immorality of slavery— though it disgraced the name of the "generous Free Briton, who knows the Value of Liberty," unlike the Spanish (naturally cruel), the French ("a slave himself") or the Dutch (who would "sacrifice everything to Gain")—necessitated parliamentary action because planters were happy to work their slaves to death for immediate gain. They were the sort of people who bought stock in the South Sea Company at "exorbitant prices," hoping to sell out quickly at great profit, even though they knew the situation could not end well. "This is at present pretty much the Case of many a *Jamaica* planter," he noted, "and the End must be the same [as the South Sea bubble] if some strong Remedies be not applied."[41]

The remedy he suggested—the abolition of the slave trade—was never a viable option, given the support for the trade in all areas of the imperial government, but his pamphlet shows how imperial statesmen as early as the mid-1740s were associating the abolition of the slave trade with security in ways Fergus notes was key for William Pitt and George Canning in the 1790s.[42] But his other remedies—greater imperial defense, paid for by increased taxation—were adopted by Moore and the Jamaican Assembly in their reconfiguration of security after Tacky's Revolt was finally put down at the end of 1761 with the capture of the last conspirator, Simon, in the mountains of St. Elizabeth. For Trelawney, fixing Jamaica's security had to be at the forefront of parliamentary attention, with only paying off the public debt and stopping "a Parcel of *Highlanders*, with the Scum of the *Lowlands*, . . . as they list through a once warlike Land with Impunity," being more urgent topics of public interest.[43] Linking the pacification of Jacobitism in Scotland with the security concerns of Jamaica illustrated the ways in which events in one part of the empire intruded into other parts.[44]

That Trelawney was an especially astute observer of Jamaican slave management practices in the middle of the eighteenth century can be seen in a brief analysis of Thistlewood's slave management practices in the difficult year of 1756, four years before the revolt broke out.[45] Thistlewood was essentially alone in dealing with a restive enslaved population. Jamaica had slave laws, but practical help in bringing the resources of the state to bear on slave management was in short supply. Before Tacky, slave managers in Jamaica were unconstrained in how they punished slaves, treating them with remarkable violence, in ways calculated to breed slave resentment. But legal mechanisms for disciplining enslaved people outside the plantation barely existed. Thistlewood, for example, brought an enslaved person before a jury of local planters in 1752 after he had barely escaped with his life just after Christmas when a runaway slave called Congo Sam attacked and nearly killed him. Thistlewood fought with Sam, overcame him with some belated help from slaves watching what was going on, handcuffed him, threw him in the bilboes, or stocks, and took him to jail to await trial, which happened on January 6, 1753. To Thistlewood's amazement, the three magistrate planters acquitted Sam, drawing the acerbic response from Thistlewood that they were "old women." Thistlewood was a new overseer; Sam belonged to a prominent planter; the magistrates did not want to annoy a peer by mutilating or killing without adequate recompense a valuable slave.[46]

The year 1756 was a difficult one, with provisions short and enslaved people hungry and discontented. Thistlewood was at his cruelest and least forgiving in this year, devising gruesome if inventive punishments such as "Derby's dose," when a slave defecated into the mouth of a slave and the slave who was then gagged. His slaves were starved, overworked, excessively punished, and desperately unhappy. One sign of that unhappiness was frequent running away. In 1756, Thistlewood lost 665 days from enslaved people who were runaway. Four persistent runaways accounted for 534 days in total. He mentioned on 104 occasions a slave who was a runaway or a slave who had been returned to the plantation. Thistlewood retrieved his slaves almost entirely by himself with some dubious help from his other slaves, used as a police force. Only once did Thistlewood record that a slave was sent home by another overseer. Otherwise, the return of slaves was subcontracted to confidential slaves.[47]

These methods used by slave managers in the 1750s to deal with runaways, methods that were based on individual action and collaboration with enslaved people, were not particularly successful. It was allied with another policy, which involved horrific punishment, as can be seen in how the eponymous Egypt (named after the plantation on which he labored) was treated after he ran away. He ran away once for thirty days, once for eighteen days, once for eleven days, and once for three days. His first absconding took place on 9 April 1756. He returned, presumably voluntarily, three days later and was whipped, "pickled" with lime juice rubbed into his wounds and had a "cart chain" locked around his neck. The rough treatment did not work. Egypt ran away again on 16 April, staying away for some time before being brought home on 1 May by "one of John James' Negroes." This time he had a pot-hook put about his neck (painful and put on in ways that made it difficult to sleep at night) after being whipped and sent to work in the fields. His pot-hook was taken off on 13 May and he was chained together with Quacoo, also a persistent runaway. Being chained together did not prevent the two slaves from running away, which they did on 5 June. He was away for a month before being returned to the plantation, without Quacoo, who had broken his chain and got loose from Egypt. Egypt was whipped and forced to work in chains for another five days before his chain was taken off. He died on 12 September 1756.

Before Tacky's Revolt erupted in late May 1760, Thistlewood followed old methods in dealing with runaways. He sent slaves out with tickets to search for runaways and then punished slaves himself when they were captured or returned of their own accord. But from midyear of 1760, these methods changed abruptly. On 30 July, Thistlewood noted how "several runaways and Addison's Cuffee belonging to Colin Campbell were flogg'd and mark'd at Savanna-la-Mar." The colonial state was starting to take responsibility for dealing with runaways, in the same way that they were devoting much time to the punishment of rebels. Thistlewood adapted quickly. On 9 August he gathered together his white under-ling, John Groves, his driver, Mason Quashe (the man usually sent out to look for runaways), four enslaved men, and one enslaved woman and set out on an expedition, "armed with 5 cutlasses, guns etc," to flush out "Achilles and his companions." His team found "many huts," food, drink, and belongings but no runaways. Thistlewood ordered the huts to be burned and "stood ambushed in the morass, arm'd with my gun and pistols in case any had come that way." The shock of Tacky galvanized white managers to take enslaved discipline more seriously, though Thistlewood thought still more could be done. He lamented on 9 November 1760, "But one Negro come to me for a ticket, the rest go without no person questioning them!" Thistlewood thought that there were "500 Negroes on the Road to Leeward every Sunday with plantanes etc," but "few had tickets." One sign that whites were taking security more seriously came a month later. On 11 December, Thistlewood's employer, John Cope, told Thistlewood that the Westmoreland Vestry was following up on an order from the government that the militia would patrol on both horse and foot for several weeks around Christmas "to allow no meetings of Negroes and no Negroe to be off his plantation without a ticket."

—

This is not my first evaluation of Tacky's Revolt as a key breaking line in Jamaica's eighteenth-century history. In previous accounts, I outlined how the revolt developed in Westmoreland; have used Thistlewood's diaries to explore day-to-day events in the rebellion; have examined how the revolt was received in Britain (badly) and the role that it may have played in the early development of British abolitionism; and have discussed how the successful resolution of the conflict by the Jamaican state affected vital areas of colonial politics and society such as the development of policies that constrained the possibilities of free colored people acquiring wealth and political influence. I also deal with the consequences of Tacky's Revolt in Chapter 4 in this volume and the evolution of the Jamaican state between 1739 and 1795 in the Epilogue. It would be tedious in this chapter to repeat what I have written before.[48]

What is important to note, however, is that silences and absences in the archives compromise our understanding of this conflict. It is a strangely and incompletely documented event about which, while we have some idea of "what" happened, we are left to struggle with understanding "why" the principal actors acted as they did. Long's account in *The History of Jamaica* is the only full account of the insurrection, and it is highly polemical even when it appears to be an accurate retelling of events. The only extensive contemporary description of the event is in Thistlewood's diaries. To these major sources can be added some fragmentary letters and official documents, most of which have been usefully collected in a website created by Vincent Brown, and newspaper accounts in Britain and North America, surveyed exhaustively by Maria Alessandra Bollettino.[49] The evidentiary record on Tacky is thin and confusing, even in comparison to other eighteenth-century slave rebellions. Trying to understand Tacky's Revolt highlights a problem endemic to scholars of eighteenth-century slavery, which is that almost everything we know about slave rebellions comes from biased and compromised sources, in which the voices of enslaved people are erased in favor of the voices of their oppressors.[50]

Having Wager and Henry Moore as the principal players in this account allows us to reflect a little on the historical archive and how it privileges some actors over others.[51] That Wager is largely invisible in the historical record is unsurprising: he was, presumably, illiterate or at least left no records about his life and opinions for historians to access. He was tortured to death, but the Jamaican authorities do not appear to have interrogated him about his motives before putting him in a gibbet, in which he starved to death over a period of days. It is one of the more curious facts about the repression of Tacky's Revolt that white authorities showed virtually no interest in finding out why it occurred, who was its leadership, what the links were between leaders and followers, and how enslaved people on isolated plantations whose movements were meant to be strictly controlled by masters were able to communicate with each other over vast distances secretly and effectively. White Jamaicans' main concern instead was in establishing spectacles of terror and awesome displays of state power, intended to frighten any slave who might be contemplating future rebellion.

Thistlewood told a revealing story about how whites prevented knowledge of the origins of the revolt from being known. On 2 September 1760, while in a party to track down runaways, his employer, John Cope, told him that when he was watching a slave called Cardiff being "Burnt at the bay," Cardiff told him "and many others present that Multitudes off Negroes had took Swear that if they failed off success in this rebellion, to rise again the Same day two year and advised them to be upon their guard." Cardiff, according to Cope, "was going to make further discoveries by accusing Col. Barclay's Tacky and others." Having his slave accused, however, put Barclay "in a rage." He "Call'd the marshall Villain for not making a fiercer Fire." Cardiff died quickly thereafter without naming others, which Thistlewood thought "a Sad Mistake."[52]

If it wished to prevent knowledge of the revolt from emerging, then the Jamaican state succeeded remarkably well. We know little about the revolt and its origins while we have much more knowledge about how it was put down. As John Garrigus and I have argued, the main significance of Tacky's Revolt was that it was the emblematic slave rebellion of the mid-eighteenth century. Colonists in the Greater Antilles interpreted its repression as showing that if they and their state were resolute, determined, and ferocious in how they dealt with rebels then they could survive whatever trouble enslaved people threw at them. The reception in Britain of the ferocity of the Jamaican state in destroying slave rebels was less positive for colonists. Britons were horrified that their fellow countrymen could sink so low as to act with such barbarity. It helped start a small abolitionist movement in Britain in which the reaction to the Jamaican slave rebellion was seen in Christian terms—planters as savages and rebels as Christian martyrs. These actors were abstract figures, however. There was no apparent interest in investigating who Wager was as a person (or Tacky or Jamaica, the leaders in St. Mary, or Simon, the last leader to be apprehended, in St. Elizabeth). The rebel represented an abstract ideal, either the relentless villain of Edward Long's description or the noble but brave savage, dying in torment but doing so without showing any awareness of pain.[53]

We know less, for example, about Wager than we do about other slave rebels, notably those who failed to put their plans into execution. Authorities intensively interrogated slaves involved in the 1741 plan to burn down buildings in New York; the slaves who sought insurrection in Antigua in 1736; the slaves who were falsely accused of poisoning whites in Saint-Domingue in 1757; and Denmark Vesey and his coconspirators in South Carolina in 1820.[54] They killed Wager without, as far as we are aware, any investigation into his motives and aims. What little we know about him comes from snippets of information, mostly from Thomas Thistlewood. Here is the sum of our knowledge about him. He may have been a "Prince in Guinea," and a slave trader and warrior, tributary to the king of Dahomey. He "was supriz'd while hunting, and Sold for a Slave, brought to Jamaica and sold to Capt. Forest." Thistlewood thought that Wager had "come to this country 6 or 7 years ago," or in 1753 or 1754. Thistlewood was told by Stephen Parkinson that Apongo sailed with Forrest during the Seven Years' War, including going in the ship *Wager* to England, "from whence he had his name."

He may or may not have seen service in Guadeloupe during the Seven Years' War, but he probably knew other slaves who belonged to Forrest who had been in Guadeloupe. Long alleged that "these men were the more dangerous as they had been in arms at Guadeloupe, and seen something of military operations, in which they acquired so much skill." Wager was clearly a remarkable slave, whose qualities even white Jamaicans noted. Thistlewood told an amazing story that Wager "used when a Slave Sometimes to go to Strathbogie to See Mr. Cope, who had a Table Set out, a Cloth laid, &tc. for him, and would have purchas'd him and Sent him home had Capt. Forest Come to the Island." John Cope was the father of Thistlewood's employer and in the 1740s had been governor of Cape Coast castle in the Gold Coast. This story seems to confirm that Apongo/Wager was Coromantee, a fact noted in a report in the *Spanish Town Gazette*, reprinted in the *Pennsylvania Gazette*. That report noted that Wager was "their Chief Coromantee Commander" and was taken up by government forces on 3 July. Thistlewood noted this in his diary entry for that day, commenting that "he was King off the Rebells: but deposed off late since wounded." Thistlewood repeated on 12 July 1760 a rumor: "It is said, Wager and his wife had a Quarrill in the Negroe ground, the Sunday they began, and that She threatened to discover the Plot, was the Occasion of beginning that Evening, other ways not to have been till the Shipping had Sail'd." Dr. Gorse told Thistlewood that Wager told him when he was being tried (of which no records exist) "that the Night" the revolt broke out "he advised Coming directly to the Bay, but the others were afraid too much." His end came in early August. On 29 July Thistlewood noted that Wager was "to hang in Chains for three days then be took down and burnt." Perhaps fortunately for Wager, he "died before his three days expired," presumably on 1 or 2 August 1760. On 1 August Thistlewood commented that according to Dr. Miller the rebels' plan was to "kill all the Negroes they can, and as soon as dry weather come fire all the plantations they Can, till they force the Whites to give them Free like Cudjoe's Negroes."[55]

Surprisingly, we do not know much more about his antagonist, Henry Moore, despite his prominence as a colonial governor involved in two major imperial crises, his literacy, and his high status as a baronet in an imperial regime devoted to knowledge and information.[56] It does not appear that he left any manuscripts or letters outside a small number of official letters collected in the colonial office records for Jamaica, 1756–62, and for New York, 1765–69.[57] Certainly the amount of material relating to him would not be sufficient for a biography. Unlike Wager, he has an entry in the *Oxford Dictionary of National Biography*, but that too is brief. A native-born Jamaican whose planter grandfather had moved from Barbados to Jamaica in the later seventeenth century, he was born on February 7, 1713, and was educated at Eton and at Leiden University. He married, aged thirty-eight, Catherine Maria Long, the elder sister of Edward Long, with whom he had a daughter and a son, Sir John Henry Moore (1756–80), a poet and Cambridge graduate who died unmarried aged twenty-three. Moore returned to Jamaica in the 1730s and after serving as an assemblyman in the 1740s became a member of Council in 1752 and quickly moved up the ranks to become lieutenant governor between 1756 and 1759. After a brief respite he returned after Governor Haldane's death in 1759 to run the island until being replaced by William Henry Lyttleton in 1762. It was a successful governorship, with Moore retaining a healthy relationship with the legislature, unlike his predecessor, Charles Knowles, and his successor, Lyttleton.[58] His major achievement, of course, was overcoming Tacky. As a reward for his services, he was created a baronet on 28 January 1764 and appointed governor of New York, where he served between July 1765 and his death on 18 September 1769. He arrived in New York in a period of crisis, during the Stamp Act furor, but he handled this crisis more adroitly than his predecessor (and successor), Colden Cadwallader. Joseph Tiedemann calls him a "shrewd politician, who adeptly governed the province during a tense period in imperial relations." His obituary celebrated him as a statesman who had governed "with such a Degree of *Wisdom* and *Temper,* as to gain the Approbation of his Sovereign and the Esteem of the People committed to his care."[59]

—

If we consider Moore and Wager as the principal protagonists in the great slave rebellion of 1760, then in the context of Jamaica the clear winner of the contest was Sir Henry Moore.[60] His resoluteness and quick action, as well as the fortuitous presence of two regiments (the Forty-Ninth and Seventy-Fourth) and a naval squadron on the island, enabled him to defeat rebels easily in St. Mary's and with some more difficulty in Westmoreland. He utilized British regulars, sailors from the British navy, local militia, Maroons, and free black rangers against the rebels to great effect. His actions were praised effusively by Admiral Thomas Cotes, who reported to Secretary of the Admiralty John Clevland about how the rebellion in St. Mary's had broken out and how it had been put down, largely owing to Moore declaring martial law and sending "two strong Parties of Regular Troops to support the Militia and free Negroes, who have had two smart Skirmishes with the Rebels."[61] The British regulars—battle-hardened troops—were especially effective, being in "the Van" of attacks and behaving with "great impetuosity" in overcoming rebels. They and Maroons and free blacks killed at least seventeen rebels in St Mary's on 14 April, including the rebel leaders, Tacky and Jamaica, and at least 130 rebels in Westmoreland on 2 June, for the loss of one soldier. Leonard Stedman, writing on 4 June, reckoned the rebel losses in Westmoreland as about two hundred and commented on the ferocity meted out to prisoners—"what they bring in alive we burn and some we hang in Gibbets."

The rebels were quickly out of ammunition, and Stedman claimed they were so desperate that "we find daily their Women and Children hanging to prevent there coming in." The Assembly noted in a "List of rebels taken and killed" on 27 June 1760 that "from the best accounts a great number died of their wounds, [and] many have destroyed themselves, several drove over Precipices and at least 150 stole [back] into their various Estates." By 14 July, Moore was able to report to the Board of Trade that the revolt was over "by the vigilance of the Officers . . . and the Principal Offenders taken up, try'd and punished." His actions, he claimed, had squeezed the rebels between two forces so that "the daily advantages gained over the Rebels were so considerable that they were soon reduc'd to great distress."[62]

Moore went quickly to work to effect laws that would make rebellion more difficult. Some of these plans involved improving plantation discipline, making local slave patrols more effective, and more rigorously enforcing slave laws. In addition, Moore "introduced a proper discipline among the Militia." He ordered barracks to be built in each parish and made sure that British regulars were kept on the island on a permanent basis. He even instituted a lottery "to provide regular payment for His Majesty's forces." These reforms to support the military and improved policing measures were not cheap. They led to large increases in imperial expenditure and in local taxation, with colonial taxes jumping from £20,167 Jamaican currency in 1750 to £71,248 in 1760, £53,571 in 1761, and about £35,000 per annum until the mid-1770s. In addition, local parishes spent about £35–45,000 per annum on barracks for several years.[63] White Jamaicans willingly paid such taxes to keep themselves safe.

The lessons learned by the Jamaican state following Tacky greatly improved security on the island. These improvements can be seen in how slave managers were supported in dealing with *petit marronage*. After Tacky, instead of having to rely on the assistance of slaves when apprehending runaways, white supervisors received assistance from soldiers, black and white, who were available to go on slave patrols after runaways. Planters were more careful in making sure that their slaves had tickets when sent out on the road. If a slave was found on a road without a ticket, whites were entitled to have them seized and sent to jail, no matter the inconvenience this might cause for plantation slave management. And the courts took slave discipline more seriously. Justice was less likely to be spontaneous and variable and more likely to follow rules of procedure. On 14 June 1765, when Thistlewood hauled a persistent runaway, Plato, before the Westmoreland slave court, he was met with a quite different response than in 1752. Plato was "Sentenced to have 100 lashes at 4 difft Places on the Bay (25 at each place) and to have his right Ear cut off, which was immediately executed." Thistlewood was on the bench himself on 20 March 1776, to hear a case about a "runaway Negro belonging to Mr. Beckford." He joined others in acquitting the runaway, "he not having been 3 years in the Country." But the chief magistrate, Dr. Panton, over-ruled the jury, ordering the runaway to have "39 lashes under the gallows." In short, collective white unity replaced a highly individualized response to prob-lems of slave management. In the aftermath of the foiled Hanover plot of 1776, for example, the parish of Westmoreland on 16 April 1777 held "a meeting off the Justices and Vestry" and "resolved to be on our guard and to inspect their behav-ior Narrowly." The authority of white power was supported, and enslaved people were kept cowed and terrorized.

Terror was insufficient to stop enslaved people from rebelling, but it was enough to ensure that slave rebellions failed. Between 1760 and 1776, there were at least three conspiracy scares in Jamaica, two of which were started but quickly put down, in St. Mary's in 1765 and in Westmoreland in 1766. Another conspir-acy, in Hanover in 1776, followed the withdrawal of the Fiftieth Regiment to fight in British North America on 3 July 1776. The revolt did not come to fruition, but its planning suggests it may have been as likely to have been as extensive as the 1760 rebellion.[64]

The rebellion in St. Mary's in 1765, like that in 1760, came as a surprise to white residents. The events of the rebellion can be followed in the correspon-dence of Moore's successor as governor, William Henry Lyttleton. Nevertheless, whites were paying more attention to what slaves were doing in 1765 than five years earlier, as seen in the already cited letter from David Miller to Brigadier General William Lewis written five days before the rebellion.[65] Lewis himself wrote a revealing letter to Lyttleton after the St. Mary's rebellion had been put down that outlined at length efforts taken to prevent rebellion in Westmoreland. He noted that "sometime last year I discovered that a Negro Smith [owned by] Mr. Crawford had made several painted Irons to fix on the end of launces for one of Capt. Forrest's Negroes." Lewis caused "the fellow to be taken up" and interrogated. His overseer declared he was innocent so "the fellow" was released. In April the overseer told him that "the Negroes of Forrest's and Midgeham Estates were in a Conspiracy to rise and kill the white people." He thought the rumor unfounded but ordered British regulars to the estates "to preserve the tranquility of these parts." He was annoyed, however, that the trustees for each estate "positively refused to allow any soldiers to come into their estates" and asked the governor for "the authority to push the matter." Hearing about the rebellion in St. Mary, Lewis "used every precaution necessary to strike Terror in the minds of the Slaves," including banning Drumming, enacting a curfew, and calling out the militia.[66] The result of such actions was to keep the rebellion of 1765 confined to St. Mary's.

A key protagonist in the St. Mary's revolt was Zachary Bayly, an extremely rich Kingston merchant and politician with large sugar estates in the parish. He had also been a central figure in Tacky's Revolt in 1760 where he had been kept "six weeks closely employed" in what he described as the most "dangerous or troublesome Affair" he had ever been engaged in.[67] These events seem to have prepared him for the revolt that broke out on the morning of 25 November 1765. Writing to the governor on that day at 11 A.M., he described at great length how "Coromantee slaves" from Whitehall plantation had murdered the planter, Mathew Byndloss, and tried to kill the owner of Whitehall, Mary Beckford and her overseer. Bearing in mind Bayly's narrative was self-serving in regard to portraying him as active and vigilant from the start, he described mustering all the white people in the district and chasing after the slave rebels with twenty-one whites and twelve blacks, sending for support from the Maroons in St. Georges and Portland, and promising to "take what people I can muster . . . [to] traverse round the plants and provision grounds in the Neighbourhood," where he believed that the rebels were "skulking."

He asked the governor to send sailors and soldiers and believed that if this was done the governor would not be "under the disagreeable necessity of putting the Island under Martial law." These requests were quickly acceded to, leading Bayly to thank the governor for his "Vigilant Care" in getting soldiers to Bayly within a day. He declared that "I shall not leave this parish while one of those villains are out," noting that already five rebels had "destroyed themselves," after realizing that they had little support, according to Bayly, within the enslaved community. The revolt did not spread. William Patrick Browne wrote on 30 November that the slaves in St. John "are very quiet and by appearance not disposed to be otherwise."[68]

Unlike 1760, Bayly and the magistrates in St. Mary's were very interested in finding out why the revolt occurred. A slave called Creole Cuffee testified in December that it was a Coromantee plot to "fight Backra." Bayly agreed, although he found the testimony given by different slaves to be conflicted, unbelievable, and self-serving. Quickly, all the Coromantees in the parish were seized, meaning that Bayly found "about 100 negroes in confinement, and most of the white people in the parish employed in guarding them," even though there was "no evidence to be found against any of them except two or three of the Frontier negroes" (from his own property). Despite this, Bayly urged for executions to be started immediately, "lest they should by any means escape." Between 2 and 19 December 1765, three rebels were burnt to death, and ten were hanged. Another thirty-three slaves were transported, and eleven were acquitted. Sholto Douglas, a local planter, whom Bayly praised for "his zeal and assiduity," investigated the revolt intensively and "discovered a conspiracy in which about thirty male and female of his own and neighbours' slaves are said to be concerned." The ringleader was determined to be Abruco, alias Blackwell, "head boyler and a very sensible fellow," who was "apprehended the first morning, tried and burnt on Wednesday 28 November."[69] According to Long, he "had been tried formerly on suspicion of being concerned in the rebellion of 1760, and [was] acquitted for want of sufficient evidence."[70]

The measures put in place in 1760 also worked to quell slave rebellion in 1776, a rebellion, Sheridan notes in his extensive article on the planned revolt, that was not a purely Coromantee-led revolt but that "indeed the most valuable Negroes in general, on the Estates to which they belong, Creoles, Coromantees and Eboes are principally concerned." Reverend Dr. John Lindsay reported a conversation with a servant who told him that the plan was extensive, that it involved poison and a "General Massacre of Whites throughout the island," and that "even the Creole Negroes, who were the savers of their Master's Mistress in the Rebellion [of] 1760 were now engaged against them."[71] It appears to have been a carefully planned and wide-ranging plan, involving enslaved people from forty-seven estates containing 8,618 slaves, of whom perhaps hundreds were directly involved in planning.[72] It was discovered on 15 July, a week before it was meant to be put into operation. Governor Sir Basil Keith sprung quickly into action, declaring martial law and an embargo on shipping. He dispatched a company of light infantry to the parish to assist the mustered militia and ordered the sloop *Atalanta* to Hanover. The parish was equally active, seizing and examining at length the ringleaders, and proceeding immediately to executions and transportations. Eventually, of 135 rebels tried for conspiracy, seventeen were executed (mostly by hanging), forty-five were transported, eleven severely whipped, and sixty-two acquitted. The savage punishments were accompanied by a strengthening of the military defense of the island, as Keith detailed in a letter to the British government on 6 August.[73] Keith insisted that everything was now in order but admitted that there was an "apparent Spirit of insolence among the Slaves, over the whole Island. . . . We are now in the most imminent Danger and the most pressing Necessity."[74]

Keith's actions were praised by white legislators, who congratulated him on his vigilance and activity. They quickly passed six laws to improve security through strengthening the militia and providing for "the better order and government of the Maroons (there was some suspicion of Maroon encouragement of the planned revolt, though Superintendent Colonel John James denied this comprehensively), as well as supporting planters to retrieve runaway slaves through raising raiding parties to range through the countryside. Whites in general took the planned revolt very seriously and greatly increased their surveillance of enslaved people. A Hanover planter in May 1777 declared that a rebellion was impossible as there was "always a strong guard of white people at all suspected places." The expense of such activities was heavy, however, and was eventually borne by the Jamaican government rather than by Hanover Parish.[75]

The transformation of the Jamaican security state began in 1739 with the conclusion of a treaty between the Jamaican government and Maroons. The treaty signed in that year and the year following became an essential part of Maroon identity and the foundation of a lasting relationship between the Maroons and white society.[76] The Maroons were not numerous but were a formidable fighting force, whose efforts were crucial to white success in 1760, when they fulfilled the terms of the treaty of 1740. Edward Long described with some amazement a dance they made in honor of Governor Lyttleton in 1764, full of "wild and warlike capers," which, Kathleen Wilson argues, both "pledge[d] their skills on behalf to the powers that be" and "invited speculation about the consequences of the withdrawal of such consent," as demonstrating both "militarism and subjection."[77] Through the treaties, Maroons agreed to hunt down and return runaway slaves and to aid in the suppression of internal and external attack. In return, the Jamaican state agreed to give them their own land for cultivation and a superintendent under whom they were supposed to fight. The cost of appeasing the Maroons was substantial, around £2,000 per annum by 1769, spent on white superintendents, rewards for returning runaways, and bribes to confirm their allegiance.[78]

White colonists occasionally muttered about Maroons as unreliable allies and as becoming "too formidable." The worst year for white relations with Maroons prior to the start of the Second Maroon War in 1795 came in 1774. In ways typical of Jamaica, a small incident flared up into something more serious. Windward Maroons were called out to capture a runaway at Old Harbour in St. Catherine Parish and, possibly intoxicated, got into a dispute with a sea captain, who thought they were about to seize his enslaved property. Words turned into action, and gunfire was exchanged, with the Maroons killing an enslaved man after the sea captain had fired and missed their leader, Captain Davy. The killing of an enslaved man encouraged another sea captain to start to give the Maroon killer a whipping (showing how whites considered all blacks, free or enslaved equivalent to slaves). After repeatedly warning the captain that was giving a whipping that he did not want to hurt him but that he needed to stop, the Maroon, Sam Grant, shot the sea captain and escaped to the Maroon town of Scott Hall. It "provoked a general Alarm throughout the Island, as if the maroons had revolted, as many weak and wicked Reports were raised on the occasion." Some white Jamaicans wanted to change the balance of power in the interior by hemming in the Maroons into their settlements and trying to make them conform to the colonial legal system. That was not the general opinion, however. The Jamaican state made sure that the fracas at Old Harbour did not result in Sam Grant being prosecuted.[79]

In general, however unreliable the alliance between the two parties might have been, it was an enduring partnership. It was also a peaceful relationship, in marked contrast to normal colonial relations with indigenes, where violence was an everyday occurrence. Helen McKee stresses how nonviolent Maroon-white relations were between 1763 and 1795, a tendency more striking given the inherent violence endemic to Jamaican society.[80] The reason for the harmony that largely existed between whites and Maroons is that wealthy white Jamaicans were aware of what an effective security force the Maroons were and how relatively easy and cheap it was to keep them happy. Whites came to rely on them for their own safety—Long claimed that planters often made special deals with Maroons, offering them protection money in return for Maroons keeping restive enslaved people under control. In 1763, for example, Thistlewood related how Cudjoe, the Maroon leader in western Jamaica, chased and captured eleven runaways who had murdered some whites and then disappeared, killing three and bringing the other eight to Savanna-la-Mar, where they were burnt and hanged. They gave intelligence of rebellion to whites on occasion and were active in attacking and subduing runaway slave bandits, the most notable being the infamous Three Fingered Jack in 1780–81. They occasionally worked within the plantation economy and acted as a local police force. Colonists came to depend on Maroons "for protection as mutual recognition of the benefits of the alliance emerged."[81]

—

There are many reasons why Jamaica did not become the fourteenth colony to declare independence from Britain in 1776. Jamaica was instinctively loyal and content with the imperial relationship it had forged in the settlement with the Maroons in 1739 and in the reconfigurations of security arrangements on the island that were hammered out by Moore and Lyttleton with the Jamaican Assembly, white Jamaicans, and Maroons after 1760. These arrangements led to considerably higher rates of taxation in Jamaica than in British North America but to greater security for white colonists. The year 1760 is an important one for the internal history of Jamaica and for considering Jamaica's role in a British Atlantic World starting to come apart following the Stamp Act in 1765. In short, the issues that concerned British North America about the imperial relationship—taxation without representation and increasing levels of taxation, parliamentary sovereignty, and preserving mainland colonies from Native American attacks and possibly from slave insurrection—did not matter as much in Jamaica. White Jamaicans were happy to pay high taxation and in the booming years after the end of the Seven Years' War could afford to pay those taxes. Parliamentary sovereignty was, as noted above, less important than debates over prerogative, debates largely solved by judicial decision in 1774.[82] Talking about declaring independence from Britain in a society in which nine out of ten people were enslaved was foolish.[83] In addition, Jamaica's position in the Greater Antilles, sandwiched between colonies belonging to France and Spain, meant that Jamaica did not have the freedom that North American colonies imagined they had to strike out as independent entities unconnected to European powers.[84] White Jamaicans knew from their near-death experience in 1760 how invaluable it was to have the imperial state guaranteeing their security. They were prepared to pay the sums necessary for protection. Importantly, those sums were large but not exorbitant and affordable in a rapidly growing economy.[85]

The irony of the American Revolution in British North America, of course, is that the hugely expensive and destructive war that led to independence could be funded only by a massive increase in state debt, by enormous increases in taxation, and by the creation of a larger and more intrusive state than had previously existed in the mainland colonies.[86] The fiscal reforms of the West Indian–born U.S. treasury secretary, Alexander Hamilton, in the early 1790s and the fortuitous godsend of the French Revolutionary Wars, which initiated steady and high economic growth rates in the United States between 1793 and 1807 following a calamitous collapse of gross domestic product between 1776 and 1789, solved the problem of government revenue—the single biggest problem that the Founding Fathers needed to address in the crucial First Congress of 1789.[87] As was the case in Jamaica between 1739 and 1795, statesmen in the United States found that the population was willing to accept increased rates of taxation if they could see that the purpose for which such taxes were used contributed to the national welfare and if the system of taxation agreed on was transparent, contestable, and controlled by representatives of the people rather than distant and arbitrary powers.[88]

Jamaica was another story. The French Revolutionary Wars were as much an economic godsend to Jamaica as to the United States as it profited greatly from the destruction of plantation agriculture in its major competitor as a result of the Haitian Revolution.[89] But the shock of slave revolt in Saint-Domingue, and the expense of a global war in which the Caribbean played a crucial part, forced Jamaican slaveholders into a new political and economic dependence on Britain.[90] Planters fell in line with imperial dictates. Not only was the level of taxation raised to much higher levels than before 1793; the imperial government was firmer in how they asserted political control in the Caribbean. It is significant that despite a great deal of territory being added to the British Empire in the West Indies in the 1790s, Britain refused to allow any new West Indian colony a legislative assembly and enforced new legislation that forced planters to modify their slave management practices.[91] Unlike 1760–61, when white Jamaicans were able to reinvigorate local security systems to prevent further slave rebellion affordably, the cost of security became increasingly problematic from the mid-1790s, posing an "unprecedented burden" on white Jamaican society. Jamaica had to cope with putting down a new and very expensive Maroon War in 1795, which nearly bankrupted the state, just as it was being asked to pay large amounts of money to an imperial state embarked on a costly campaign in Saint-Domingue.[92] Aaron Graham estimates that government spending ratcheted up to £420,000 in the late 1790s, and public debt, which had reduced from £114,608 in 1782 to £11,657 in 1793, grew massively to £457,801. That meant that the tax rate per capita doubled in the late 1790s, peaking at 6–8 percent in 1802.[93]

It was the French Revolutionary Wars and the Haitian revolution that transformed imperial thinking about Jamaica and that led to the destruction after the Second Maroon War of a security system that had served white Jamaicans well for half a century. These events, more than the consequences of the American Revolution and the rise of abolitionism, both important matters, were what contributed most to the decline of the Jamaican planter class in the 1790s and 1800s, mainly because the radicalization of the Atlantic World through the Haitian and French Revolutions thrust Jamaicans into a maelstrom of geopolitical considerations that compromised their ability to control their own politics.[94] The defeat of Tacky, Jamaica, Wager, and Simon in 1760–61 convinced wiser heads in the assembly (those wealthy planters like John Tharp who expended considerable effort in trying to preserve harmonious relations with Maroons in the 1780s and 1790s and who disagreed strongly with the Jamaican government's decision to transport Trelawney Town Maroons to Nova Scotia) that the settlement of 1760 made Jamaica safe from all attacks, internal and external. The settlement depended on increased surveillance of the enslaved population and unwavering support of slave managers in their management policies; a rapid and extraordinarily violent response by the Jamaican state to any sign of enslaved unrest; and the pacification of Maroons through annual subventions and using them as an internal police force. The settlement had been foreshadowed by Edward Trelawney's treaty with the Maroons in 1739 and confirmed through Henry Moore's resolute actions after Tacky in 1761–62. Trelawney and Moore were notably successful governors.

Another very successful governor was Archibald Campbell, a general in the British army, who governed Jamaica in its next period of crisis, in 1781–82, when French invasion was likely.[95] In 1783, after Admiral Rodney's victory at the Battle of the Saintes and the destruction of most of the French troops destined to invade Jamaica by disease in Saint-Domingue, Campbell wrote a long memoir explaining how he would have defended the island through retreating to the interior districts of Jamaica, an area that "in the last twenty years . . . have undergone immense improvement" and which were naturally healthier than the coastal regions of the island.[96] Campbell did not mention Maroons, though their presence within the interior as a police force was implicit in his analysis. In a communication with Lord Shelburne he admitted that "no lasting support can be reasonably expected from the Militia of a sugar colony, but that it must derive its permanent and sure defense from Regulars alone."[97]

It might seem that Jamaica, with four men-of-war and other vessels, eight thousand seamen, 3,014 regulars, 6,801 militia (of whom 1,417 were free people of color), one hundred "stout negroes to Drag guns," and five thousand black pioneers would be overwhelmed by an invading force of thirty thousand French soldiers, of whom twenty-four thousand were regulars and fifty "sail of the line." Campbell, however, had great confidence in his ability to see off the enemy through abandoning Kingston to invaders. He planned to retreat to the interior, where he believed that an army "posted in the healthy mountains" and with adequate provisions could hold off the French for two years. The French, would be "necessarily exposed to the sultry plains of the Coast, deprived of every shelter, except what Tents could afford, and which in case of rain would prove ruinous to their army." In short, he believed "that it is certain that the British Forces would possess every advantage which could possibly fall to the lot of an inferior army in defense." Indeed, he argued that "the Island is capable of being defended with infinitive advantages against a very superior number." The French, he predicted would have tried "to get inhabitants to defect" (the common response by white planters in most Caribbean islands to invasion) and would have encouraged slaves to "think the times favourable for insurrections." But white Jamaicans, he averred, were in 1782 "cheerful, ready for service and in proportion as their danger increased, inspired with a degree of military endeavor, which seemed to indicate hopes of improvement." They and their black soldiers would have offered "gallant resistance" if the French had invaded. He was certain that "the combined forces of France and Spain in 1782 and 1783 would have been completely discomfitted in their attempts against Jamaica."[98]

If some governors were successful, others fitted Edward Long's contemptuous epithet of Jamaica's governors as "hirelings, tools and sycophants, men of narrow souls and mean prejudices," mostly interested in self-advancement, "whose conceit and vanity, on being elevated to a vice royalty, will pervert what little understanding they possess."[99] Alexander Lindsay, Earl of Balcarres, who started an unnecessary and vicious war against the Maroons in 1795–96, which wrecked the delicate balance of security in Jamaica, was one such man overwhelmed by "conceit and vanity." The outbreak of the Haitian Revolution spread panic throughout the island, like what had occurred in 1760 and 1782.

Balcarres, however, unlike Campbell and Moore, did not cool the panic but exacerbated it. Despite an almost complete lack of slave unrest following the outbreak of revolution in Saint-Domingue (Governor Sir Adam Williamson commented in January 1792 that enslaved people were "well-disposed and obedient" and that they wondered why the white people were so busy), a quietude that lasted throughout the 1790s, Balcarres was heavily influenced by the planter-historian Bryan Edwards, whose account of a visit to Saint-Domingue in 1793 was highly inflammatory and who had always an animus toward Maroons. He convinced himself that Maroons were in league with the rebels in Saint-Domingue and that they planned to arm Jamaican slaves and make themselves the masters of the island.[100] It took only the smallest incident—an event like that which had occurred in 1774, the flogging of two young Maroon men—for fighting to break out. Balcarres favored a policy of genocidal extermination rather than treating with Maroon leaders. When invited to meet with six Maroon captains in early August 1795, he had them all arrested and imprisoned. He even imported fighting dogs from Cuba to tear Maroons to pieces. His actions divided the white community. The residents of St. James and Trelawney formed a "violent opposition party" under the leadership of John James, the well-regarded ex-superintendent of the Trelawney Town Maroons. Balcarres responded with the outrageous accusation that James was a traitor who wanted to replace him in office and whose plan was that "the Maroons with himself in their head and supported by the negroes were to give law to this country."[101]

Balcarres won a hard-fought and pyrrhic victory. He boasted to the Duke of Portland that because of his efforts "the most perfectly tranquillity is restored to the Island: the Slaves on every Plantation are obedient, contented & happy. . . . Thus, has ended the Nation of Trelawny Maroons, a People . . . [who] were not to be overcome but would ultimately acquire the Dominion of this Island." He collected a £700 reward from the Jamaican Assembly for his actions, which included a cynical violation of the treaty, which Major General George Walpole had negotiated with Maroon leaders so that Maroon leaders were flogged and six hundred transported to Canada and then to Sierra Leone.[102] Walpole was disgusted with Balcarres's perfidy, resigned his commission, refused to accept a £600 Assembly grant, and became an abolitionist.[103]

The financial cost of victory, as we have seen, was substantial and meant that henceforth Jamaica depended on public credit as the gap between income and expenditure was always negative. More important, Balcarres's folly broke down a workable system of security, which had served the white inhabitants (but alas, not its black citizens) very well since 1739. White Jamaicans were no longer as clearly in charge of local politics as they had been since 1760. That Balcarres was able to act as he did in northwest Jamaica when planters resisted his actions vigorously, leading him to complain that there "was an imperium in imperio" in Trelawney and St. James, with residents who "opposed and thwarted everything that was done," showed that the imperial state was beginning to place wider geopolitical interests over local concerns. In this respect, 1795–96 marked the end of the American Revolution in Jamaica. Or at least it marked the end of a period in Jamaican history where planters had a disproportionate influence in British politics and were largely left in charge of politics and Jamaican colonial spending. William Pitt the Elder declared in the mid-eighteenth century that sugar planters should be thought of in the same way "as the landed interest of this kingdom." It was a "barbarism," he thought, "to consider them otherwise." His son, by contrast, was not a friend of the planter interest, always being sympathetic to abolitionist arguments in favor of ending the slave trade. He was convinced that reforming the British slave system was the only way to secure the colonies from slave rebellion or the disaster that had befallen colonists in Saint-Domingue. He wanted Britain to legislate to ensure that these aims succeeded.[104]

In short, by 1796 there was a new political relationship between Jamaica and Britain in which Britain broke with the principles that had shaped its eighteenth-century colonial system in favor of a more coercive approach to white colonists. Britain could not leave the West Indies to determine its own security arrangements because it was paying so much attention to the geopolitics of territorial expansion in the Caribbean at the expense of revolutionary France and beleaguered Saint-Domingue. This was a different ideological age, a new age of revolution from below that shattered the self-confidence of planters, amplifying their fears about personal and collective safety. Planters turned to the protection of an assertive and powerful imperial state, which insisted that the West Indies submit itself to the British legislature. Jamaica was part of a new global world that preoccupied leading statesmen in new and unusual ways. In 1792–93 Britain believed that the conflagration in Saint-Domingue allowed it the opportunity to carve out a vast new West Indian colony to replace its North American colonies and to complement its new possessions in India and Australia. But within two years, these plans had turned to dust. The British invasion of Saint-Domingue was a terrible disaster, with great armies destroyed by disease and little military progress being made. The year 1795 saw not just Balcarres's folly against the Maroons but also the largest slave rebellion in the British West Indies since Tacky's Revolt, Fedon's Revolt in Grenada. The British government doubled down in 1796, sending its largest ever expedition to leave Britain to Saint-Domingue and taking from the Dutch and Spanish colonies in the southern Caribbean and northeast South America.[105]

In such a febrile environment, what Jamaicans wanted was of little importance to a government seeing the West Indies as a theater for global domination. As Michael Duffy eloquently puts it, "the impact of the French Revolution on British attitudes to the West Indies was thus to raise them to an importance beyond all that had gone before, and then to deflate that importance and divert British attention." He continues: "the efforts of the 1790s thus constitute an attempt at a turning point around which the British empire failed to turn."[106] The success of Toussaint Louverture and others in Haiti showed Britons that the lesson learned by Jamaicans in 1760—that absolute repression of any slave revolt and vigilant surveillance of the enslaved population kept slavery secure and white colonists safe—was not inviolate against vigorous revolutionary resistance. Events north of Jamaica between 1791 and 1804 showed that the American Revolution was well and truly finished in the Caribbean and that a new phase in Jamaican history had begun.

Chapter 5

The Ambiguous Place of Free
People in Jamaica

Jamaican society divided along two great axes. The first separated the small number of free persons from the large mass of unfree persons. The second divided the population by race, with whites accorded the full privileges of civilized society and blacks excluded from civic participation. The correlation between conditions and color was significant: to be free was to be white; to be white was to be free. But the fit between color and status was not perfect. Some whites—servants, initially, and Jews throughout the seventeenth and eighteenth centuries—had limited freedom, and some blacks and colored were not slaves. White Jamaicans found it easier to resolve the former anomaly than the latter. Although distinctions of rank never disappeared and although anti-Semitism was both marked and virulent, both Jews and dependent white men were gradually incorporated into civic society. The historian Bryan Edwards claimed that a remarkable equality prevailed between all ranks of white men, a category that increasingly came to include assimilated Jews, many of whom had prospered in Jamaica.[1]

Incorporating free people with African blood (henceforth free people) into civic society, however, proved much more difficult, especially as a caste system based on a rigid definition of color became ever more entrenched. Racial orthodoxy ruled that African blood consigned a person to slavery. All slaves were black, and all blacks were slaves. The former proposition was undeniable: except for a minuscule number of slaves in the seventeenth and early eighteenth centuries who were Amerindian,[2] all enslaved people in Jamaica were African. But the latter proposition became untenable over time as a small but growing proportion of people of African descent left the ranks of the unfree and joined the ranks of the free. In doing so, these people violated a central tenet animating Jamaican life—that whiteness equaled freedom while blackness meant slavery. In the seventeenth century, when freed Africans and their children were very few, white Jamaicans pretended that they did not exist. That changed in the eighteenth century as, in part due to the gender imbalance in white society in the seventeenth century white men sought out black women as partners and had children with such women, leading to a large population of color and a somewhat diverse colonial society in which nonwhite free people occupied an ambivalent status. As Daniel Livesay notes in the most comprehensive study to date of this population, free people of color included free people, "whose plights were often not much better than the enslaved," and free people, who "endured a wide range of experiences." Only a small proportion, the object of his impressive study, inherited sizeable sums of money and "lived rather comfortable lives."[3] This chapter, in contradistinction to Livesay's analysis, which is mostly focused on free people at the elite level and is especially interested in mixed race people who traveled to Britain for an apprenticeship or education, mainly deals with those free blacks and free mixed-race people within Jamaica who eked out a living as artisans or small-scale planting or lived on the margins of society.[4]

———

Free blacks were not enumerated as a separate category in the censuses of either 1662 or 1673, and no laws were enacted specifically about free people until the second decade of the eighteenth century.[5] But during the eighteenth century both the absolute number of free people and the proportion of the total free population that were free people rose dramatically. By the last decade of slavery, the freed population had reached perhaps forty-two thousand, over two and a half times the percentage of the white population and 11.4 percent of the total population.[6] Well before the freed population had reached such proportions, however, the growth, composition, and character of the free people population of Jamaica had become an issue of compelling concern for white Jamaicans. The free people community were a problem and an indictment. Its existence forced a reordering of the simplistic categorizing of people by race that governed white Jamaicans' thinking. How to keep free people from becoming overmighty, yet still a support for a "weak and feeble" regime,[7] periodically exercised white authority, most noticeably in the aftermath of Tacky's Revolt of 1760 when white Jamaicans systematically rethought racial policy on the island. Whites were faced with an uncomfortable paradox: as free people became increasingly essential as a support to white authority over enslaved people, they felt that it was necessary to ensure that free people be confined to an intermediate position below whites but above slaves.

Little sustained attention was devoted to free people until the passage of major legislation limiting freed rights in 1761. Before that date, white Jamaicans alternated between increasing and constricting free people's rights and privileges. Policy was enacted on an ad hoc basis, depending on political exigencies and the changing circumstances of individual free people. It was also subsumed, as Daniel

Livesay, insists, within a discourse focused around the central issues in white society—promoting white settlement in the face of declining white population and security threats from a resurgent Maroon population. In this period, when white Jamaicans harbored a strong belief that their society was destined to become a tropical Albion settled by Britons and when the number of free people was still small, free people were less a problem than an irritation. The presence of free people was a blot on Jamaica's reputation, the fruit of Jamaican immorality. If free people were barely visible in the body politic, few whites worried about them.

But as free people became more visible, both in numbers and as people with wealth potentially equal to that held by whites, they became a political problem. After 1761, whites attempted to solve the problem through regulation. Their intention was to lower the pretensions of free people by confining them through restrictive legislation to a limited and circumscribed place in society. Free people were to be poor but "useful," a group with limited freedom who neither mingled with enslaved people nor consorted with whites. Thus, the ability of free people to "pass" into white society, their opportunities to acquire a propertied estate, and their freedom of movement were severely restricted by statute.

White attempts to restrict free people's progress were handicapped by free people's desire not to be confined to so narrow a place in society. Moreover, as white men continued to be infatuated by freed women, continued to breed a "spurious progeny," and continued to advance free people's economic prospects, white legislators could not keep free people within their prescribed caste position below whites. In addition, free people became militarily increasingly indispensable. By 1788, freed men accounted for over a quarter of ordinary troops.[8] By 1813, the percentage of rank and file who were free people was 46 percent.[9] Free people's importance as military allies became clear in the 1790s when Jamaica faced the twin threats of a Maroon rebellion and the Saint-Domingue Revolution.[10]

By the late eighteenth century, whites were forced to incorporate free people into civic society, as the number of free people grew. Gad Heuman suggests that the freed population more than tripled in the three decades after 1790.[11] As early as 1793, Bryan Edwards predicted a future ascendancy of free people, stating that "the West Indies are destined to be governed by the children of its soil," and recommended the extension of full civil privileges to free people.[12] By the 1820s, John Stewart was writing that it was folly to continue to deprive free people of the privileges of whites.[13] Nevertheless, despite the twin crises of growing abolitionist attacks on the slave system and increasing economic adversity, whites were hostile to any attempts to improve free people's social position. Concessions to free people were made grudgingly and unwillingly. Whites proved unable to overcome deeply rooted color prejudices to cultivate the most rapidly growing sector of the free population. Consequently, they alienated free people from white interests and forced them to ally themselves instead with the large black enslaved population.[14]

In retrospect, free people were whites' escape hatch. Demographic disaster ensured that white population remained low and that it comprised mostly newly arrived immigrants.[15] The demands of the hugely profitable Jamaican economy ensured that enslaved people would continue to vastly outnumber whites. If whites had been able to incorporate free people fully into free society and had been able to gain their strategic support, white authority would have been greatly enhanced. Possibly, whites would have been then able to put up a stronger defense against the abolitionist assault of the early nineteenth century. Certainly, the transition from slavery to freedom after 1834 would have been less traumatic for the white ruling class. If whites had been able to see the rise of a class of free people as the "whitening" of an African society rather than the "blackening" of a European culture, they may have been able to preserve their privileged position for longer than was to be the case.[16]

Readers familiar with the limited historiography on free people in Jamaica will note that this argument downplays the uniqueness of Jamaica's treatment of mixed-race people among British American jurisdictions. Winthrop Jordan high-lighted Jamaica as the only major English colony to give legislative countenance to the social ascent of mixed-race people. He outlined a method adopted by the legislature in 1733 whereby people of color would be considered white if they were "Three Degrees removed in a Lineal Descent from The Negro Ancestor exclusive." Jordan noted that Jamaica was "unique in its practice of publicly trans-forming Negroes into white men." He attributed Jamaican uniqueness to the colony's shortage of white women. Arnold Sio argued that the special exemptions accorded to coloreds in Jamaica resulted from specific military and political con-ditions as well as the severe shortage of whites in Jamaica. Nevertheless, closer examination of the reasons why the legislature passed the 1733 act defining who was and was not white reveals that it occurred after a member complained that wealthy, light-skinned mixed race people had voted at an election. The house was acting purely in relation to one family, the Goldings of Vere, who appear to have been able to pass as whites. While the legislature allowed the Goldings the rights and privileges of Englishmen, including the right to vote, they were careful to insist that the Goldings' case was exceptional. Possibly the Goldings may have been the only family of mixed-race people able to define themselves as white. The legislation that Jordan believes allowed light-skinned mixed-race people to become whites was passed to prevent all but a very select group of free people from claiming the privileges of whites. In this respect, Jamaica resembled the mainland plantation colonies in explicitly restricting the suffrage to white men, contrary to traditional English understandings of suffrage and naturalization. How they restricted the suffrage to whites, however, was designed so that a few wealthy mixed-race people, such as planter William Cunningham (who in 1738 was specifically granted by the Jamaican Assembly the right to vote, against the wishes of the Council of Jamaica), were able to continue to enjoy their accepted white status. Theoretically, free people could always become white if they could prove that their "negroe stain" had been extremely diluted by white blood. The definition of what constituted a white person enunciated in 1733 was retained in later legislation. But the number of free people who were able to claim all the privileges of whites was limited to a select few, each of whom gained such rights early in the eighteenth century. As Samuel and Edith Hurwitz have pointed out in their careful study of private bill legislation for free people, of 128 bills only four provided free people with all the rights of white men, including suffrage and office holding.[17]

Livesay treats privilege bills at length and argues that some Jamaicans, such as Governor Edward Trelawney, who argued in 1746 that "having a due Proportion of Freemen (of one Colour or another, white, black or yellow since white Men cannot, at least immediately be got)," supported the limited but feasible ways in which wealthy free people were able to gain privileges that made them near white people in status through economic as much as phenotypic attributes.[18] He examines the Golding case at length, concluding that class position (the Goldings were at the top of the island's brown community) was in this period able to be seen as synonymous with class privilege. Although he notes, as argued above, that it was extremely hard for people with "corruption of blood" to become white, that a few people did, Livesay argues, had significant consequences. Jamaica, he contends, unlike other British American colonies, did not follow a "one drop rule" of racial categorization, making it virtually impossible for free people to ever become white. One could become white after four generations. Jamaica, alone in British America, made concessions towards free people that were slightly akin to an Iberian model of racial toleration, although I argue that such concessions were halting and countermanded by legislation less favorable to free people. The long-term trend was to maintain Jamaica as a segregated colony.[19]

He emphasizes, as I do not, the extent to which the mere possibility of becoming white over several generations opened possibilities for the very wealthiest free mixed-race people to have all the privileges of whiteness and none of the civil impediments of being considered a "mulatto." He stresses that privilege bills proved an "aspirational decision that saw individuals of color as a potential seedbed for a new, robust white population and showed "that a strong settler society trumped any ideological zealotry about conceptions of blood purity" and "legitimized the traditional interracial relationships dominating the island." Livesay, in short, allows for the possibility that mixed-race Jamaicans could become legally white. He argues that while for the first half of the eighteenth century most free people saw their political rights eroded, for an elite few who could claim economic and cultural capital the ambiguity in Jamaica's complex set of laws governing the stats and behavior of free people allowed them the possibility of joining and augmenting a failing white settler society. He concludes that "at the end of the 1750s, a generation's time span of allowing mixed-race individuals to become legally white, or at least exempt from most race-based laws, created a new and uncertain racial dynamic . . . [where] lineal descent away from African ancestors had fully enfranchised certain individuals, enough so that some visitors noted them as white."[20]

Nevertheless, the evidence that Livesay adduces in support of his argument is counterbalanced by the abundant evidence he also provides that this desire among Jamaican elites to foster and make white a small elite group of wealthy mixed-race people, many of whom were connected to white elite men through familial ties, was fiercely contested by non-elite whites. It might be that kinship between free people and white assemblymen smoothed the passage of privilege bills, but it bred great resentment among less privileged whites. They developed a hierarchy of racial markers that they adhered to religiously and in which the primacy of the white race over free people, tinged with African heritage, and the enslaved was paramount. Poor and middling whites bristled at the notion that free people might hold the same rights as themselves and as privilege bills came increasingly in front of the Assembly, so too the complaints from outside the Assembly became more frequent and more vociferous.[21]

That free people might join or even supplant whites as colonial rulers contradicted two of white Jamaicans' most firmly held beliefs. The innate superiority of Europeans over Africans made it unthinkable that a person tinged with African blood could consider himself or herself equal to a European: the "distinction requisite and necessary to be kept in the island between white persons and Negroes" needed to be maintained at all costs.[22] But it was more than just racism that prevented whites from fully accepting free people as equals. Accepting free people as equals, white Jamaicans knew, was an admission of Creole degeneracy. Racial orthodoxy had it that free people were a corrupt, mongrel breed who retained within them the barbarism characteristic of their African forebears.[23] A racial Gresham's law operated: blacks' "bad" blood drove out the "good" blood of whites. Free people could not be British because being British meant being white. White Jamaicans wanted to believe that their society was becoming both more British (and thus whiter) and more settled. Given their strongly ethnocentric understanding of British identity, increasing acceptance of free people could be accomplished only at the expense of increasing Britishness. Therefore, if whites believed that Jamaica would eventually become a settler society populated by people of British stock, whites insisted that free people had to be excluded, as non-Britons, from claims to civic membership.

What were the characteristics of this distinctive group whom whites found so difficult to incorporate into civil society? The first freed persons of notice were John Williams and Emanuel Bartholomew, who were granted special privileges through private bills passed by the Jamaican legislature in 1707.[24] In 1730, free people numbered 1,010—nearly 11 percent of the free population but just 1.2 percent of the total population. In his 1774 history, historian Edward Long counted 3,408 free people in Jamaica in 1762, a 337 percent increase over 1730. Further growth saw the number of free people reach 4,093 in 1774 and 7,605 by 1788. By the latter date, free people accounted for nearly 30 percent of the free population and 3 percent of the total population. If the population figures for 1774 and 1788 are accurate, the numbers of free people were increasing by 234 people per year in the 1770s and 1780s. The rapid rise in the rate of freed population increase after midcentury can be seen in the parish registers for St. Andrew, St. Catherine, and Kingston. Figures on free people derived from parish registers need to be interpreted with care as ministers varied greatly in whether they included or excluded free people from church ceremonies. Nevertheless, freed baptisms in all three parishes doubled, and freed funerals increased four-or five-fold between 1740 and 1780, suggesting an explosion in the numbers of free people on the island.

The geographical distribution of free people was uneven. Like whites, free people clustered in towns, especially Kingston and St. Jago de la Vega. In 1730, over half of all free people dwelt in the parishes of Kingston and St. Catherine. Kingston was home to over a quarter of all free people. Kingston became more of a magnet for free people over time, claiming 43 percent of free people by 1788. Over two-thirds of free people lived either in the four parishes surrounding St. Jago de la Vega or the three parishes adjoining or including Kingston. Free people were more likely to be urban than either whites or slaves were: comparable figures for whites and slaves living in these seven parishes were 47 percent and 22 percent.

As well as being urban rather than rural, free people tended to be female rather than male, children rather than adults, and as likely to be colored as to be Negro. The census of 1730 provides the only breakdown of the freed population by sex and age. It shows that whereas adult men were less than 16 percent of the freed population, adult women made up 37 percent and children a further 47 percent of free people. This bias toward women and children was maintained over time. Women and children accounted for 75 percent of freed deaths in Kingston between 1753 and 1774. Freedwomen outnumbered freed men in St. James Parish in 1774 by over three to one. Most free people in St. James in 1774 and in the Kingston burial register, 1753–74, were colored.[25]

Free people were probably healthier than whites and may have been able to establish naturally increasing families. Data is scarce, but a partial reconstitution

of one of the very oldest freed families in Jamaica, the Bartholomews of St. Catherine, indicates more successful family formation than among whites. The first
Bartholomews to be freed, Emanuel and Whannica, had nine children between
1691 and 1706, of whom at least three married and had children. Three grandchil-
dren of Emanuel and Whannica survived to marry and produced a minimum of
eight children, ensuring Bartholomew survival.[26] But natural increase accounted
for only a small proportion of the growth of the freed population. Most slaves
were freed either by will or by deed of manumission.

Filial affection inspired some testators to lavish wealth on mixed-race children, as the Jamaican Assembly noted with great concern in 1761. It noted fourteen men who bequeathed large sums to mixed-race children. In the case of the
very wealthy William Foster, the owner of a sugar plantation, two pens, and
several houses, as well as over £40,000 sterling in personal possessions, an
extremely large estate passed to a single heir or heiress who was a free person.[27]
That few free people experienced such generosity from their parents did little to
allay white Jamaicans' concern about the wisdom of allowing any free people to
acquire such opulent fortunes. As shall be discussed below, the Assembly was
determined to ensure that testators would restrain their generosity when making
bequests to free people.

More slaves may have been manumitted in their owners' lifetimes than at
their owners' deaths. Until the last decades of the eighteenth century, annual
manumissions were moderate. Between 1748 and 1759, ninety deeds of manumission were recorded each year. By 1773, the number of deeds of manumission
had jumped to 161. After 1761, free people were required to register with local
authorities. The register of free persons for Kingston has survived and shows a
similar pattern of moderate manumissions before 1765, increasing sharply in the
1770s and 1780s. In the 1760s, 302 free people (thirty-four per annum) regis-
tered at Kingston. Between 1770 and 1777, the annual number of free people
registering at Kingston shot up to nearly fifty-eight per annum, increasing further
between 1778 and 1783 to seventy-three annually.

The deeds of manumission preserved at the Jamaican Archives offer only a
bare summary of how manumissions occurred and for what reason. They do
note, however, the names of both owners and free people. Although the reasons
given for granting freedom tend to be formulaic—"for true and faithful service," for example—some deeds allow more precision about the reasons for
granting freedom. An analysis of seventy-eight manumissions in which one
hundred slaves were freed in 1748 and fifty-nine manumissions freeing eighty-
seven slaves in the first eight months of 1767 outline a variety of reasons for
manumission, as well as allowing conclusions to be drawn about the identity of
manumitters. As we might expect, manumitters tended to be of high status:
thirty-nine were designated as esquires or gentlemen, nine were merchants,

and eighteen were planters. Nevertheless, not all manumitters were high-status men. Two fishermen, six carpenters, and a mariner, as well as four free persons and eleven women, manumitted slaves.

A manumission was occasionally merely a confirmation of arrangements in a will. Thus, one deed noted that Michael Price of Westmoreland had left by will an enslaved woman named Manning to his son for life, after which she was to have her freedom. More commonly, slaves were manumitted because, as George Galbraith noted of York, they had given "long and faithful service" to their owner. Sometimes, freedom was given for services rendered to the state: Mingo, Marlborough, and Tom were all freed because they had "deserted from the enemy," presumably from the Spanish. Freedom could be gained through a slave's own effort, or through the efforts of relatives. One deed described how Jeremy Garland had provided in his 1744 will for his Negro woman Mimbo and her child Billy to have their freedom once they paid his estate £60, which they had done by 1748. In 1767, at least seven manumissions were clearly self-purchases where slaves bought their freedom from their owners. By far the most common reason for manumission, even if seldom directly mentioned in deeds of manumission, was because slaves were either the children or, less frequently, the mistresses of white men. Thus, Philip Nottage, William Cartwright, and Henry Green all freed mixed-race children whom they noted were their children by Negro women. Almost all the nearly 50 percent of manumitted slaves who were children would have been white men's children.[28]

Manumission by will or deed gave slaves freedom but freedom of a very limited kind. Enslaved people who were freed gained rights to their bodies, to the fruits of their own labor, and the ability to pass on their freedom to their children. But they obtained none of the other rights of whites, such as the right to sue or to give evidence against whites in a courtroom. The children of free people, who did not suffer the stain of having once been enslaved, were granted some legal privileges, such as the right to a trial by jury, and could give evidence in cases involving other free people. But even they faced legal restrictions, inspiring some of the more affluent to petition the Assembly to pass private bills granting them legal exemptions.

At first glance private bill legislation for free people appears to have provided free people with an escape from blackness. Bills specified that petitioners were henceforth to be granted all the rights and privileges of Englishmen. Yet, as Samuel and Edith Hurwitz have carefully documented, such rights and privileges were strictly limited. To avoid any misinterpretation, the Assembly made sure to state that except for the privileges specifically granted in each private bill, the petitioners "shall be and remain to all intents and purposes in the state and circumstances of a free Negroe as before this act, any law, custom or usage to the contrary, notwithstanding."[29] Moreover, legislators made it clear that granting free people some of the privileges of white men occurred only in special circumstances where free people had "considerable property in land and slaves," were Christian, and had enough education "to enable them to earn their bread in a reputable manner."[30]

These restrictions were sufficiently rigorous that before 1761 only twenty-nine acts had been passed granting privileges to ninety-one freed people. Even such a limited number of blacks passing some way into whiteness disturbed whites. The initial granting of private bills in 1708 led the Assembly to decide that no more petitions ought to be heard in that year. A surge in private bill legislation in 1747, when seven acts were passed involving twenty-five freed people, caused the Assembly to once again declare a moratorium on hearing free people's petitions and led it to pass an act that clarified the legislature's understanding that if a freed was granted privileges it did not mean that he or she had "the liberties of whites."[31] Yet whites contrived to fear that the passing of private bills granting free people enhanced judicial rights might suggest to free people that they were equal to whites. At the same time as it passed legislation restricting what free people could inherit, the Jamaican Assembly began to reword private acts. The rights to which free people were entitled were enumerated but so too were the continuing restrictions that they suffered. From 1761 until 1813, privileges granted by private act were fewer and less generous than those customarily granted before 1761, denying to most free people the ability to testify against whites in civil action and excluding free people from voting or attaining public office. Free people continued, however, to press for private bill legislation, even if the privileges they gained were strictly limited. Between 1762 and 1780 a further forty-nine private acts were passed involving 191 freed people.[32] Livesay argues that the specifics of the 1761 law still meant that there were frequent private exemptions of mixed-race people. He admits, however, that the 1761 act embodied a major shift in policy towards free people, "putting a significant kink in the flow of wealth from affluent whites to offspring of color" and "worked against the previous generation's strategy to incorporate a select, but not insignificant, number of mixed-race colonists fully into white settler society."[33]

—

Almost every advance in free people's privileges was countered by a legislative restriction on free people's rights. Legislation was driven mostly by free people's needs and by white reaction to free people's assertiveness. The granting of the first private acts to free people in 1707 was followed in 1711 by an act making it an offence to employ free people in public office, an act a year later prohibiting free people from acting as navigators or as drivers of carriages for hire, and in 1715 by legislation designed to exclude free people from supervisory positions on plantations.[34] Restrictions on free people could result from whites' reactions to the assertiveness of individual free people, as in the treatment of Francis Williams, the youngest son of John Williams, the first free Negro granted a private act. Williams was a man of talent and education, whom Edward Long tells us was educated in England at the expense of the Duke of Montagu, who wanted to see whether a Negro was as capable of learning as a white. Long devotes a large section of his history to the career of Williams, principally to denigrate Williams's literary achievements.[35] There is no reason to accept Long's portrait of Williams as an excessively vain man, "severe with slaves and children . . . who looked down with sovereign contempt on his fellow Blacks," given that Long's account is marked by malice and was written expressly to show that Africans were inferior in mind to Europeans.[36] Yet Williams was clearly a proud man unwilling to adopt a subservient attitude. In 1724, Williams became involved in a passionate argument with colonial official William Broderick. Broderick claimed in a petition to the Assembly that Williams insulted him by calling him a "white dog" and that Williams had struck him while declaring that he was "as good a man as ever stood on Broderick's land." In addition, Broderick claimed that Williams had asserted that he was exempted from "such trials as other negroes."[37]

Broderick's petition was predictably self-serving—a Committee of the Home found that Broderick began the dispute by calling Williams a "black dog." But the committee rejected Williams's desire to present a petition of his own against Broderick, even though Williams claimed his right to do so because he had been given a certificate of general naturalization in England in 1720 and had taken oaths of loyalty in 1723. Williams's behavior, the committee declared, "is of great encouragement to the negroes of the island in general and may be attended with ill consequences to the white people thereof." His pretensions that he was as good as a white man because he had certificates of naturalization could not be tolerated. The Assembly immediately brought in a bill reducing Williams to the same legal status as other free Negroes. Further, the bill required not only Williams but all free people to wear badges of distinction, prohibited them from wearing arms or using swords except on militia duty, barred them from living in the towns without a license, and prevented free people from working as goldsmiths or silversmiths, or from keeping shop. If a free person offended against the act three times, he or she would lose their freedom and be transported. The bill did not pass without opposition. Williams protested vehemently against a bill he believed was designed explicitly to ruin him—"it being notorious that no free Negro wears a Sword and pistols besides himself." He presented a petition to the Board of Trade. The Board of Trade's counsel, Francis Fane, agreed, arguing that Williams should be exempted from the act. It is unclear whether Williams did gain an exemption, but the act stood. The actions of an individual freedman led to significant restrictions for all free people.[38]

What bothered white Jamaicans about Williams was that his claim to be Broderick's equal was plausible. Williams owned an estate that he calculated was worth over £20,000 and was well educated and genteel. Edward Long went to considerable lengths to pour scorn on Williams's character and to demonstrate that his poetry was execrable. But his tone was defensive: Williams's very existence contradicted the ideology of innate white superiority. Long castigated Williams as a "white man acting under a black skin," which all but admitted what Long most feared: that a black could emulate white behavior so successfully as to be the intellectual equal of a white man.[39]

White fears about blacks emulating or passing as whites led to one of the most significant (and retrogressive) acts passed in eighteenth-century Jamaica: the 1761 Act to Prevent the Inconveniences Arising from Exorbitant Grants and Devices to Negroes. This act heralded the end of legislation expressly designed to advance white settlement and marriage in favor of legislation intended to limit the growing wealth and power of free people. The act was thus a futile effort to halt the eventual "browning" of Jamaica. But it was also passed to keep free people in their place, a place increasingly premised on free people's color rather than on their wealth and status. As in nearby Saint-Domingue, from the 1760s onward whites intensified efforts to segregate their societies according to color and attempted to establish more precise definitions of individuals' racial status. This redrawing and solidification of racial boundaries was not as complete in Jamaica as it was in Saint-Domingue, where African ancestry, however distant, became accounted as a permanent stain that justified barring a man from full civil privileges. Moreover, the denial of civil rights to people with African blood in their lineage could be done retrospectively. Men who had previously been accepted within white society were increasingly excluded on grounds of race.[40]

The color bar in Jamaica grew more rigid after 1761, but it was never impermeable: the richest and the lightest-skinned mixed-race people, especially those who had been accorded the privileges of whites in the first half of the eighteenth century, occasionally could pass as white. In addition, Jamaican legislation, even in the restrictive act of 1761, always allowed "persons of the fourth degree (from a negro), when granted the rights and privileges of whites" an exemption from regulations applying to free people. Nevertheless, the way this clause was written into the 1761 legislation significantly restricted customary understandings of how a person could become defined as effectively white. Not only was the method of "passing" moved out one generation further than had been previously accepted (to four degrees rather than three degrees from an African ancestor); such transgression of racial boundaries was allowed only to the very small number of free people who had been already granted before 1761 private bills guaranteeing them the rights and privileges of whites.[41]

Whites found it conceptually difficult to categorize someone like Anna Petronella Woodart, the mixed-race heiress of the wealthy planter William Foster, who not only had an opulent fortune of well over £50,000 but also was a Christian who had been given a "most liberal education."[42] She was very well educated and lived in England, which made her less of a threat to Jamaican politics, than someone like John Golding. Livesay argues that she epitomized the eradication of color differences between white and black, contrary to the ideological position of most whites, and "embodied the successful transition in to whiteness for which so many officials strived over a generations; worth of lawmaking."[43] Within this system of racial classification, free people were supposed to be poor, and freed women were assumed to be debauched. Edward Long was highly dismissive of wealthy mixed-race people, claiming that the wealthiest deserted Jamaica for England (as Anna Woodart indeed had done), thus impoverishing rather than enriching the colony. But poor free people, he asserted, were potentially very useful members of society. In the strictly racially divided Jamaica that he envisioned, free people (whom he pictured as entirely comprising colored people rather than blacks) would hold an intermediate position, above slaves but below whites. If they were "retrieved from profound ignorance . . . [and] instructed in Christian morals and obliged . . . to serve a regular apprenticeship to artifices and tradesmen" they could become "orderly subjects, and faithful defenders of the country." He saw free people as forming "the centre of connexion between the two extremes" and thus allowing for a three-tiered caste society "dependent on each other, and rising in a proper climax of subordination, in which the whites would hold the highest place."[44]

Not surprisingly, Edward Long was a great supporter of the provisions of the 1761 act where "mulattoes" were prevented from buying or inheriting property worth over £2,000. He agreed with the major aim of the act, which was to eliminate that class of opulent mixed-race people whom Long felt to be a great drain on society. But Long was also concerned that mixed-race people (all of whom he believed deserved to be manumitted) should not be a burden on the public. Any free person who did not have land or a trade to follow, Long argued, should, as in Antigua, be obliged to serve as domestics to white families.[45] As Long argued, the purpose of the 1761 act was to keep free people from becoming as wealthy as whites. Free people were to be poor but useful—peasants and minor trades people, providing services at reasonable cost to whites, who as a class formed a barrier between whites and slaves. The makers of the 1761 act largely succeeded in their aim but at considerable cost. Restrictions on what free people could receive and, more important, on what protection they could receive under the law, prevented free people from attaining any considerable position or wealth. It made their position permanently precarious.

Few free people were as wealthy as the freed people who so greatly concerned the legislature in 1761. The wealthy mixed-race planter William Cunningham, who died in Westmoreland in 1762 with a personal estate of £16,780 sterling, including 160 slaves, was atypical. Mary Tower and Sarah Williams, free black women who died in Kingston with estates worth less than £20, were more typical. Two censuses of parish population confirm the humble position of free people. A 1754 census of St. Jago de la Vega and St. Catherine lists 405 free people in the parish of whom 114 were independent household heads. The great majority of freed householders were in lowly occupations. Women were seamstresses, and men were tradesmen, especially carpenters. Just three free people in St. Catherine were landowners. The three landowners were small landowners, owning around sixty acres each.[46] A listing of male household heads for St. James Parish in 1774 shows a similar wealth pattern. Of sixty-two freedmen able to bear arms, three were small planters, forty-six were tradesmen, of whom half were carpenters, with a further nine fishermen and four servants.[47]

The legal disadvantages placed on free people made them vulnerable to white depredations, even before 1761. One of the key points made by Francis Williams in his 1730 petition to the Board of Trade was that without the legal privileges of whites he was vulnerable to "ill disposed People" who might "spirit up the Minds of their Ignorant Slaves to Assault and waylay him" when he "sues for the recovery of his own."[48] In practice, free people could be assaulted at will by whites with little chance of redress. In 1756, for example, Benjamin La Cruize, a freeborn Negro prisoner in St. Jago jail, petitioned the Assembly to complain about his treatment. Arrested and jailed in February 1752, La Cruize was at liberty by December 1755 when he had his eye "thrust out . . . without provocation" on Christmas Day by Deputy Marshal Charles Baker. To add insult to injury, Baker prosecuted La Cruize for breaching the peace, and La Cruize was imprisoned for a term of three months. Five months after the end of his sentence, however, La Cruize remained in jail, despite being granted a writ of habeas corpus at the last court and despite no sentence being produced to detain him. La Cruize was between a rock and a hard place: he wanted to be released but feared the wrath of Baker who had sworn that La Cruize would be in danger of his life when he was released. The Assembly granted La Cruize's petition but insisted that he be conscripted without pay into the Royal Navy.[49]

For their part, free people demonstrated little attachment to slaves' interests and indeed made considerable efforts to distinguish themselves from enslaved people.[50] Most appear to have aspired to join the society of whites rather than the company of slaves. White observers noted that mixed-race women notoriously preferred the company of white men to that of colored or black men. The best description of their attitudes and behavior comes from John Stewart, who wrote, in the 1820s, one of the better accounts of Jamaican life. What he describes seems to reflect customs of long standing in Jamaica, customs that probably also existed in the eighteenth century. Free people, he noted, "feel a kind of pride in being removed some degrees from the negro race and affect as much as possible the manners and customs of the whites." "Most of the females of color," he continued, "think it more genteel and reputable to be the kept mistress of a white man, if he is in opulent circumstances, and can afford to indulge their taste for finery and parade, than to be united in wedlock with the most respectable individual of their own class."[51]

—

Mixed-race women gained many advantages in becoming the mistresses of wealthy white men. Some idea of what mixed-race women gained from concubinage can be seen in the histories of the mixed-race daughters of St. Andrew planter John Augier. Given their freedom and a small share of their father's estate in his 1722 will, Augier's five daughters augmented their fortunes by becoming the mistresses of Kingston merchants. Mary Augier was kept by William Tyndall, by whom she had two children; Jenny Augier's partner was Theophilact Blechynden, by whom she had a child; Susanna Augier was the mistress first of Peter Caillard, then of Gibson Dalzell, both of whom are noted in 1761 as having given mixed-race children sizeable bequests; Frances Augier had three children by William Muir and Samuel Spence; and Jane Augier was the mistress of John Ducommon, by whom she had four children. Mary, Susanna, and Jane each acquired considerable property; they were granted additional privileges through private acts enacted in the second quarter of the eighteenth century.

But the act of 1761 made such "passing" more difficult. Both Jane Augier and her quadroon son William, for example, were denoted by color rather than status in the Kingston Parish register when they died in 1766, whereas their relative Mary had been described as a widow, not a mullata, when she died early in 1761.[52] By the late eighteenth and early nineteenth centuries, free people were no longer welcome in polite white society, and formal racial segregation was widespread. Free people who identified with their white rather than their black ancestry found themselves denied access to white society.[53] Culturally, free people were in no-man's-land. They were, as the mid-nineteenth-century colored writer and politician Richard Hill evocatively put it, "barkless, branchless, and blighted trunks upon a cursed root."[54]

Unable to join white society, free people had difficulty identifying with or being accepted in slave society. Links between free people and slaves, of course, always existed, especially among poorer free people, and these links gave slaves the potential to undercut white authority. The *Weekly Jamaican Courant*, for example, carried an advertisement in 1726 in which it was suggested that a runaway Creole Negro woman had been "intic'd away by Ishmael, a free Mulatto

Fellow (who is her husband)."[55] Yet the relationship between free people and slaves was more often tense than cooperative. Part of the tension arose from how free people obtained their freedom. Slaves were unlikely to look kindly on slave women consorting with white men and even less likely to think fond thoughts about slaves such as Sarah, who was granted her freedom and an annuity of £10 for life, as a reward for killing a rebellious Negro, Ancouma, with whom she had lived in the woods for five months, or applaud the actions of sixteen slaves freed "as a reward for their faithful service in the late rebellion" of 1760.[56] Enslaved people probably viewed these freedmen as turncoats and traitors. Slaves' contempt for slaves who gained their freedom by compromising other slaves can be seen in a remarkable petition of 1745. Thomas Edwards had been the slave of Thomas Fuller, a wealthy planter in St. John Parish, until 1744, when he revealed a proposed slave revolt to white authority. In gratitude for his foiling of a slave plot, the Assembly ordered Edwards to be set free from December 21, 1744, granting him in addition £30 and an annuity of £30. Edwards, however, was a marked man, "in daily danger" as Edwards put it, "of being Murder'd by the Negroe slaves who often assaulted him." Edwards fled to England, but even there he was met by great hostility from Negroes from Jamaica who had clearly heard of his treachery. In an abject, almost whining, entreaty to the Board of Trade, Edwards complained that threats from blacks from Jamaica, "whether free or slaves," prevented "his endeavouring to get into business." Moreover, Edwards's wife in Jamaica, an enslaved woman suffering "the Miseries of slavery," had to endure "the additional Misfortune to be daily abused by other slaves, for your petitioner's fidelity."[57]

But free people's aspirations were not reciprocated by whites. Whites continued to doubt free people's constancy, believing them potential rebels in disguise. Not surprisingly, such condescension caused resentment. Immediately following the passing of the 1761 act, for example, Governor William Lyttleton noted to Secretary of State Egremont that he was finding it very difficult to persuade free people to join the militia, despite offering a large bounty to recruits. Lyttleton thought that free people did not enlist because "most of them have beneficial Trades and find a comfortable maintenance here with their wives and families," but free people's dissatisfaction with legislative restriction on their economic advancement may have also made them reluctant to support whites in their quest for security.[58]

Whites had only themselves to blame for the eventual unwillingness of free people to "naturally attach themselves to the white race as the most honourable relations, and so become a barrier against the designs of the Blacks."[59] Throughout the eighteenth century, whites made no attempt to improve the legal status of free people. Indeed, with the 1761 act, they declared their determination to whittle away what rights free people did have. Every push by free people to advance their own position was treated by whites as a dangerous attempt to break down what they understood as rigid racial barriers and was met by a tightening of the restrictions that made free people's lives so precarious.

—

White resistance to allowing free people greater civic participation arose in part from their negative views of free people's moral character. In *An Essay on Slavery,* the anonymously authored tract of 1746 probably written by Governor Edward Trelawney, for example, free people were described as "a lazy debauched Race, a very Nuisance to the Community."[60] Edward Long had similar views. He condemned mixed-race women as leeches, bleeding their white keepers of money and possessions, and portrayed mixed race people in general as lascivious, vain, extravagant, and born thieves and liars.[61] Even those writers who believed it obvious that browns rather than whites would become the permanent masters of the island could not shake off racial assumptions about free people's degeneracy. John Stewart believed that free people's dominance of Jamaica was inevitable but still found it amazing that Africans did not realize "that the finest Grecian contour is more beautiful than their large or gross features," and he castigated free people for living "in idleness and vice." Free people, he asserted, were "indolent, dishonest, and vicious, leading a wretched and precarious sort of life."[62]

Just as important as racist assumptions in shaping white views on free people and in preventing whites from giving free people greater rights and privileges was whites' continued belief that Jamaica was soon to become a more settled country, full of Europeans and their legitimate progeny. Whites found it difficult to accept free people as useful members of free society, entitled to the privileges of whites, because whites' vision of what Jamaica ought to be explicitly excluded free people. Whites envisioned Jamaica as settled and improved. The eighteenth-century white Creole understanding of "improvement" was that society came to resemble ever more closely British society.[63] Britain was the very model of an improved, settled, thriving, and civilized society. White Jamaicans were proud of the extent to which Jamaica was a valuable addition to British imperial power through its extraordinary economic vitality. They stressed also how Jamaica was becoming ever more British, both in landscape and in moral and social character. With increased settlement, Jamaica was taking on the appearance of an established agricultural society with cultivated fields and improved landscapes in the countryside and with well-furnished and sumptuous multistoried brick homes in the towns.[64] The white owners of these dwellings lived in ways that ever more closely approximated British standards. As Long noted, "in manner of living" white Jamaicans had the "fine ambition of initiating the manners of . . . Europeans on every point." They succeeded so well, Long argued, that they "differ[ed] not much from their brethren at home, except in a greater profusion of dishes, a larger retinue of domestics, and in weaving more expensive cloaths."[65]

But, as Long realized, white Jamaicans were not Britons yet. Britain's demographic makeup was not that of Jamaica, where 90 percent of the population were black slaves and where a growing proportion of the free population were not white. To be improved meant becoming British, but to be British was to be white. Thus, Jews could eventually be regarded as citizens. Indeed, Jamaica preceded Britain by several years in granting Jews full civil privileges.[66] But to include free people in this expanded definition of who was allowed membership in civic society would push the boundaries of who was entitled to the privileges of Britons too far. Whites could hardly include free people in their vision of an improved Jamaica when their opinion of African character was so overwhelmingly negative. To become more British, Jamaica needed to have a free population that was whiter rather than browner.

Fear of Africanization was at the nub of white fears about increasing free people's participation in civic society. If the thriving society of white Jamaica could be seen on the one hand as becoming ever more demonstrably European, ample evidence existed on the other hand whereby it was possible to see white Jamaicans as becoming Africanized. Suckled by black domestics, Creole whites imbibed African modes of behavior with their nurses' milk and became "habituated by Precept and Example, to Sensuality, Selfishness, and Despotism."[67] In their language, their food, their beliefs, and their manners, whites, brought up from birth in "constant intercourse" with blacks, copied "insensibly" the habits of Africans. Their Africanization was shown visibly in the "Drawling, dissonant gibberish" that Creole women used with their "sable hand-maids" and in their adoption of Africans' "awkward carriage and vulgar manners."[68] The vices that characterized white Jamaicans were those that they also associated with blacks, notably, a tendency toward indolence, an inordinate lack of self-restraint in satisfying animal passions, moral degeneracy, and addiction to vice. In their attachments to colored women, it was as if Creoles had "suck[ed] in the *Affection*, or rather *Infection*, with their Nurse[']s Milk, and the rest of the Inhabitants acquire it by Custom and Habitude."[69]

The disease metaphor was apt, in a land and among a people devastated by continual epidemiological disaster.[70] Africanization and yellow fever were culturally similar. Yellow fever and other diseases prevented whites from establishing a viable settler society full of flourishing families of Europeans and the descendants of Europeans. The few whites who survived the assault of disease were confronted by another type of "infection"—African influences—that just as surely as disease, whites knew, doomed European settlement and eventually European control of Jamaica. In the seventeenth and eighteenth centuries, whites could do little to counter the effects of disease, given their very limited understanding of epidemiology. But by limiting what free people could do and by excluding them as much as was possible from white society, whites could at least halt the spread of one "infection": the gradual spread of African ideas, values, and behaviors into the white population.

—

Their efforts to stem the tide toward the Africanization of Jamaica proved futile, however. By the early nineteenth century, as ideas of Jamaica as a place of extensive white settlement had faded, it was becoming clear that Jamaica was not to become an extension of Europe in the tropics but was to become instead, as David Eltis notes was the case of New World colonization before the mid-nineteenth century, an extension of Africa into the Americas.[71] British migrants to Jamaica in the seventeenth and eighteenth centuries were undone by their own economic success. The great wealth generated by tropical plantation agriculture enabled whites to import enormous numbers of slaves from Africa into Jamaica. These importations did more than transform the racial balance on the island. Africans brought with them biological pathogens to which whites had little resistance. Disease thus continued to ravage the white population. Whites remained too few to be able to transform Jamaica into a tropical Albion. Instead, Jamaica evolved after over 150 years of British occupation into little more than a giant sugar factory where a small number of white managers, mostly sojourners rather than permanent residents, lorded it over a host of enslaved black people.[72]

By 1823, observers like John Stewart were willing to accept that extensive white settlement had failed, that whites were indeed just birds of passage in Jamaica, and that the eventual controllers of Jamaica would not be whites but their colored descendants. Respectable free people, he argued, should have more power as their numbers were increasing and "it is in vain that . . . laws and provisions are thrown in the way of this people's acquiring an ascendancy in this country."[73] Stewart's assumption that Jamaica would soon become a brown man's country were shared by the anonymous author of *Marly; or, A Planter's Life*, a supposed novel that was in fact a series of polemics draped around a formulaic plot, designed to show the essential humanity of slaveholders and the necessity for emancipation to be gradual rather than immediate. In a curious passage in the book, a passage that would have made Edward Long's blood run cold, the author suggested that a great improvement would happen if every "mulatto" child was freed at birth. Long had proposed a similar policy toward mixed-race people as a way of creating an intermediate class between whites and blacks, but the author of *Marly* turned Long's reasoning on its head. As black women naturally preferred a white partner to a black man, *Marly*'s author argued, mixed-race children would frequently be born. Over time, it was suggested, slavery might disappear as blackness became diluted. Dilution of blackness was desirable as a brown population was preferable to a black one. Eventually, the author speculated, the population would be "whitened," provided Europeans retained political control of the island, through the missionary efforts of white men spreading their Caucasian blood throughout the non-European world.[74] By the 1820s, as the author recognized, white Jamaicans could no longer pretend that their society was becoming Europeanized. It had clearly been Africanized, instead.

Chapter 6

The *Somerset* Decision and the Birth of Proslavery Arguments in the British West Indies

William Murray, Earl of Mansfield, was a legendary chief justice in eighteenth-century England, perhaps the most important judge in British history, and the presiding officer over any number of famous cases. Probably his most famous decision was the one given in *Somerset v. Steuart* (1772). The case was initiated by a group of English humanitarians, led by Granville Sharp, Britain's most important early abolitionist, who challenged the legality of allowing the slave James Somerset's master to order Somerset back to the colonies. It staggered through eight hearings between December 1771 and June 1772, gathering public attention along the way, mainly because of Sharp's determined advocacy. West Indian interests were also very concerned about the possible outcome of the case, funding the defense costs of Somerset's owner, Charles Steuart.[1] Mansfield gave a statement to a packed courtroom on 22 June 1772 that demonstrated the rare ability to be oracular and gnomic at the same time. He found that Somerset, an enslaved man who had arrived in London from Virginia via Boston and who dwelled there with his master, could not be forcibly removed to Jamaica.[2] The most extreme reading of what Mansfield decided in 1772 was that he provided the legislative means whereby slavery was set on the road to extinction. He produced strong words about how slavery was not an institution that did credit to Britain, declaring that "the state of slavery is of such nature that is incapable of being introduced on any reasons, moral or political, but only positive law.... It is so odious that nothing can be suffered to support it but positive law."[3]

That was not a reading, however, that Mansfield encouraged. His finding suggested a fundamental break in the colonial understanding of how slavery should be treated, but it was a judgment given reluctantly. This explains the reasoning underlying the case about slavery being mostly a local issue never having been adopted as policy by the British state. Mansfield was very aware of the value of the plantation colonies and the slave trade to Britain and had close friends with interests in the West Indies, like his old school friend London merchant and politician William Beckford, an absentee planter who was among the largest slave owners in Jamaica. He was enough of a political animal to worry about the consequences of a judgment that attacked the legal basis of holding Africans in chattel slavery. Thus, he was disturbed at what would happen if he was forced to rule on the matter under debate.[4] He urged the two sides to settle so that he would not have to make a decision.[5] Just a year earlier, adjudicating on a petition by Thomas Lewis, a black man freed largely through Granville Sharp's efforts, he declined to agree to a general principle that slaves were entitled to habeas corpus, declaring that "I don't know what the consequences may be, if the masters were to lose their property by accidentally bringing their slaves to England. I hope it never will be discussed; for I would have all masters think them free, and all Negroes think they were not, because then they would both behave better."[6]

His judgment in the end was short (it lasted not much more than a minute when presented in court), and he refused to elaborate on what he meant. It may have been a momentous decision, but it did not have a great deal of practical legal consequences for enslaved people. Benjamin Franklin was not too far from the truth when he derided the importance of *Somerset* as a judgment that merely meant that one slave only could not be made to board a ship bound for Jamaica. For Franklin, what Mansfield's ruling mainly showed was the moral arrogance of the English people. He responded to Mansfield with a piece in a London paper that he described to Pennsylvania abolitionist Anthony Benezet as "Remarks on the Hypocrisy of this Country . . . for promoting the [slave] Treade, while it piqu'd itself on its Virtue Love of Liberty, and the Equity in its Courts in setting free a single Negro."[7]

Lord Mansfield had no intention of making law that referred to anyone except James Somerset. He later argued that his ruling "was a determination [that] got no further than that the master cannot by force compel him [Somerset] to go out of the kingdom."[8] He said this to dispel the popular impression left by *Somerset* that his decision had ended slavery in Britain. We can see the limited actual impact of the case by the fact that slave cases continued to come before British courts after 1772, that newspapers continued to carry advertisements for slaves, and that even Mansfield's specific ruling in *Somerset* was regularly flaunted by slave owners taking slaves back to the West Indies and North America. Even though *Somerset* did give slaves the potential for freedom in allowing them access to habeas corpus, slave owners learned how to circumvent the restrictions placed on them by doing such things as forcing slaves to sign or mark an indenture of service. That indenture only occasionally allowed servants access to wages and allowed, even after the abolition of the slave trade in 1807, for blacks (defined as workers, not slaves) to be shipped wherever slave owners wanted them to go.[9]

Moreover, Mansfield's judicial successors were aware of how unwilling Mansfield was to move away from the bare facts of the case to more general statements about the legality of slavery. They tended to interpret the law as narrowly as possible.[10]

Yet such comments made after the event were misleading.[11] Lord Mansfield understood he was doing more than just making a narrow adjudication on a single case. Mansfield knew what he was doing when he declared that slavery could be instituted only by positive law. He was making a deliberate effort to demolish the legal justification for slavery on any other basis not just in England, but in the colonies. His ruling meant that slavery existed only within those jurisdictions that had passed laws specifically to protect it. Slavery was thus circumscribed and fatally weakened as an institution, if only in England.[12] Even if the effect of the case was merely to make colonial slave owners more wary about bringing slaves into England, as the ruling seemed to suggest, and that their rights over enslaved property were more limited than they thought, it was still a momentous decision.[13] However hedged Mansfield's dicta on slavery were, they heralded a decisive shift in the struggle between masters and servants in Britain. The opinion was rightly seen by England's black population of ten thousand people as a substantial broadening of their rights under what the law termed was a condition of slavish servitude. Reputedly, when Mansfield concluded his brief statement that allowed Somerset to go free, large numbers of black Londoners in the gallery made a bow of respect to the judge.[14]

Because slavery had to be positively sanctioned, it could be selectively altered or abolished, as indeed happened in Pennsylvania, Massachusetts, and New York during and after the American Revolution.[15] This also raised substantial issues about compensation if Parliament ever chose to abolish slavery. Such a prospect lay well in the future. But *Somerset* meant that slaves could sue for freedom if they were removed from England. The ruling was also likely to lead to increased slave runaways, as slaves came to realize that they could become free or protected from excessive force both in England and in the colonies. The practical effect on slaveholders, therefore, was that they had to be careful about moving slaves from one jurisdiction—*Somerset* was mostly a conflict-of-laws case—to another jurisdiction, especially if they intended to dwell in Britain.[16]

Slave owners, at least in the British West Indies, were aware of the larger implications of the case, even if colonists in North America were not especially concerned that *Somerset* marked a turning point in metropolitan understandings about the morality and efficacy of the chattel slavery of people of African descent in British America.[17] West Indian planters, on the other hand, were certainly exercised about *Somerset*. One sign of their concern was that three of the strongest intellectuals in the West Indies quickly wrote pamphlets denouncing Mansfield's rulings: Jamaican Edward Long, Barbadian Samuel Estwick, and Antiguan Samuel

Martin.[18] There were strong practical reasons why West Indians would be more concerned than North Americans about the possible ramifications of *Somerset*. Unlike British North America, especially the Chesapeake, where the enslaved African population had experienced natural demographic growth for a genera- tion, thus making the external slave trade largely unnecessary, the enslaved popu- lation of the British West Indies declined year on year, making planters very dependent on new additions of labor from the Atlantic slave trade.[19] Even the slightest hint that there was any threat to this lucrative and vital trade from the machinations (as they saw them) of English abolitionists made West Indians very uneasy and quick to defend existing arrangements.

This chapter analyzes the West Indian context of this famous decision. It argues that much of the impetus behind the case arose less from abhorrence about the institution of slavery than from a distaste for what the effects of being slave owners were on the character of white West Indians. Thus, the early abolitionist Granville Sharp made a relatively minor case about the right of a slave resident in England not to be moved away from the country into a cause célèbre. This chap- ter examines the reasoning behind the decision itself; how the decision helped galvanize both antislavery and proslavery discourse after 1772; and how the reac- tion to *Somerset* helps explain why British West Indians took a different direction to that taken by British North Americans in the lead-up to the American Revolution.

The decade of the 1760s, a decade when British West Indian prosperity reached new heights and when West Indian planters and merchants fondly believed that they were people of inestimable value to the British empire owing to the vast wealth they produced for imperial coffers, was one in which the image of the West Indian slave owner changed decisively, and changed decisively for the worse as far as white West Indians were concerned. As the West Indies became more prominent within the empire and as the personalities of white West Indians became better known to Britons, especially the cruelty with which they treated enslaved people and the arrogance by which they presented themselves in public discourse, metropolitan Britons started to distance themselves from people they increasingly viewed as not so much Britons living overseas as much as un-British tyrannical despots.

The *Somerset* case made clear to a generally unsuspecting West Indian audi- ence that a growing number of Britons found what white West Indians did so disagreeable that they were prepared to start agitating that the Atlantic slave trade—the lifeline keeping the British West Indian economy going—was immoral and that it should be stopped. West Indians protested how they were being increasingly demonized in metropolitan thinking. They developed proslav- ery arguments, utilizing the languages of commerce, security, and imperialism, to counter attacks on them by Britons who used the languages of humanity and natural justice to attack slavery. But they were handicapped in how they could respond to what they considered bewildering metropolitan attacks on a vital part of their world because their insistence that they were true-born Britons and the fellow countrymen of British people living in the British archipelago forced them to reply to abolitionist attacks on them within a discourse that abolitionists had largely themselves shaped. This discourse increasingly suggested that some people who claimed to be British could be disqualified from that status if they did things that marked them out as fundamentally non-British, according to criteria chosen by metropolitan opponents of West Indian planters and merchants, rather than by West Indians themselves.

In addition, the *Somerset* case advanced understandings, especially in the northern colonies of America, that there was a palpable difference between West Indians and Americans, seen most graphically in the cruelty of West Indians toward Africans, and even more so in the willingness of British North Americans, but not West Indians, to transcend their sense of themselves as British subjects in order to embrace a new form of belonging, in this case to the new nation of the United States. Here there was an irony. The decision in *Somerset* should have shown to British North Americans the extent to which Britain was willing to legislate for Americans even in matters where Americans especially claimed the rights of local autonomy, such as in the ability to make laws concerning slavery. But except for a few farsighted people, Americans did not connect what Sharp and Mansfield were doing regarding James Somerset to larger conceptions about American constitutional rights.[20]

West Indians were far less sanguine about the *Somerset* decision. They knew that, however guarded Mansfield's remarks, his decision had the potential to harm the slave trade and perhaps even the institution of slavery itself. That they recognized the danger of Mansfield's judgment and saw that *Somerset* marked a change in the relationship between Britain and the West Indies is clear from the quality of the responses made by Long, Estwick, and Martin to Mansfield's decision. Nevertheless, *Somerset* had less impact in the West Indies than might have been expected. No West Indian legislature, for example, protested their opposition to the decision to Parliament. Moreover, the case did little to encourage West Indians to think about joining their American cousins in dissent and then rebellion against imperial power.[21]

The reasons for this relative lack of outrage about *Somerset* in the places where it was most likely to affect colonists arose from four interconnected factors: the diffidence by which Mansfield declared his judgment; the fact that the case was adjudicated at a time of maximum prosperity for planters in the Caribbean; the inability of abolitionists to seize on the decision as a mean of galvanizing interest in their cause; and the incipient loyalism of West Indians and their willingness to believe that Parliament would always act sympathetically to fellow Britons (as they saw themselves).

—

The single most important reason why West Indians did not get especially agitated about *Somerset* in 1772 was that this was an especially propitious period in the history of the plantation world in the British Caribbean. The decision arrived at a time of maximum prosperity for West Indians and at a time when they felt more confident than ever before about their value to the British Empire. In this respect, therefore, it was a decision that greatly surprised West Indians, just as in 1788 planters and merchants in the West Indies were caught unaware by the sudden realization that a major abolitionist movement was on the verge of success in Britain.[22] West Indians felt happy in their position in the empire. A couple of events in Jamaica from around 1772 suggest just how comfortable life was for planters and merchants and how pleased they were about their place in the imperial universe. The years between the Seven Years' War and the American Revolution were, as George Metcalf concluded, "a brief golden age for the plantocracy."[23] Plantation profits were high, the place of Jamaica within the empire was secure (and much more secure than was the case for British North America and India), and enslaved people were temporarily peaceful. These were years of plenty and peace. In addition, the harmony of the planter class in this period came from its satisfaction that it had won their most important constitutional battles and that both governors and the imperial government had to respect its power and privileges.[24]

Matters were so good that in 1772 the Jamaican Assembly made the rare gesture of making a governor the subject of commendation. William Trelawney—for whom a new parish was to be named—was praised, a month before his death, in fulsome terms: "We cannot but acknowledge with the greatest Justice that the present Harmony and general Tranquillity of this Country are the natural effects of Your Excellency's steady and impartial Administration." For Edward Long, the planter-historian, such praise (and the gift of £1,000 sterling for "giving his remains an honourable interment") was signal evidence that "there are no people in this world who exceed the gentlemen of this island in a noble and disinterested munificence." Planters, he argued, might be unjustly accused of "wilfully seeking occasions to quarrel with their governors." But this was true only for governors they did not respect. When, as with Trelawney, planters thought a governor was willing to accept their constitutional positions, they displayed "liberality and a just deference to those governors who have deserved well by the mildness and equity of their administration."[25]

In that same year, the triumphantly successful speaker of the Assembly, Sir Charles Price, made a baronet at the recommendation of Trelawney and acclaimed with the moniker "the Patriot" by his compatriots, also died. Edward Long eulogized him as "endued with uncommon natural talents with a truly patriotic attachment to his country." By patriotism, Long meant that Price, "though ever ready to assist and facilitate administration, while conducted for the great principle of public good," was "always the steady, persevering, and intrepid opponent to illegal and pernicious measures of governors." In fact, however, Price was a man of less integrity than Long presupposed. He had used his long tenure as Speaker to craft highly advantageous land deals for himself. His immense slave and sugar empire, one of the greatest in the world at his death, was mostly a mirage. He was heavily indebted to London merchants, and his estates were mortgaged and remortgaged. Price died in 1772 before his estate and reputation unraveled.[26]

There were other West Indians of stellar reputation whose fortunes were more secure than that of Price and who had real clout within the British Empire. William Beckford, born in Jamaica in 1709 but permanently resident in London after the late 1730s, radical Lord Mayor of London in the 1760s, and close friend of William Pitt the Elder, was lampooned as a rich and sexually suspect philistine, but he was an important figure in British political life. Moreover, he was envied for his wealth much more than he was derided for the means—slavery—by which his wealth was derived. When he died in 1770, reputedly the richest man in England, he was the owner of Jamaican property worth £114,269 including 1,356 slaves. He was honored by his grateful fellow Londoners with an ornate memorial in the Guildhall in London.[27]

Other West Indian planters, including men who chose to stay in the islands rather than migrate to Britain, were also highly lauded. Colonel Samuel Martin (1693–1776) of Antigua died aged eighty-three on the eve of the American Revolution, admired inside and outside his native island as an exemplary planter. The owner of 304 enslaved people in 1768, he wrote the most important work on plantation management published in the African period of slavery in the Caribbean, *An Essay upon Plantership* (1754). In this work, he advocated the kind treatment of enslaved people as economically beneficial. One reads such comments from West Indian planters with a suitable degree of skepticism. Nevertheless, Martin appears to be the very model of a West Indian planter, the real-life equivalent of Sir George Ellison, lauded by Sarah Scott in her 1766 novel of the same name. Like the fictional Ellison, Martin was a high-minded gentleman who could be described as a man whose "sentiments are . . . noble, generous, humane and [which] ought to be graven in the heart of every West Indian planter." The Scottish traveler Janet Schaw agreed. Meeting Martin in 1774, she gushed effusively about the old man's politeness and amiability, describing him as "the loved

and revered father of Antigua to whom it owes a thousand advantages, and whose
age is daily employed to make it more improved and happy."[28]

White West Indians therefore were self-confident, self-assured, and increasingly integrated into metropolitan British society from the Seven Years' War
onward. Some West Indians even merged into the British aristocracy, including
members of the Beckford family; and a few, such as the Lascelles family of Barbados and Yorkshire and the Pennant family of Jamaica and North Wales, even
became ennobled themselves.[29] The extent to which West Indians in Britain had
become part of the established order is heralded in a famous anecdote, occasioned when the carriage containing George III and William Pitt the Younger was
overtaken near Weymouth, Dorset, by another, much grander carriage, belonging
to a West Indian absentee planter. The monarch turned to the prime minister and
inquired, "Sugar, sugar hey? All that sugar! How are the duties, hey Pitt, how are
the duties?"[30]

The anecdote relies for its interpretative power on the association between
the West Indian planter and great wealth. West Indians were convinced of their
value to the empire for two reasons. They gloried in the enormous wealth that
sugar and the plantation system brought them, and they believed that they demonstrated in their behavior, values, and political attachments that they were proud
British subjects of an imperial monarch. The wealth that the West Indies brought
to the empire was the more substantial contribution. Historians debate whether
the money that cascaded into planter and merchant pockets, and which allowed
West Indians to buy the elaborate carriages that embarrassed George III by their
grandeur, was overall beneficial to the empire or whether it was sizeable enough
to alter the path of economic growth in Britain. There is no doubt, however, that
the mid-eighteenth-century West Indian economies were highly profitable, that
some of those profits flowed to Britain, and that this wealth gave West Indians an
outside influence in imperial counsels, especially when it came to West Indian
policy.

The annual rates of return from sugar in the eighteenth century were very
healthy, with profits especially high from the 1740s through the 1770s. J. R. Ward
suggests that British Caribbean agriculture exhibited constantly high returns of
over 10 percent on capital in the half century before the American Revolution,
with a downturn in the 1780s more than matched by high profits in the 1790s.
David Ryden has refined these estimates to show that productivity improvements
after 1763 were more than sufficient to justify rapidly rising prices of slaves and
land with the average gross returns on slave investment between 1752 and 1807
being 17.8 percent, with rates of return exceeding 20 percent per annum during
the Seven Years' War and in an extended period of prosperity between 1787 and
1798. Only in 1780, when the west of Jamaica was devastated by the largest

hurricane in its history, did rates of return fall to close to nil. The overall conclusion must be in support of Seymour Drescher's famous conclusion that the British government committed "econocide" by deciding to abolish the slave trade. Slavery made plantations in the West Indies profitable until the very last years of the Atlantic slave trade.[31]

What this meant for planters is that they were as a group very rich, whether they were resident in the West Indies or whether they moved, as a small minority did, to Britain, where they transferred their money into various forms of conspicuous consumption.[32] That wealth, West Indians believed, gave them especial prominence as people who counted, and who needed to be consulted, in imperial matters. West Indians formed the most powerful colonial lobby in London. Their opposition to the Stamp Act, for example, was influential in encouraging British opponents of that act, like the Marquess of Rockingham, to increase their agitation against government policy. They were also able to make successful pleas for preferential treatment from the home government. During the 1760s, Andrew O'Shaughnessy notes, "British colonial policy increasingly discriminated against the North American colonies in favor of the British West Indies." Lord North was known to say of West Indians that "they were the only masters he ever had," while Benjamin Franklin complained that the "West Indies vastly outweigh us of the Northern Colonies."[33]

Not all Britons, however, were convinced that West Indians were quite the ornament to the nation that West Indians thought they were.[34] One way of reading the debate around the meaning of *Somerset* is that it was a public airing of metropolitan unease about how West Indians behaved. West Indians thought their wealth insulated them from public criticism. They believed that their claims to be as British as any other set of Britons were self-evident and well warranted. The debate over *Somerset* showed, however, that such assumptions were unjustified. Many Britons had always disliked West Indians as nouveau riches philistines. Nevertheless, the reasons for metropolitan distaste for West Indians had deeper and more substantial roots, grounded in a view of West Indians that saw their behavior to be so cruel as to be un-British and their character as so shaped by an arrogance bred by having dominion over hundreds of enslaved people that they were a disgraceful, rather than useful, addition to British identity.[35]

The end of the Seven Years' War brought Britain to new heights of imperial grandeur and European power.[36] But the accession of new territories and the unprecedented attention suddenly given to events in far distant places like Bengal and the Ohio valley, which had seldom come to the notice of the British public, brought what happened in imperial places under much greater scrutiny than in the past. And many Britons did not like what they saw when they had revealed to them the true character of British behavior overseas. They were increasingly

deeply engaged with empire, were proud of how it demonstrated their commitment to Protestantism, commercial growth, and liberty, but were also concerned
about how rules of European civility did not apply overseas. Even British nation-
als acted in ways that were unacceptable in Britain and indeed acted in ways that
were shameful, embarrassing, and unworthy of a great nation and a great people.[37]

As Jack P. Greene notes, this critique of empire was "not the product of some
general reassessment of empire by a few major analysts" but "was a loose bundle
of separate critiques by quite different and often unrelated groups arising out of
attempts to diagnose, understand, and resolve specific problems associated with
particular areas of the overseas empire."[38] The East and West Indies were especially the focus of such critiques, as they were assessments of colonies that were
profoundly influenced by the language of alterity. In short, metropolitans who
wanted to criticize various aspects of empire concentrated on specific examples of
outrageous colonial behavior that showed that colonials were demonstrably un-
British and the very opposite of the "kin" or "brethren" that colonials and some
imperial officials alike thought they were. As Greene concludes, these critics suggested that "notwithstanding claims to the contrary by the objects of their scorn,
the overseas portion of the British Empire was British only in the sense of being
subject to metropolitan Britain and emphatically not because Britons abroad
exhibited, much less shared, the same civil and humane values as Britons at
home."[39] As a number of writers, such as Richard Bourke, Hannah Weiss Muller,
Dana Rabin, and Brooke N. Newman, have begun to explore, following on from
work such as that by H. V. Bowen, P. J. Marshall, Jack P. Greene, and myself, the
1760s and 1770s saw the start of wider debates about the place of nonwhite
subjects in the empire, the status of black subjecthood, the rights of Britons based
overseas, and how the behavior of such Britons toward nonwhites ought to be
evaluated and sometimes condemned.[40] *Somerset* was not the only event in metropolitan politics in 1772 with imperial dimensions. Another one of these imperial
issues also involved the West Indies. The start of the First Carib War initiated
ferocious debate in Parliament, in which the venality of West Indian planters, the
language of humanitarianism, and the extent to which nonwhite people were
entitled to the benefits of British subjecthood were all disputed.[41]

The changing views held by Britons about West Indians and about their place
in British life can be seen in the positions put forward from 1769 by Granville
Sharp. Sharp was an extraordinary and indefatigable reformer, who turned his
attention to a whole range of ills that he thought were debasing the moral character of Britain, leading to political corruption and the disenfranchisement of ordinary Britons. He was the grandson of an archbishop of York and the youngest son
from a very large family of an archdeacon. He had been forced by his family
position as youngest son to work his way up in society (although aided by much
wealthier older brothers), starting as an apprentice to a linen draper. Perhaps

because he was aware from his own life about how circumstances beyond an individual's control could affect one's life's chances, he campaigned tirelessly about all manner of injustices, including many domestic ills. He was not just an antislavery advocate but a wide-ranging reformer who believed strongly in two things: the need to bolster the moral authority of the established church, an institution he always thought had the potential to redeem both America and Britain from its many sins; and the desirability of restoring to Britain the ancient law that had characterized the theocratic commonwealth of ancient Israel.[42]

His outrage about how colonial slavery had come to be accepted as legal in Britain emerged from his reading of works on medieval England. He was convinced from his lengthy studies into the history of villenage and the creation and implementation of colonial laws about slavery that slavery had been illegal for centuries. He was also motivated by a deep desire to allow the Church of England, of which he was a devoted evangelical member, to be an instrument of change. He thought that a reformed church could end not just physical slavery but also "civil and political slavery, as well as Slavery to sensual appetites."[43] He had larger objects in view than just the recognition in law that West Indians could not force slavery on England. He wanted a thoroughgoing moral and political reformation of the British nation. Included in this grand plan of reform was a strong xenophobic streak. It was foreigners, and foreign innovations, such as bringing colonial slavery to the metropolis, that, he argued, were corrupting the British people. Long before such views became commonplace during the American Revolution, Sharp believed that southern slave owners and even more so West Indian planters were not fellow countrymen so much as foreigners. He believed that they were determined to introduce their tyrannical and essentially un-British ways of life and their alien political beliefs into a nation in ways that would compromise the freedom of Britons, especially poor Britons.[44]

His anger over the reality of slavery existing in England because of slaveholders bringing African servants to England was intensified by his concern for the poor and by a xenophobic belief that Britain was being overrun by blacks. Slavery was inherently dangerous, he argued, because it was hard to limit it to just one set of people. If Africans could be enslaved, so too could Britons. He noted that "the practice of slaveholding is now only in its infancy among us." But it could easily spread if considered legal because if "such practices are permitted much longer with impunity, *the evil will take root; precedent and custom will too soon be pleaded in its behalf.*" The result eventually would be the introduction, once again, of villenage. Sharp believed that new forms of villenage would be hard just to confine to blacks but would become institutions "in which the poorer sort, even of the original English themselves, might in time be involved."[45]

As Dana Rabin has argued, these explorations of villenage were an important step toward envisioning whiteness as essential to freedom and in explaining how whiteness was constituted as race by both proponents and opponents of villen-age.[46] The history of villenage played a crucial role in the arguments put forward on both sides of the case. Proslavers believed that villenage provided a precedent for enslavement. Initially, Sharp thought that villenage regulations were still in force and that West Indian slavery "is justly entitled to succeed it and to be established on that obsolete foundation." He came to this view when trying to understand a case that was an important predecessor of *Somerset*, though, strangely it was not raised by either side in that case except to be denied by Somerset's attorneys as not relevant as it concerned marriage. Mary Hylas was a slave brought to England by her master, John Newton, who then married a slave owned by Newton's sister-in-law. Newton ordered her to be returned to Barba-dos; Mary's husband, John, sued; and Justice Wilmot had to decide between the laws of slavery and the laws of marriage. He found in Mary and John's favor and told Newton to either return Mary or face a penalty. Sharp thought of Mary as a *neif*, or female villein. As Katherine Paugh argues, "by sticking to older scholarship on villenage and the laws of marriage, Sharp was able to avoid contemplating the complexities of contract and consent for enslaved, married women."[47]

By 1772, however, Sharp insisted that villenage had become obsolete in a nation devoted to freedom. The lawyers for Somerset argued that villenage did not set a precedent in any case, because villeins were white English people who by race and residence were entitled to privileges to which no African could lay claim. For Sharp, it was crucial that Mansfield needed in his decision to make clear, at least implicitly, that planters could not use the historical precedent of villenage to try and intrude slavery as part of English law. English law, he believed was irreconcilable with plantation chattel slavery. The center could give but never receive law. Villenage was defunct, because, as Sergeant Davy argued, "the English Constitution could not bear that any Set of Men under any Circumstances should be put in such a State in this Country." If by some strange chance a person "confessed himself a Villein in England," then "it is impossible that any judge would hear more of it—they would be told that all England would revolt at it—that the Genius of the People would not suffer it." As Rabin concludes, "Mansfield's decision perpetuated a legal apparatus that ensured that human rights in Kingston and London would be relegated to different jurisdictions and tied to local rather than universal definitions." This separation fitted into assump-tions that connected villenage to whiteness and, with the abolishing of villenage in the mid-sixteenth century, linked whiteness in turn to freedom.[48]

The advocates representing Somerset in court warmed to this xenophobic theme. Sergeant Davy ominously declared that if Mansfield decided wrongly then nothing would stop a West Indian owner of an English estate from "stocking his farm with negroes" and that "instead of the farmers who now drive his plough, he would say call out a hundred of my fellows and set them to the plough, and go to the ironmongers for half a score of tortures to make them do it better." Sharp had given Davy the day before the case started an iron muzzle used to discipline slaves in the West Indies to "show that men are not to be entrusted with an absolute authority over their brethren."[49] Davy made sure that he displayed it, somewhat gratuitously, to a shocked courtroom. He called "in God's name" for "an Act of Parliament to prevent the abominable Number of Negroes being brought here by those West Indian planters . . . before we have an Importation of them . . . to prevent the abominable practice of bringing them over in such Numbers." If there was no such law, he warned, race mixing would result, with dire consequences: "I don't know what our Progeny may be I mean of what Colour . . . . It would not be a pleasant Sight nor would be endured but would occasion a great deal of Heart Burn." Thus, it would be best "if these Gentlemen did not bring over so many of them," because he worried for the country if a planter resident in England became, as in Jamaica, "the Grandfather of half a Score of Slaves."

Even if allowing blacks to be held in slavery in England did not lead to the enslavement of the English poor, it was still a bad thing as it introduced an alien presence into the country. This group of aliens were growing in numbers and were contributing to growing levels of crime. Even worse, what West Indians were doing was bringing in a group of lowly paid servants who were bound to displace white workers from their positions. There was already "a dangerous increase of slaves in this Kingdom" and because blacks were "already much too numerous," Sharp thought that "the public good seems to require some restraints on the unnatural increase of black subjects."[50] Davy hinted that allowing more slaves into England to work for absentee planters would also lead to rapid population increase.[51] Sharp later celebrated the result in *Somerset* as preventing the arrival of slaveholders from the West Indies "bringing with them swarms of Negro attendants into this island." Mansfield's decision meant that Britain was not "overrun with a vast multitude of poor wretched slaves" who would inevitably "engross the employment and subsistence of the free labourer and industrious poor."[52]

Behind all these arguments lay a much simpler belief. That belief was that slavery was an innovation invented in the islands by tyrannical planters who wanted to bring a foreign experiment to Britain and in doing so make Britain "as base, wicked and Tyrannical as our colonies." Sharp did not become opposed to slavery because he had a philosophical objection to the institution. He became an antislavery advocate because he saw slaveholders in action and did not like what he saw. The episode that famously turned him against slavery was when he found Jonathan Strong, an enslaved man, badly hurt after his master had beaten him. Moreover, he discovered, to his shock, that Strong's owner reclaimed property rights in Strong after the slave had recovered and then decided to send his enslaved property out of the country. He could not believe that slave owners could be so cruel, but the more he studied slavery, and the more he knew about how West Indian planters behaved, the more he came to realize that violence maintained the system and that planters acted with the authority of arbitrary monarchs to treat slaves abominably.

His 1769 tract against slavery was full of condemnation of West Indians as cruel and hypocritical tyrants. He went through colonial law exhaustively to show that the planters of Virginia, Barbados, and Jamaica had illegally developed laws about slavery. He concluded about the Virginia laws that they were "the most consummate wickedness, I suppose, that anybody of people under the specious form of legislature were ever guilty of." Jamaican lawmakers, he asserted "do not scruple to charge the slightest and most natural offences with the most opprobrious epithets," introducing barbaric punishments long discarded back in Europe.[53] What they had done, he argued, made West Indian planters fundamentally un-British, because slavery by its very nature only enhanced malign characteristics like "avarice, choler, lust, revenge, caprice and all other human infirmities." It made ordinary men tyrants: "every petty planter who avails himself of the service of Slaves, is an arbitrary monarch, or rather a lawless Basha in his own territories."[54]

—

Somerset attracted much attention from the British public, including West Indians and merchants trading with Africa. Having funded Steuart's unsuccessful defense against his slave's petition for habeas corpus, West Indians in Britain were horrified by Mansfield's decision, understanding that the limited practical results arising from the case did not stop it from being a significant change in how Britain thought about the slave trade and about slavery in general. Very quickly, Martin, Estwick, and Long wrote polemics attacking Mansfield and his judgment, accompanying similar polemics from British supporters of the slave trade.[55] What these West Indian writers found especially disturbing about the *Somerset* decision and about the general trend in British thinking about the empire in the 1770s and 1780s was Lord Mansfield's determination to treat James Somerset as a subject of the realm with as many rights to British justice as his owner, Charles Steuart. As Edward Long argued in opposition to *Somerset*, "Negroe labourers [were] merely a commodity," not people with rights that could compete with the rights of free-born Britons. The white West Indian was a "freeborn Briton" or a "native subject." The African was, whether born in Africa or in the colonies, always an alien. Long insisted that "this class of people were neither meant, nor intended, in any of the general laws of the realm, made for the benefit of its genuine and free-born subjects." How, he asked, could Mansfield make any consideration about the "rights" of someone who, for the purposes of law, did not exist, but who was, rather, a "fit object of purchase and sale, transferable like any other goods or chattels," and how, Long expostulated, "could it even be considered that the rights of free-born Britons be subordinated to the rights of foreigners?"[56]

The most convincing arguments that proslavery writers made were those based on property rights and on the commercial value of the plantations and the slave trade. It was when arguments turned to notions of who and who did not belong to the political nation, and even more when the case began to turn on the character of West Indian planters, that arguments began to unravel. Their arguments in regard to slaves being property, and that destroying the slave trade as the main support of a flourishing plantation system was commercial folly of the highest order, have been well covered elsewhere.[57] These arguments were always hard for antislavery advocates to combat, much harder than the traditional justifications made for slavery being an institution with biblical precedence and with historical roots in English customs of villenage. By the mid-eighteenth century, philosophers, led first by Montesquieu and followed by numerous other Enlightenment scholars, including adherents of Scottish commonsense philosophy, had demolished the moral underpinnings of support for slavery.[58]

But it was more difficult to counter proslavery arguments that the laws of England supported the slave trade and protected slave owners' rights in their enslaved property. Long and Estwick were able to cite multiple sources to show that property in the form of slaves was protected under a multitude of laws and customs. Furthermore, they were able to show just how valuable the slave trade was and by extension were the plantations that this trade maintained. In addition, both writers made strong cases that the plantations were very valuable possessions of Britain, producing goods that otherwise Britain would have to get at high cost from elsewhere.[59] Moreover, they argued insistently that Britain and its plantation colonies were enmeshed together and that it was impossible to treat them as separate entities. Antislavery advocates wanted to posit a break between what happened in Britain and what happened in Britain's American colonies or in Britain's trade with Africa, drawing distinct geographical boundaries around the mother country and her peripheries and denying that what went on elsewhere had anything to do with them. By contrast, proslavery advocates had a larger vision, seeing the West Indian colonies as integral parts of a British empire that could not be divided between colony and metropolis.[60]

They viewed the world just as a cosmopolitan businessman would: as an interconnected global world in which trade and commerce could not easily be confined to just one place. Moreover, they made strong and effective arguments that Britain not only had orchestrated the beginnings of the slave trade and had implemented laws and made policies that protected this trade, but also were the principal beneficiaries of the wealth that was produced by slaves in British plantations. As Samuel Estwick argued, "whatever the state and condition of Negroes, it is Great Britain and not America that is responsible for it: that this is therefore a British, and not an American question." He concluded that "whatever property America might have in its Drugs, it is Great Britain that receives the essential oyl extracted from them."[61]

West Indian proslavery advocates, however, were undone by their racism, their lack of sensitivity to the interests of the English working class, and their inattention to those parts of Sharp's critique and Somerset's defense that appealed to the xenophobic side of English opinion. Martin, Estwick, and Long were all committed racists, and each made mention in their pamphlets of how Africans could be characterized as inferior people owing to their savagery and brutality. Martin argued that the slave trade did Africans a favor, redeeming them "from the most cruel slavery to a milder and more comfortable state of life than in their own country." Estwick speculated that Africans were different not "in kind" but even "in species." One way by which he justified that speculation was by observing African culture. He was struck by how "there is no sort of plan or system of morality received by these tribes of Africa." He lamented Africans' "barbarity" toward their children, which "debases their nature even below that of brutes"; commented on their "cruelty to their aged parents"; and ridiculed their religion, which he thought "seems the effect only of outward impressions, and in which neither the head nor heart have any concern."[62]

These racist attacks raised an immediate question that proslavery writers were unable to answer satisfactorily: if Africans were closer to brutes than humans, then why would Britons want to have them come into their kingdom? Moreover, in a question that did not seem to occur to either Martin, Estwick, or Long, if Africans were flooding into the country, as all agreed was the case, who was responsible for this infestation of alien people? The answer was that the only people who wanted blacks living in Britain were devilishly proud West Indian planters and merchants, selfishly wanting to bring into England their items of property that they could treat worse than any white servant.

But England was a land, Sharp pointedly noted, that had no need for slaves and that had a surplus of perfectly capable domestic servants able to do everything that slaves could do and more besides. There was no reason for West Indians to debase England by bringing in people everyone agreed were undesirable and unwelcome additions to a nation already suffering from letting in too many aliens. Moreover, Sharp hinted, the deeper motivation West Indians had in bringing their slaves to England was that they could reduce the lives of the English poor to as lowly a condition as that of plantation slaves.[63] Proslavery writers had two responses to this mostly unanswered question. First, they insisted, as Long argued, that Britons were able to take their property wherever they liked within the British Empire. Responding to junior counsel James Mansfield's quip— "where is this mighty magic air of the West Indies that by transporting [slaves] for a while there, they should become our absolute property here?"—Long argued that "the pretended magical touch of the *English air*" could not "like the *presto* of a juggler, turn" West Indians' "gold into counters" without taking away property rights "solemnly guaranteed by the consent of the nation in Parliament."[64]

Second, rather lamely, they advocated that Parliament make a positive law banning West Indian slave owners from bringing their enslaved property to Britain. This argument, of course, contradicted their first argument, that slaves were property transferable throughout the empire. It implicitly contradicted the main argument that proslavery writers had, which was that metropolitan governments had to recognize the validity of colonial laws. It also flew in the face of reality, which was that colonial slave owners insisted on bringing their enslaved property with them when they went to Britain. Both Estwick and Martin, however, made this argument. Martin combined a hope that "Parliament will expel Negroes now here" with a nod toward fears of miscegenation by adding that "prohibiting" slaves coming into Britain would "save the natural beauty of Britons from the Morisco tint."[65]

Long did not make such an argument or, more precisely, did not make an argument in these terms. But his lengthy disquisition on what he thought about the presence of large numbers of blacks in Britain seriously undercut his more considered arguments about *Somerset* and provided fresh ammunition to British antislavery advocates who supported the xenophobic element of Sharp's condemnation of Africans and West Indians. Long was not typical in his thinking about race in late eighteenth-century Britain. Indeed, in his obsession with transgressive sexuality and in his tendency toward theories of polygenesis—as shown in his notorious argument that as orangutans were sexually drawn to the higher species of Africans, so too Africans wanted to copulate with more highly regarded specimens of the human race, including English women—he was on the extreme end of scientific racial thought in 1772.[66] He combined, in addition to his extreme views on race, a startling condescension toward the English poor that was at odds with Granville Sharp's clear and consistent concern about the state of England's poor and excoriation of the English upper class for its various immoralities.[67] Long's racism and his class prejudice undermined his more cogent arguments by suggesting not only that Sharp was right in fearing the consequences of "swarms" of black servants on the conditions of white servants and the working poor but also that the natural propensity of black men to mate with white women and the natural immorality of lower-class white women would lead to a miscegenist nightmare.

He became derailed halfway through his response to Mansfield by his need to relate at length his views on African character and on the English poor, views he was starting to elaborate so as to publish in his great three-volume history of Jamaica, published in 1774.[68] His belief in the innate inferiority and lack of intelligence of African slaves was so strong, as evidenced in his later history in his lengthy digression intended to prove that the black poet Francis Williams was a fraud, that he found it impossible to even contemplate the idea that any black had the industry capable of being a worthwhile naturalized subject. The only way that he thought a black person might become rich in England was if he won the lottery and thus, as he wrote in a passage that he surely intended to evoke horror in his readers that such a thing might be possible, could become a wealthy landowner. But this flight of fancy was clearly intended to be no more than a worst-case scenario because it was a matter of conviction for Long that it was legally impossible for a slave from the colonies to ever become a naturalized subject. He found it difficult even getting his head around the idea of Africans that Sharp insisted on, which was that Negroes were humans, not beasts or merely property.[69]

Not surprisingly, given his highly negative view about blacks living in Britain, whom he considered "a dissolute, idle, profligate crew," Long agreed that there should be some restraint on "the unnatural increase of *blacks* imported into it." He wrote approvingly of the custom in France where "no Negroe slave can be brought from the colonies . . . except to be bound apprentice to some handicraft."[70] He admitted also that there was no reason for any slave owner to bring slaves to Britain (though he defended resolutely their right to do so) as they were "retained in families more for ostentation than for any *laudable* use." This thought led him onto what can only be called a rant, about how "worthless" blacks were and into a fantasy about the dreadful possibilities of miscegenation, in which he castigated blacks for their immorality and where he made wild and highly derogatory aspersions about what poor women in England wanted sexually from black men.[71]

What worried him was how the worst kind of whites seemed drawn inexorably to blacks in Britain, many of whom he thought, because of misguided actions such as that envisaged by Mansfield and Sharp, had "eloped from their respective owners." These black "renegades" fell "into the company of vicious white servants and abundant prostitutes of the town." They turned to crime, and, even worse, to marriage, being thoroughly "debauched in their morals." The people that black runaways wanted to marry were lower-class English women who, he asserted, were "remarkably fond of the blacks, for reasons too brutal to mention." He made clear what he was referring to by following up this calumny on such women by suggesting that "they would connect themselves with horses and asses, if the law permitted them." But the physical assets of black men did not make them faithful, and they quickly moved on to other women and "then abandon their new wife and mulatto progeny." The results, he averred, were likely to be disastrous if unchecked. It would lead to extensive miscegenation and "in a few generations more, the English blood will become contaminated . . . until the whole nation resembles the *Portuguese* and the *Moriscos* in complexion of skin and baseness of mind."[72]

Having raised the nightmare, as he saw it, of inevitable miscegenation if the number of blacks in Britain were not curbed, he compounded his fearful remarks by claiming that the mulatto children abandoned by irresponsible black fathers would inevitably become a burden on local parishes and would become part of a larger body of poor people, both native born and also from "the swarms of needy dependents continually pouring in, from the foreign states around us." Long's warning was an eighteenth-century version of the traditional xenophobic lament that foreigners were coming in to seduce local women, steal jobs from natives, and live off the state as a criminal class. Unlike Sharp, who showed genuine sympathy for the English working poor, Long moved quickly from some anodyne and formulaic concerns for how blacks and foreigners were taking poor people's jobs to contempt—most blatantly in his outrageous comments about poor women's fondness for well-endowed black men but also in an assumption that poor whites and free blacks would combine to increase crime and create disorder. The apocalyptic result, he argued, would be miscegenation on a wide scale, which would be "a venomous and dangerous ulcer, that threatens to spread its malignancy far and wide" with an infusion into Great Britain of "lazy, lawless Negroes, living in a state of nature" allying themselves in crime and disease with "an inconsiderable remnant of white subjects, unable to provide themselves with a mere support of existence."[73]

But none of this, Long believed, was the planters' fault. They were not to blame "that the nation begins to be emblazoned with the African tint."[74] They should instead be applauded, he thought, for their contribution to the growing wealth of the nation. What he missed, however, was how easy it was for his readers, few of whom had the blind optimism in planters' innate goodness that Long exhibited in his history of Jamaica, to read things differently.[75] It was easy to see that the doomsday story Long told all stemmed from the selfish desires of corrupt and tyrannical West Indian planters to bring their slaves to England just for purposes of ostentatious display. Moreover, everyone knew, including Long, who lambasted white West Indian men for "rioting in the goatish embraces" of their colored mistresses rather than taking white women into matrimony, that the principal practical promoters of miscegenation were not amorous poor white women but were rich planters and merchants.[76]

As Sergeant Davy had pointed out in his advocacy, what was at stake if *Somerset* was decided in Steuart's favor was the ability of foreigners, Russian or West Indian, to bring female slaves to Britain whom they could rape with impunity or, as James Mansfield noted, who they could whip mercilessly in English public squares.[77] Long's argument also confirmed Sharp's contention that allowing black slaves into the country would eventually lead to whites being reduced to the status of blacks rather than the blacks improving to the status of whites. And it seemed to many readers that the fault of the social disaster that Long hinted at was due to the planters he lauded. These men came to Britain loaded with money, determined to introduce the customs and horrors of the Caribbean into polite English society, such as enslavement, and continually lusting inappropriately after black women.

In retrospect, the wealth and constitutional strength of the West Indian planters, even in 1772, was not as impervious to outside attacks as planters imagined. The Seven Years' War was not the beginning of the end—it was the start of two decades of unprecedented planter power and enormous prosperity. It was the beginning, however, of metropolitan rethinking of the West Indian image and, most important, of metropolitan doubts about white West Indian claims to be freeborn Britons. White West Indians were the first in what became a long line of loyalists abandoned by Britain.[78] The speed and effectiveness of the abolitionist campaign against the slave trade generally, and against planters specifically, took white West Indians very much by surprise. Their lack of awareness of their personal unpopularity in Britain was, in retrospect, a major tactical mistake.[79] The campaign against slavery and the slave trade was motivated not just by humanitarian concern for the poor Negro. It was given additional force by abolitionists' strong personal distaste for the planter personality and the planters' way of life. As Kathleen Wilson summarizes, "The fabulously wealthy Caribbean planter that emerged in fact and fiction came to represent West Indian uncouthness, backwardness and degeneracy that inverted the acclaimed standards of civility and culture."[80]

Proslavery advocates' delay in fighting the representational battle over the image of the West Indian planter meant that antislavery advocates had a head start of many years. They were able to implant in the British mind the image of the West Indian planter as cruel, avaricious, and decidedly un-British. By the end of the American Revolution, that image had become a powerful prejudice, held equally by opponents and supporters of the British slave trade and the plantation interest. It took very little effort for antislavery writers to be able to draw on the highly negative representation of planters so powerfully expressed in the writings of Granville Sharp and in the legal strategies developed by Sergeant Davy and Francis Hargreaves in defending James Somerset in 1772. Antislavery writers drove home to the British public that the West Indies was an alien and frightening place where people claiming to be British did things that disgraced the name of Britain. Crucially, this discourse did not rely just on the denigration of planters and merchants as fundamentally un-British. It also relied on a deep residue of xenophobic dislike of foreigners and contempt for alien "infestations" into the body politic. Proslavery writers by the 1780s had developed a rhetorical strategy that emphasized both a more inclusive idea of nation, which insisted that West Indians belonged to a large British empire that transcended the narrow bounds of the British archipelago, and also a well worked-out and pernicious idea of race, which connected race explicitly with "color" and with developing notions of scientifically justified racial difference.[81]

Their assumptions ran up against another discourse, however, that concerned the relative rights of citizens against subjects. This debate became urgent from the mid-eighteenth century as disputes over the citizenship ideal moved from being political in nature (who created laws) to legal (who belonged to the nation and what rights did they have). In this move, the concepts of "citizen" and "slave" became interchangeable as both citizens and subjects were deemed obedient to the laws instituted by community and both pledged allegiance to and sought protection from that community. It was under this new equation of citizens and subjects that Jews in Britain could be thought to have some belonging to the British nation, leading to the Jew Bill of 1753–54. Nevertheless, whether subjects—people bound by laws they did not necessarily make—and citizens—people who participated in the making of laws and who became part of the political nation—were in fact indistinguishable remained a highly contested notion. The relative rights of various kinds of people in New World settings—Catholics, Jews, and free people of color—often, as with *Somerset*, raised significant legal and political issues, both in Britain and in France.[82]

In *Somerset*, however, the question of to what extent people of African descent could become British subjects was sidestepped. Slavery, it was argued by Davy, could not exist in England as a successor of villenage because villenage no longer existed, and even if it was supposed that it could exist, only white Englishmen and not Africans could be villeins. Villenage was racially determined—it was a historical experience of white English people. What status people who were Virginian slaves had in England did not need to be discussed, as the matter at hand was narrowly about whether the laws of England allowed slavery to exist. Davy argued, and Mansfield agreed, that the laws of Virginia had no force in England. Laws could flow from empire to colony, not the other way around. Mansfield made his decision on very narrow grounds, affirming the supremacy of English municipal institutions and conventions without nullifying colonial laws. Somerset was thus an alien according to British law, coming from a different country. It was English law that "a Foreigner cannot be imprisoned here on any Law of his own Country." Mansfield expressed no opinion on whether the laws of Virginia and Jamaica had legal force in their home places. He held that "the power of a master over his servant is different in different Countries, in some it is more, in others less extensive." Therefore, he concluded that a master's power over a servant "must be regulated according to the law of the place it is exercised." But under the laws of England, a servant could not be sent abroad by a master against his will.[83] What remained unclear is whether a black servant and his descendants could ever become British or whether they would be permanently alien residents. It is revealing that we do not know what happened to James Somerset after the case was decided in his favor. Did he or his children become English? Did he even stay in the country?[84]

The assertions of proslavery advocates—that an inclusive idea of nationhood allowed colonists to be thought of as Britons while racism provided the mechanism for excluding "inferior" races—proved a powerful idea, laying the foundation for scientific racism in the nineteenth century.[85] It became the foundation for the expansion of British liberty overseas into settler colonies in which English traditions of liberty and the rule of law were asserted as the inheritance of white male settlers while being something that could be denied on racial reasons to nonwhites and perhaps also to women, including white women.[86] But regarding the specific issue that proslavery advocates were fighting—stopping the abolition of the slave trade—it was a rhetorical strategy that backfired. Granville Sharp and the advocates for James Somerset were able in 1772 to successfully demonize West Indians, both black and white, as alien presences in a land that needed to erect a *cordon sanitaire* to prevent being infected with the sin of slavery. They also drew attention to what they saw as the problem of increasing numbers of culturally inferior and sexually deviant black people. Daniel Livesay notes that this theme was very prominent in the arguments made by Sergeant Davy in support of Somerset, which envisioned England overrun by blacks corrupting the already suspect urban lower classes.[87] Edward Long had inadvertently stoked these xenophobic concerns about England being flooded by Africans through racist diatribes that hinted darkly at miscegenation and improper cultural mixing, implying that anything to do with the West Indies reeked of cultural debauchery and sexual excess.[88] The latter message, combined with a belief that the West Indies was a corrupt, exotic place where Britons degenerated rather than thrived, drove out counter-messages that the West Indies, like the East Indies (where a similar and even more pervasive tale of Britons corrupted by Asian vice was being told in the mid-1780s during the dramatic trial of Warren Hastings, ex-governor of Bengal), was a modern, progressive place that was a cornucopia of riches augmenting British commerce.[89]

White West Indians were slow learners. They began to deal belatedly with the question of their image in Britain only in the mid-1780s, especially in response to the first serious polemic directed against them, written by James Ramsay in 1784, a cleric with experience in the West Indies. Ramsay was motivated, like Sharp, to become an antislavery writer as much from a detestation of the proud and irreligious West Indian planter as from concern for the West Indian slave.[90] Proslavery writers tended to stress other things, such as the importance of the plantations to British commerce, the inviolability of property rights, and the complicity of Britain in recognizing slave owners' legal rights to the ownership of African slaves. They emphasized how if enslaved people were given their freedom, then there would be a reversal of natural order, with black people enslaving whites rather than the other way around.[91] In short, they concentrated their rhetorical power on areas where they had strong arguments but neglected responding to the most powerful arguments made by their opponents.

By the time that abolitionism became a real force in British politics in the 1780s, the image of the West Indian planter as being as cruel and as barbaric as the savages he owned was firmly implanted in the British mind. Even those people who might naturally have supported hedonistic West Indians, and who were likely to turn against do-gooders like William Wilberforce, had begun to believe not only that Africans were racially inferior but that white West Indians had degenerated from the behavior expected of Britons and were themselves subject to significant, and distasteful, degrees of Africanization.[92] Caricatures of the West Indian planter, as in James Gillray's powerful "Barbarities in the West Indies" (1791), emphasized that the "West Indian plantation system produced English-men who were so depraved that they assumed the debased, animalized forms of the very "savages" they brutalized."[93]

In retrospect, West Indian writers were too successful in their assertions that what *Somerset* entailed was the opening of Britain to the horror of miscegenation and cultural corruption. They thought that Britons would be able to distinguish between the noble, upright planters who brought Africans into Britain and the grotesque and disturbing Africans who were arriving in England in ever-greater numbers during and after the *Somerset* trial. They were wrong. Just like Granville Sharp, metropolitan observers elided together as similarly alien and undesirable not just African slaves but also their owners—sources of the corruption that seemed about to engulf a previously uncorrupted land. West Indian planters and merchants had lost an important battle—their image in metropolitan discourse—before the real war over the fate of the British slave trade had even begun. *Somerset* was thus not just a significant law case, even with minimal direct effects. It was a major cultural moment in the reconfiguring of West Indians in the British mind. It was the start of a process by which the most ostensibly loyal of Britain's subjects in the Americas were abandoned by the mother country.

Chapter 7

The *Zong,* Jamaican Commerce, and
the American Revolution

Only one slave ship in the Transatlantic Slave Trade Data Base goes by the name of the *Zong.* This single vessel, however, is probably the most notorious of all the slave ships that crossed the Atlantic in the long period of the transatlantic slave trade. It is famous because of events that happened onboard in late November and early December 1781—the deliberate murder through throwing overboard of African captives in order to later claim insurance on these captives as lost cargo—that became the subject of an infamous court case in 1783.[1] The abolitionist Granville Sharp tirelessly promulgated to a shocked public both the callous conduct of the crew of the *Zong* in murdering, on three separate occasions, 122 Africans and the even more shocking ways in which the British legal system treated a case of what abolitionists thought was mass murder as a routine, if legally complicated, case of maritime fraud.[2]

The *Zong* was an unlucky ship on an unlucky voyage. It had an inexperienced captain—Luke Collingwood, a surgeon by trade who had never commanded a slave vessel. It also had a small and quarrelsome crew, who were at loggerheads with each other by the time that the *Zong* had reached the Caribbean after an unusually long Atlantic crossing. First, after leaving St. Kitts in mid-November 1781 the crew discovered that their water barrels were leaking, meaning that water supplies became short by the time that the ship neared Jamaica (though no sailor or captive was put on short allowance). Second, someone made the disastrous decision of mistaking the western end of Jamaica for Cape Tiburon in eastern Saint-Domingue. This navigational error meant that by 29 November 1781, the *Zong* was hopelessly off course, becalmed somewhere off the southwest of Jamaica, far away from its intended landing point of Kingston.[3]

Adrift, with water running low, and fearing slave insurrection, the crew, according to their testimony, later contested by Granville Sharp, consulted and decided to force overboard African captives to preserve water for survivors. They killed the Africans in three batches—fifty-four were killed on 29 November;

forty-two were murdered on 1 December; and sometime in early December (and after rain had fallen, making the problem of dwindling water supplies less acute) a further twenty-six were thrown overboard. On the final occasion, ten slaves leapt into the sea, committing suicide rather than being murdered. The *Zong* finally arrived in Black River, southwest Jamaica, on 22 December 1781, where it sold its remaining 208 slaves.[4] The ship (renamed the *Richard*) returned to England on 26 October 1782 and the owners, the Gregson syndicate, claimed the losses of the murdered slaves at £30 per head, under an insurance contract that covered the death of slaves aboard slavers due to "the perils of the sea," and the loss based on the average cost of the surviving 208 Africans sold into slavery at Black River in January 1782.[5]

The *Zong* case became a pivotal moment in the development of a humanitarian sensibility.[6] Lord Mansfield made a notorious comment adjudicating in the case when it came before the Court of King's Bench on 22 May 1783. He stated that "the Matter left to the Jury was, whether [the mass murder arose] . . . from necessity [,] for they had no doubt (tho' it shocks one very much) the Case of Slaves was the same as if Horses had been thrown over board." His further comment, that insurers had to pay up for dead slaves killed in an insurrection "just as if Horses were kill'd" but that insurers did not have to pay up for slaves dying naturally just as "you don't have to pay for horses that die a natural death," caused consternation.[7] Initially, the owners' request for financial relief was granted, and they received £3,660. The insurers appealed; Mansfield agreed with the insurers and ordered a new trial, which does not seem to have taken place, meaning that the owners of the *Richard*, previously the *Zong*, did not get their money.

—

Usually, the story of the *Zong* starts either in Liverpool, where the owners of the vessel lived, or in London, where the trial of the *Zong* occurred on 5 March and 21–22 May 1783, or in Anomabu in West Africa, from where the *Zong* departed for Jamaica. It is instructive, however, to see how the *Zong* case looks if the starting point is southwestern Jamaica, where the 208 survivors of the calamitous voyage were sold on or about 9 January 1782. By late November 1781, when the *Zong* first arrived near Jamaica, the island was in a sorry state. The entry of France into the American Revolutionary War had raised prices on most imported goods, had increased shipping costs dramatically, and had halted a golden period of near continuous economic growth that had lasted from the end of the Seven Years' War until the start of the American Revolution.[8] Plantation profits slipped dramatically. J. R. Ward estimates that plantation profits declined from an average of about 10 percent throughout the eighteenth century to 3 percent in the early 1780s. Sugar production fell from 51,218 casks shipped to Britain in 1775 to

30,282 shipped in 1783. Inventoried wealth declined, and indebtedness increased. Transatlantic commerce suffered. Between 1774 and 1781 charges on a hundredweight of sugar increased from 14 shillings to 37 shillings, and insurance rates rose from as low as 2 percent to as high as 28 percent. The price of provisions skyrocketed—if planters could find any.[9]

All these problems, however, paled in the face of natural calamity. On 3 October 1780, darkening skies heralded the arrival of a destructive hurricane. Southwestern Jamaica bore the brunt of the storm.[10] Thomas Thistlewood, a small planter and inveterate diarist living in Westmoreland Parish, thought that seventy whites and at least five hundred slaves perished. More died from famine and disease. One account thought that slaves "were every day perishing in numbers, partly from diseases (chiefly dysenteries) brought on by unwholesome food and partly from absolute starvation!" Savanna la Mar was inundated with water, and houses were destroyed. Reports estimated property damage in Westmoreland Parish alone at £678,571. Jeremiah Meyler, a merchant at Savanna la Mar, gave a vivid account of the destruction. He claimed four hundred whites died in the town. He estimated in 1783 that "I suffer'd by the hurricane only upwards of £20,000 sterling in 300 hogsheads of sugar & 150 puncheons rum, & the whole of my estates levell'd, with loss of Negroes & stock, the very support of my Negroes only, exceeded the neat proceeds of two following crops."[11] Damage was also severe in the western regions of St. Elizabeth, where Black River was, and in parts of St. James and Hanover Parishes.[12]

Thistlewood related how terrible the hurricane was for ordinary Jamaicans. The hurricane of 3 October 1780 destroyed his house, wrecked his outstanding garden, and caused, Thistlewood declared, "sad havoc all through the countryside." He calculated his damages at £1,000 and noted in his first diary entry for 1781 that he had made no money in 1780. His entries at that time bear testimony to the depression that gripped him and everyone else in the area. He noted that "Mr. James Robertson declares he is afraid to fall asleep, as such dreadful hurricanes and confusion present themselves to him, as far exceed the real one. Just so with myself and several others, the nerves so affected." On 31 December 1780, summing up the year, he noted, "The hurricane has made everybody look ten years older than they did before, and the healthiest show a great dejection in their countenances—nothing looks pleasant or agreeable since" because it was "as iff a dissolution of nature was at hand." He even tried to sell his property but had no takers, meaning that he had to undertake the rebuilding of "the ruins of a dwelling house" by himself.[13]

The early 1780s also saw the Atlantic slave trade at a low ebb. Planters stopped stocking their plantations with new slaves from Africa, despite mortality rates for slaves staying very high. But slaves who died from overwork and malnutrition were not replaced. In the West Indies the slave trade declined by one-quarter between 1776 and 1778 with twelve leading Liverpool slave companies deciding to get out of slave trading. Thus, when the Gregsons, the merchants who bought the *Zong* as a prize, decided to reenter the slave trade after a three-year hiatus they found themselves in a trading environment with significantly fewer competitors. They also were involved in a seriously diminished slaving system. In the three years before the war, 231 slaving ventures landing 60,480 slaves arrived in Jamaica. Between 1777 and 1782, however, only 95 ships with 34,526 slaves arrived in Jamaica. The year 1780 marked the nadir, with just 3,763 slaves arriving on the island. Prices for African slaves were not encouraging. In 1775, the *Thames* arrived in Lucea in western Jamaica with 341 slaves from the Gold Coast and sold these slaves for an average price of £56.66 sterling. But in 1779 the 258 slaves from the Gold Coast onboard the *Rumbold* fetched just £32.75 each. It is hardly surprising that trade dried up the next year even if prices improved a little in 1780: 368 Biafran slaves on the *Hawke* sold at Kingston for £37.85, and 295 slaves from Sierra Leone on the *Mars* sold for £43.59 apiece.[14]

Slave traders faced other problems besides low prices, high insurance, and weak demand. Trade conditions became more difficult and more expensive on the African coast. American and French privateers added to the risks of trade in the Middle Passage.[15] The revolutionary war forced slave traders to restructure their trade. They looked increasingly to the Royal Navy for support and tried, wherever possible, to travel in convoy. In the case of the *Zong*, for example, its departure from the Gold Coast in June 1781 came after the *HMS Champion* mounted an action against Dutch settlement at Commenda, with the *Zong* providing the warship with gunpowder. The seaways clear, the *Zong* left Cape Coast in convoy with the *Adventure* and *Lord George Germain*. Both ships had recently been to Jamaica, and both had captains—John Muir for the *Adventure* and William Thorburn for *Lord George Germain*—with recent experience of wartime slave trading conditions.

War also opened opportunities, however, such as the ability to seize ships flying under enemy flags. Seizure was how the Gregsons got the *Zong* in March 1781. The original plan was to send Richard Hanley in the *William* to Anomabu in December 1780 to buy slaves and transport them to Jamaica. Just as Hanley arrived, however, Captain William Llewellin of the *Alert* received letters of marque to be used against the Dutch, now at war with Britain. By late February 1781, he had captured three Dutch vessels, one of which, the *Zorgue*, was sold to Hanley for an undisclosed sum, along with 244 slaves and goods to purchase more African captives. The *Zorgue* was renamed the *Zong*. Hanley thought the acquisition a golden opportunity. Instead, it was a poisoned chalice. Having to buy the ship and its cargo at a time of economic downturn strained the Gregsons' finances.[16]

Slave traders dealt with declining returns and rising costs by increasing the size of slave cargoes sent across the Atlantic. Before the start of the American Revolution, very few ships carried more than six hundred slaves. The average number of slaves carried in slave ships going to Jamaica between 1774 and 1776 was 235. The only ships carrying more than six hundred slaves to Jamaica before 1776 were the *Liberty* in 1763 (also a war year), which loaded 741 Africans and landed 605, and the *Britannia* in 1774, with 626 slaves embarked and 609 landed. But ships carrying enormous numbers of slaves became more frequent after 1778: two in 1779, one in 1780, two in 1781, four in 1782, three in 1783, and four in 1784. The arrival of ships carrying very large numbers of slaves changed the trading dynamics at the places they landed. It was easy for the market to become swamped, especially in smaller ports, if a ship carrying many slaves arrived suddenly. In late 1781, two ships carrying over six hundred slaves arrived in Jamaica—the *Jane* with 620 slaves, arriving at Montego Bay on December 5, and the *Champion* with 691 slaves, landing at Kingston on Christmas Day. The arrival of the *Jane* in Montego Bay may have influenced Collingwood's decision to head for Black River. It meant that 2,148 slaves were crowded into western ports in December 1781, almost guaranteeing a buyer's market, especially a month before hurricane donation monies were released to planters.

Thus, the *Zong* departed for Jamaica at a particularly inopportune moment. The hurricane of 1780 and the travails resulting from over five years of warfare in the Americas meant that 1781 was probably the hardest year that Jamaicans had experienced in their lifetime. The hurricane meant that even those planters who had weathered the storms of previous years made little to nothing in 1780 and 1781. As well as economic difficulties and the ongoing problems caused by the October 1780 hurricane, Jamaicans suffered another hurricane in August 1781. This hurricane was less serious than the one a year earlier, but it led to losses of life and property. The navy suffered particularly, losing several ships, including *HMS Pelican*, under the command of Cuthbert Collingwood (seemingly no relation to Luke), later a hero of Trafalgar and Lord Nelson's close companion.[17] The loss of naval ships was particularly worrying because national affairs were going very badly. In November, news of Cornwallis's catastrophic defeat at Yorktown reached the Caribbean, as did news that a powerful French fleet was headed to the eastern Caribbean and then to Jamaica, probably likely to arrive in December 1781 or January 1782. According to British newspapers, the French flagship, the *Ville de Paris*, carried fifty thousand pairs of handcuffs and fetters to carry off slaves from Jamaica to French colonies.[18]

But there was some glimmer of light for white Jamaicans devastated by the events of 1780 and 1781. After much delay and wrangling, an enormous amount of money—£40,000 sterling—had arrived on the island late in the year as a gift from the British Parliament for hurricane relief.[19] It was the first such gift granted for natural disaster in British America since a much smaller grant had been sent to South Carolina in 1740. The great majority of the money went to afflicted planters in western Jamaica, with the planters of Westmoreland and St. Elizabeth getting the lion's share. Reflecting the racial dynamics of the island and the determination to make whiteness a key political category, the monies were directed entirely to the white population with free people of color, many of whom had suffered greatly in the hurricane, receiving next to nothing in compensation. Free people of color petitioned the Assembly that when they filed their claims the commissioners for distributing money had told them that "the donation did not extend, nor was it meant to relieve people of color." The petitioners noted that this was unfair, as they paid taxes, served in the militia, and were "good citizens," and they lamented the fact that "the humanity and charity of the British nation could mean to exclude a large body of subjects . . . because their skins are brown." They argued they should get some of "the generous gift of the parent state . . . as subjects who ought to be relieved." Their protests, however, fell on deaf ears.[20]

British politicians stressed the humanitarian aspects of this relief allocation, but they also wanted to reward the loyal West Indian colonists at a crucial time in the American Revolution. The motion to give £80,000 to Barbados and £40,000 to Jamaica passed Parliament in late January 1781 with little debate. The money arrived in Jamaica in late May. The Jamaican House of Assembly decided to confine all the payments to white people who had suffered losses in the hurricane and who were residents of western parishes. Claims were ordered to be filed by October 10, and a joint committee of the Assembly and Council approved a plan for redistribution in November. Money started to trickle to recipients late in the year, with £25,426 going to residents of Westmoreland and adjacent areas of St. Elizabeth. Some indication of the timing of how the money got into circulation comes from Thistlewood's diaries. He received his money late in December and sold his share of the hurricane relief on 9 January 1782, "for . . . £140 2d."[21] Thistlewood sold his share the exact day that 208 slaves went on auction from the *Zong*.

Such was the changed commercial environment that the Gregsons faced in reentering the slave trade business in Jamaica in 1780–81. The easy profits of the mid-1770s had disappeared to be replaced by more challenging times and different modes of operating. The advent of general European war, with Britain at war simultaneously with three slaveholding European powers—Spain, France, and the Netherlands—meant that when Richard Hanley purchased the *Zong*, slave traders faced especial difficulties. These difficulties may have encouraged Hanley in his purchase.[22] He sent word back to Liverpool that he had bought the vessel, and as of 3 July the *Zong* was insured for £8,000. It was an unusual insurance for an unusual slave ship. The Gregsons and their partners in the *Zong* syndicate arranged for a "valued policy" in which they insured their slaves not on the cost of their purchase in Africa (which in 1781 was £14.60 per slave)[23] but on their future value when sold in Jamaica.

As Ian Baucom points out, "the genius of insurance and its contribution to finance capitalism, is its insistence that the real test of something's value comes not at the moment it is made or exchanged but at the moment it is lost or destroyed." He notes that at question in the *Zong* case was not the murder of 122 people (and suicide of a further ten people) but was a form of value. Baucom asks whether it was possible to see in "a drowned slave a still existent, guaranteed, and exchangeable form of currency."[24] Baucom's question is the right one to ask in a trade where a principal aim was to reduce people into commodities. That process took place mainly aboard ship. People became numbers in captains' descriptions of them in their logs. The role of sailors was to create the commodity called "slave" to be sold to planters.[25]

What the Gregsons had done was to contract a kind of futures contract reflecting the expected sale price of enslaved people in Jamaica. These kinds of insurance contracts tended to be contracted only in colonial commerce where it was often difficult to obtain an invoice price for goods at the port of embarkation.[26] Underwriters were happy to go along with these kinds of contracts, presumably because it gave them access to an extremely lucrative trade. Underwriters issued policies for as much as £100 million per annum for property on the sea, a figure roughly equivalent to the total annual national product of England and Wales.[27] On occasion, these kinds of contracts, where underwriters were prepared to pay out on notional future losses rather than on actual losses incurred in business, could be very beneficial for slave traders. One example was *Shawe v. Felton* (1801), where the owners of a slave ship underinsured a ship's cargo and then lost the ship after they had deposited their enslaved cargo, owing to what was agreed to be "perils of the sea." They were able to claim, however, on the lost ship and also get the full value of the insurance on their by now diminished cargo, even though they had made good money on the slaves they had sold as part of the shipment.[28]

That the type of insurance the Zong had was a "valued policy" was very important in determining why the ship owners took the action that they did and perhaps in explaining the events that happened off southwestern Jamaica in November and December 1781. The anticipated future loss of property in a "valued policy" could be worth more than the actual value of that property if sold on market (especially in a glutted market) owing to the magic of "average loss." Average loss meant that the insured got back in a valued policy the "average" of the amount that each enslaved person was valued at. In effect, if a captain knew that a slave cargo was likely to deteriorate in quality and thus fetch less money at market than it was valued at for insurance purposes, he had commercial encouragement to maximize his investment by destroying slaves when their ostensible value was the sum that they had been insured for rather than for what they might fetch in an actual market. Thus, if Collingwood knew that the captives he was carrying would fetch less on arrival in Jamaica than the £30 they were insured for while at sea, he had a strong incentive to get rid of his captives at sea before they dropped in value on land.[29]

Marine insurance was an important but arcane area of the law. It was governed by standard printed forms first devised in the late seventeenth century that were far from suitable for purpose, but which were encrusted by generations of legal interpretation hanging on every sentence. As James Oldham notes, legal scholars have always considered the forms used in the eighteenth century for marine insurance absurd and ridiculous. He cites Sir Douglas Owen saying in the early twentieth century that "if such a contract were to be drawn up for the first time to-day, it would be put down as the work of a lunatic endowed with a private sense of humour." Nevertheless, Mr. Justice Buller in 1791 explained why such a form had to be continued to be used: "a policy of assurance has at all times been considered in courts of law as an absurd and incoherent instrument: but it is founded on usage and must be governed and construed by usage." As Buller suggested, this form provided clear principles about what would and would not be covered by insurers for claims on property lost at sea. Owners could claim only on property lost through the "perils of the sea." Such perils included shipwreck, piracy, arrest, and shipboard rebellion. They did not include losses caused by human error nor the death of slaves through what was termed "natural causes," which might include famine or lack of water. But "perils at sea" clearly included risks to crew from slave insurrection.[30]

When a master was forced to jettison a portion of his cargo to save the rest, an insurance claim for recompense based on the average value of the insured items of cargo could be made. The standard text, published two years before the *Zong* case came to trial, was by John Weskett. Masters could not claim for slaves dying a "natural death," including when "the captive kills himself through despair," but "when slaves are killed, or thrown into the sea to quell an insurrection on their part, then the insurers must answer." Thus, if the counsel for the plaintiffs could prove that by throwing some slaves overboard they prevented the greater evil of imminent insurrection whereby "all the blacks wou'd have killed all the Whites" then their claim should stand.[31]

The terms of the insurance claim made it clear that the Gregsons could make money on their insurance claim if the crew of the *Zong* played their cards right in late November 1781. The Gregsons would have received £8,000 if the ship had sunk with all hands, but they would have still lost money on their investment and, more important, the crew would not be alive to take advantage of any insurance money. But the Gregsons could make money if a portion of slaves were thrown overboard. The insurance premium was £1,680, or the value of fifty-six slaves. If that number were thrown off the ship and the rest of the slaves were sold in Jamaica for at least an average price of £30, then the Gregsons could make about £10,000 gross profit from the trip, which in the circumstances that the crew of the *Zong* found themselves in around early December would be a decent result. If most of the slaves thrown overboard were slaves that were unlikely to achieve high prices in the market, then the lure of killing slaves for their insurance became even more compelling. It is noteworthy, in this respect, that the first batch of Africans thrown overboard on 29 November 1781 comprised fifty-four women and children—very close to the number of slaves needed to pay for the insurance premium.[32]

Of course, all this depended on crew members knowing the inside details of claiming under marine insurance law for losses due to "perils of the sea." We have no evidence about how much knowledge of insurance law Luke Collingwood or first mate James Kelsall had. We know that Robert Stubbs claimed that it was the captain—despite his possible delirium—who drew the crew's attention to the insurance aspects of throwing slaves overboard at the same time as he was allegedly acquainting the crew with the notion that if action was not taken immediately to secure the ship through the mass murder of slaves that they would imminently be massacred by an uprising of slaves. Certainly, enough was known about marine insurance claims to understand that a captain needed to consult with all the crew as a prerequisite to a successful claim for jettisoning a cargo in an emergency.[33] The crew, Kelsall claimed, were consulted on 29 November. Such consultation of crew was not normal practice on a slave ship and had not been normal practice on the Zong, where Collingwood, as seen in his summary dismissal of Kelsall for two weeks from his position as first mate, acted in a dictatorial rather than a consultative manner.

Consultation, however, was a formal element in the claim. It is strange that a doctor-turned-captain would have such direct understanding of complex law to have anticipated later court action by making sure that the crew were consulted. Here, the role of Robert Stubbs is of interest. He alone of the crew would have had deep knowledge of commercial practices in Britain. He had been a ship captain in the 1750s,[34] a merchant for the African company in the 1760s, a slave ship owner and ship broker in the 1770s, and until 26 January governor of the Royal African Company Fort at Anomabu. He was also not a man given to worry about moral scruples. William Llewellin, captain of the Alert, which had carried Stubbs and another Royal African Company governor, John Roberts, to the Gold Coast in 1780, refused to take Stubbs back on his ship as a passenger when returning to Britain in 1781 because he thought Stubbs had a "wicked and treacherous Character."[35] Stubbs found sanctuary with Richard Hanley on the William.

His whole life before the Zong had been one long quarrel and confusion. James Walvin describes him as a deeply flawed character, whose word was widely regarded as worthless. There is at least circumstantial evidence suggesting that he was more involved in discussions about insurance claims on the Zong than he let on. The evidence does not allow for anything more than speculation, but other people at the time, notably the abolitionist Reverend George Gregory, who had

known Luke Collingwood in Liverpool and who thought him a "liberal, benevolent and well-intentioned man," suspected that Collingwood had been scape-
goated. Gregory believed that it was highly convenient for crew members to argue
that all decision-making flowed from the inexperienced captain.[36] It was also very
convenient to place all the blame on a dead man—Collingwood, sick in November 1781, had died in Kingston in January or February 1782.

The crux of the plaintiffs' case was that the crew was under the "necessity" of
throwing slaves overboard because they faced an imminent insurrection from
water-deprived captives realizing they were facing a slow and painful death. The
argument that the counsel for the plaintiffs made was curious, because, mimicking
the "valued policy" at the heart of the dispute, it argued that actions were taken to
prevent a future event rather than in response to an event that had already happened. There was no insurrection on the *Zong*. But the crew were convinced that
if they had not acted on 29 November that an insurrection would have occurred.
Killing fifty-four women and children by jettisoning them as cargo saved the crew
and the rest of the slaves on ship from imminent destruction. Both James Kelsall
and Robert Stubbs, the only eyewitness testimony that exists concerning the
decision-making process on the night of 29 November, were insistent on this
point. Stubbs supported Collingwood's decision to murder Africans because
"according to his Judgement the Captn did what was right. . . . [There was] an
absolute Necessity for throwing over the Negroes." Kelsall claimed he objected at
first but soon came around.

Mansfield and Buller, the presiding judges in 1783, did not make much of the
question of insurrection on board the *Zong*. In retrospect, it is surprising that they
did not interrogate counsel further on this matter. Of cases dealing with slave
insurrection that came before the British courts in the late eighteenth century, it
was only in *Gregson v. Gilbert* that the insurrection under discussion was imaginary, not real. In *Jones v. Schmoll* (1785), the underwriters were forced to pay the
insurance on nineteen slaves who had been killed in what the plaintiffs called "a
mutiny." In *Rohl v. Parr* (1796), the underwriters agreed to compensate the owners "by general average" for all slaves lost from insurrection amounting to more
than 5 percent of the slave cargo. The crew on the slave ship *Zumbee* killed seven
of forty-nine slaves in putting down an insurrection off Cape Coast in West
Africa. As this was more than 5 percent of the slave cargo, the judges on King's
Bench found in favor of the owners of the ship and against the underwriters.[37] But
it was made clear in court in 1783 that the *Zong* suffered no insurrection.
Solicitor-General John Lee, acting for the Gregsons, argued that by throwing the
slaves overboard the crew avoided the greater evil, for otherwise "in a few hours
there must have been such an Insurrection all the blacks wou'd have killed all the
Whites." Counsel for the defendants, Mr. Davenport, demurred. Lee replied by
quoting testimony from Stubbs. Davenport continued, however, insisting that

there had been no insurrection. Lee agreed, noting that he did not say that there was an insurrection, merely that there might have been one if preventive action had not been taken.[38]

Was an insurrection likely? The jury at the jury trial on 6 March thought so. They awarded the Gregson syndicate £3,660 for the loss of 122 Africans insured at £30 each. Insurrections were a constant threat in the Atlantic slave trade. David Richardson estimates that up to 10 percent of all transatlantic slaving voyages were affected by revolt, leading to the death of perhaps 1 percent of all captives entering the trade and adding significantly to the costs of doing business. The costs of containing coercion were high and were perhaps highest in the second half of the eighteenth century, when the frequency of slave revolts increased and when the violence on both sides was most pronounced. The crew on slave ships were terrified about the possibility of shipboard revolt and did as much as they could to try and prevent it, including exercising fierce control over African men, the most likely protagonists in any uprising. Men were kept fettered and chained almost continuously and were separated from women and children and kept in heavily patrolled apartments, usually immediately below the main deck and as far away from the weapons room as possible. A large wooden grating covered the entrance to their quarters, designed to prevent men captives from getting out in anything other than single file.[39]

The crew had good reason to be anxious about their situation on 29 November when they realized that they were at least ten days sail from Jamaica with no more than four days' supply of rationed water for themselves and 380 captive Africans. This slave ship was "in distress." The journey was already lengthy. At seventy-nine days duration on 29 November, it was nearly three weeks longer than the average slave journey to Jamaica in 1781. Moreover, the crew had known for nearly ten days that they had lost a lot of water from leaky casks and that the first and second tiers of water butts were exhausted. The *Zong* had enough water remaining after leaving Tobago on 20–21 November for ten to thirteen days at full rations, but the captives, if not the crew, were probably already on short rations well before reaching Jamaica.[40] If so, this might help explain the higher than normal losses of captives (13.1 percent) that the *Zong* had sustained even before deciding to jettison "cargo" on 29 November. The ship was troubled in other ways, as well. The captain and first mate had quarreled, leading to James Kelsall being suspended between 14 and 29 November as first mate. The captain was sick, possibly delirious. It is likely, though no evidence directly supports this contention, that Robert Stubbs was spending some time as master of the ship, despite being a passenger and despite having not sailed a slave ship for at least a quarter century. At the very least, there was an absence of effective leadership at a very crucial moment in a troubled voyage.

Most worryingly, the ship was seriously overcrowded and seriously undermanned. The usual maximum number of slaves carried by a slaver of the *Zong's* size was around 250. The *Michael*, for example, a slaver of 132 tons going to Jamaica in 1781, embarked from Africa with 210 slaves and disembarked 195. Like the *Zong*, it left Africa with a crew of twenty. Seven crewmembers died or left en route. Its ratio of slaves to crew was 10.5 to 1 on departure and 15 to 1 on arrival. The average ratio of slaves to crew for ships going to Jamaica, excluding the *Zong*, in 1781 was 8.91 to 1 when ships left Africa and 10.18 to 1 when the ships arrived in Jamaica. The *Zong* had a ratio of slaves to crew of 23.15 to 1 (19 crew and 440 slaves) when they left São Tomé, and an extraordinary 34.5 to 1 ratio (11 crew and 380 slaves) on 29 November. In addition, only one ship voyaging to Jamaica in 1781—the *Jane*, a vessel of 242 tons with 677 slaves, a ratio of 2.8 slaves per ton—was more crowded than the *Zong*, which had 2.26 slaves per ton.[41]

Nevertheless, if the *Zong* really was at risk of insurrection the actions that Collingwood and his crew took after 28 November did not protect the ship from shipboard revolt. Indeed, their actions placed the ship and the crew in great danger. The crew had a variety of options open to them. The most obvious option was to shut down the holds and let the Africans take their chances of surviving without water and meat while the crew made haste as fast as they could to the safety of port in Jamaica. Africans starved on ships all the time, and a "sickly ship" was not uncommon. Dr. Alexander Falconbridge described how bad things could get on a ship ravaged by disease: "the deck was covered with blood, and approached nearer to the resemblance of a slaughter-house than anything I can compare it to, [and] the stench and foul air were likewise intolerable."[42] The major problem with this option was that taking it would negate all future claims for insurance. It was well established that deaths of slaves at sea from a disease that could have been contracted at land, or death from "despair" or suicide were not covered in insurance claims.[43] It would also virtually ensure a bad market for selling slaves. The last ten days of a voyage were generally taken up with preparing captives for sale—taking constraints off wrists and ankles so that sores could heal, careful cleaning, using a lunar caustic to hide sores, covering up gray hair, and rubbing down bodies with palm oil so that captives glistened and gave off a healthy glow. Such preparations would be impossible in a ship where famished slaves were locked below decks.[44]

The second option was to throw overboard as many slaves as were deemed necessary for the crew's survival and then try for landfall in Jamaica. After 132 slaves had been thrown overboard and a further thirty-six had died (with thirty left dead on the decks rather than being thrown overboard, as one would expect to have happened, if only to prevent the spread of disease), the crewing ratio of the *Zong* would have fallen into still high but not unreasonably high ratios. Ten men plus Robert Stubbs were on board the *Zong* when it sailed into Black River on 22 December 1781 with 208 surviving Africans—a ratio of 18.9 Africans per crew member. A figure of nineteen Africans per crew member gave Collingwood some confidence that he could quell a slave uprising, whereas thirty-four Africans per crew member, as existed on 29 November, signaled trouble.

The third option was the one outlined by Kelsall in his interrogatory before Exchequer. The ship could wait until the water had diminished further and hope either for rain or for a passing ship to relieve their plight. The climatic conditions pertaining in early December have not usually been considered important by historians in evaluating the case of the *Zong*, but they were important in the trial at King's Bench in March 1781. It does not appear that the question of rain had come up in the original jury trial at the Guildhall on 5 March, a trial over which Mansfield had also presided. But when Sergeant Heywood, on behalf of the insurers, after the cause was over and the verdict brought in, declared that Stubbs in his written testimony had said that rain had fallen for several days while the *Zong* was at sea, and after the first murders had been committed, Mansfield became agitated. This, he declared, was "a fact which I am not really apprized of." He reiterated his surprise shortly after: "I am not aware of that fact [.] I did not attend to it. . . . It is new to me. I did not know any Thing of it." That slaves were thrown overboard after rain had fallen for several days, thus presumably alleviating the "perilous Necessity" the members of the crew were under in a ship short of water, was, to Mansfield, "a very material circumstance," even if it was "not agreed on by both sides." Mansfield knew very well that if slaves had been killed after rain had fallen then the whole nature of the case changed. It was what led him to suggest that there be a new trial. Indeed, if rain had in fact fallen, then *Gregson v. Gilbert* was no longer an insurance trial but could be, as Granville Sharp insisted in his commentaries on the case after the verdict, a murder trial.[45]

Rain did fall. Stubbs hints that it might have fallen as early as 1 December— the day of the second tranche of killings. Sharp repeats this assertion, adding that this rain continued for a day or two, enabling the crew to collect six casks of water.[46] Kelsall states more authoritatively that rain fell between 6 and 9 December. The fact of the rain supports Kelsall's objections at the 29 November meeting of the crew that the *Zong* could have waited a few days before throwing slaves overboard. It had at least four days full rations of water. On short rations, the *Zong* could have watered crew and slaves (especially if only healthy slaves were given water) for the seven days between the first killing and the start of rain and could probably have sailed very close to the Jamaican coastline, where help conceivably might have been at hand. On 29 November, however, the crew agreed that an insurrection was imminent if action was not taken soon. Some crew then took fifty-four women and children from the hold and threw them overboard. No African man was killed in this first and especially pivotal action. Then the crew waited. They tacked against the wind and headed to the southwestern hamlet of Black River. That journey took much longer than the ten to fourteen days' sail that Kelsall estimated it was likely to have taken from 29 November to reach land. If Kelsall was right about sailing times, the *Zong* should have reached Jamaica sometime between 8 and 12 December.[47] Instead, it sailed with a diminishing slave cargo for at least a further four weeks. During those "missing" weeks, they threw overboard another seventy-eight Africans, in at least two separate batches.

It seems difficult to believe that an insurrection was imminent on 29 November when the crew threw women and children out of cabin windows. Although the *Zong* had had a long journey and had been afflicted by more than normal levels of sickness, the condition of the slaves was still good. Kelsall noted that on 29 November the *Zong* had 380 slaves, "all of them in good health and condition." The slaves thrown overboard were chosen "without Respect to sick or healthy." They were all "marketable slaves." Indeed, if the aim of throwing Africans overboard was less to preserve the ship from insurrection than to ensure that potential slaves arrived in the best possible condition, then Collingwood probably succeeded. The ship may have arrived "in great distress," according to the *Cornwall Chronicle*, but the insurers disagreed. The ship, they averred, docked "in perfect safety and with Crew And the Rest of the Slaves in good health." Getting rid of the less valuable women and children rather than high-value but potentially dangerous men, knowing that the *Zong* was arriving in a crowded marketplace, seems to make more sense as an insurance scam than as a genuine attempt to forestall insurrection.

Matters may have changed by 1 December. One imagines that the days of 30 November and 1 December were extremely fraught on the *Zong*. Robert Stubbs claimed that he heard the "shrieks" of the women and children thrown "singly through the Cabin Windows" when he was in his own cabin. So too would have the captives locked below deck. Kelsall recounted how an African who spoke English had told him that the people shackled below decks "were murmuring on Account of the Fate of those who had been Drowned" and that the African pleaded on behalf of his fellow captives that "they might be suffered to live and they would not ask for either Meat or Water but could live without either till they arrived at their determined port." How likely is it that all the captives chained below decks were willing to risk dying through starvation and dehydration? Is it not just as likely that some of the captives would have preferred to go down fighting rather than acquiesce to a slow, lingering death?

It is probable that the crew would have thought these thoughts also on 1 December. It was now two days on from the events of 29 November and water supplies were running extremely low. Slaves would have gotten very little of whatever water had been distributed. It may have been on this day that "all apprehended [the slaves] should die of want of Water if they had not thrown the Slaves overboard to preserve the rest." Here was the "Perilous Necessity." And it may explain why the crew chose to throw overboard forty-two "stout healthy Men slaves"—slaves that would have fetched a premium in the slave market but also the mostly likely slaves to overpower the crew and seize the ship. Kelsall tried to justify the killings by claiming that it was an act of perverse kindness, "the shortest and least painful Mode" of destroying them and thus kinder than "suffering them to expire by degrees." Nevertheless, the accompanying statement—that if the killings had not taken place many of the Africans "would have been seized with Madness for want of water"—may have more accurately summarized the crew's feelings on 1 December. They feared that the approaching "madness" of dehydrated Africans who knew that they too would be thrown overboard like their womenfolk and children would lead to an attack that would overwhelm the small complement of crew. They might also have calculated that the longer that the water shortage continued, the weaker that the crew would have become and the less likely that they could put down an insurrection. Taking forty-two men, one by one, from a men's quarter seething with discontent was problematic, increasing the possibilities of revolt, even if the men were heavily shackled, as male captives usually were.

And then it rained, possibly on 1 December itself and more lengthily after 6 December. Sometime after 6 December and the arrival of rainwater came the most inexplicable act of the entire journey. A further thirty-six Africans were thrown or jumped overboard. Why would the crew do this? The threat of insurrection had been quelled (if it was to happen, it surely would have happened in the period of the first two killings, when tempers were raw and water shortage acute). The arrival of rainwater had reduced the immediate danger that the *Zong* was under. The rain was the "material circumstance" that changed the verdict in the trial. In the jury trial at Guildhall, the jury did not know that rain had fallen and found in favor of the plaintiffs. At King's Bench, that it rained before some murders were committed became evident and led Mansfield to declare that there needed to be a new trial. We would like to know more about this third set of murders. Were the captives "stout and healthy" or like the thirty dead slaves lying on the deck that Stubbs said greeted Jamaicans on the ship's arrival at Black River? On what day did the third lot of killings occur? Why did the third lot of killings occur at least a week after the second murders? Was this set of killings to stop the spread of disease or an act of greed by a compromised crew to maximize an insurance payout and thus make a bad voyage marginally profitable to their employers? If a third trial of the murders on the *Zong* had proceeded, we might know some of the answers to these questions. But, then again, we might not have heard anything about the *Zong* at all if this third set of murders had not occurred. One thought is that these murders were so unable to be justified by the commercial logic of the time that they aroused the suspicion of the insurers that they were being fleeced.

A recently discovered misfiled letter in the British Library written by Granville Sharp to the Lords Commissioner of the Admiralty on July 2, 1783, suggests that

Sharp at least had his doubts about the facts of the case as presented by the crew in testimony before the courts. Sharp noted that by the fateful day of 29 November over sixty slaves and seven whites had died and "*a great number of the remaining Slave[s] on the last day mentioned were sick of some disorder or disorders & likely to die or not live long.*" Sharp made mention of this fact because "dead and dying slaves would have been a dead loss to the Owners . . . unless some pretence or expedient had been to throw the loss upon the Insurers." He believed that the "*Sickness & Mortality*" was "*not occasioned by the want of water*" as they only discovered that day that fresh water was reduced to two hundred gallons "*before any Soul had been put to short allowance and before there was any present or real want* of water." It was before water had become a problem that Collingwood, according to Sharp, told the crew of his plan and "palliate[d] the inhuman proposal" by arguing that "it would not be so cruel to throw the poor sick Wretches . . . into the Sea as to suffer them to linger out a few days under the disorders they were affected." The chief mate, Kelsall, objected to this proposal, saying "there was no present want of water to justify such a Measure." Sharp's views on how the murders had occurred were that they were deliberate from the start, doubting that the "mistake" in navigation occurred. He noted that on November 27 "instead of proceeding to some port, *either thro' ignorance, or a sinister intention,* [*Collingwood]ran the ship to Leeward,* alledging tha[t] he mistook Jamaica for Hispaniola." Sharp seems to think the actions deliberate, as he follows this statement with outlining the sickness onboard the ship.[48]

Another mystery associated with the *Zong* is what happened in the month or six weeks after the rains had fallen on either 1 or 6–8 December and after the ten to fourteen days that Kelsall estimated it would take the stricken ship to return to Jamaica. Why did the *Zong* dock at Black River rather than Montego Bay? The answer, I argue, is associated with the dynamics of Jamaican commerce and politics in November and December 1781. The *Zong* wanted to maximize its opportunities in a crowded slave market.[49] Its captain—either the very sick Collingwood or Robert Stubbs, the self-proclaimed innocent bystander—decided that it was a good idea to delay arrival so that slaves would be sold in January rather than December.

Selling slaves at Black River was a very curious thing to do in 1781. The Free Port Act of the mid-1760s had broken the monopoly of Kingston over the buying and selling of slaves on the island. An increasing number of ships began to bypass Kingston to sell slaves in the booming western provinces, where the plantation economy was expanding and where prices for slaves were higher than in the East.[50] Between 1763 and 1784, forty-three ships landed 12,361 slaves at Montego Bay. Given the *Zong*'s likely location, southwest of Jamaica, Montego Bay was the obvious place to head to after the navigational error of 27 November had taken the ship away from its likely destination of Kingston. The problem was that

Montego Bay and the nearby port of Lucea were crowded with slave ships. Both *Zong*'s traveling companions, the *Adventure* and *Lord George Germain*, had dropped their slaves in the northwest, on 16 December and 30 November respectively, and the *Jane* had landed her 620 slaves in Montego Bay on 5 December. Another ship, the *Liverpool*, may also have been in port from late November. If the *Zong* had docked at Montego Bay in the second week of December, their 250 or so slaves would have joined at least 1,430 and possibly 1,758 slaves for purchase. The *Ulysses* arrived with 390 slaves on 19 December, making the marketplace even more crowded.

It made sense, then, to try and bypass Montego Bay and to try and sell slaves further south. The obvious place to dock was Savanna-la-Mar, the capital of Westmoreland Parish. Between 1763 and 1785, fifteen ships landed 3,496 slaves at Savanna-la-Mar. The port, however, had fallen out of favor. No ship had docked there since four ships had gone there in 1768. Moreover, the town had been virtually destroyed in the 1780 hurricane. The *Zong* pushed on to Black River, around fifty kilometers by road east from Savanna-la-Mar. It was a pioneering journey, the first of only eight voyages in total and three before 1789 to the town. No other ship landed at Black River until the *Tharp* did in 1784.

There was logic in what the *Zong* did, however. Black River served as a center for planters who lived in the prime sugar territory east of Savanna-la-Mar. Thomas Thistlewood's career shows that there was a constant flow of people in the region between Savanna-la-Mar and Black River. His first position, in 1750, was at Vineyard Pen, close to Black River, before he moved to properties a few miles east of Savanna-la-Mar.[51] Planters in this region had just received a considerable share of the £25,426 hurricane relief money sent to Westmoreland and St. Elizabeth Parishes. Their slave losses from the hurricane had been significant; they had not had the money to replace their slave forces until hurricane relief money arrived; and they were eager to use their new cash on their most desperate need, new slaves. Thomas Thistlewood's decision to sell his share of the relief money in early January (he had lost no slaves in the hurricane and thus did not need to buy) tells us that money for buying slaves had suddenly became available by early 1782 when none had been available in November or early December 1781. Thus, it would have been the assistance of British imperial money that accounted for the relatively successful sale of 208 slaves. In the circumstances, clearing £7,488 gross profit, with the expectation of a further £3,660 from a successful insurance claim for 122 slaves killed to stop insurrection, was a good return from a stressful trip. If we assume that the real cost of purchasing slaves on the African coast was around £16 per slave, or £7,072, and if we add in an insurance premium of £1,680 and wages and other costs of perhaps £1,500, then the *Zong* with gross profits of £11,148 and costs of £10,252 would have made a small profit for its owners. By contrast, if it had limped into Montego Bay in early

December with around three hundred slaves in poor condition in a market full of much better quality slaves, which slavers were trying to sell to planters without much ready cash, it is hard to see how the ship could have cleared a profit at all.

The *Zong*, thus, was doing good business in late 1781 and early 1782. Of course, this statement presupposes what is difficult to presuppose: that an inexperienced ship captain and a disgraced and financially inept ex-governor of a slave fort, whose actions on a disastrous slaving voyage marked them out as highly incompetent sailors, would have been able to take commercial decisions of some complexity. Adrift at sea and without any communication with a wider world, would Collingwood, Kelsall, or Stubbs have had the knowledge of conditions in western Jamaica at the end of the American Revolution that have been outlined above? There are some reasons to suggest, however, that Collingwood or Stubbs knew what they were doing when they murdered captives for insurance, dallied for longer than they might have in the Caribbean Sea, and landed in a port hitherto unused for slave sales. Hearsay evidence presented to the Privy Council in 1791 about how sailors stranded on the Isles of Pines in 1779 had contemplated making captives walk the plank to claim insurance suggests that what was done on the *Zong* on three separate occasions was not outside the knowledge of men engaged in the slave trade.[52]

Moreover, Collingwood and Stubbs would have known something about conditions in western Jamaica from their conversations with Captains John Muir of the *Adventure* and William Thorburn of *Lord George Germain*, both of whom had been recently in Jamaica on previous voyages and who intended to sell their slaves in Lucea and Montego Bay. Captains kept a very close eye on their competitors and anxiously assessed the state of the market where they were to sell slaves for as much as they could. Samuel Gamble, captain of the *Sandown*, which made a generally disastrous voyage to Jamaica in early 1794, noted in his journal the day he arrived in Kingston that two other slave ships from Angola and Bonny had arrived at the same time as him and "that there's upward of 3000 Slaves in the Harbour for Sale." Within a week he glumly commented that "Planters & others complain that they cannot purchase what Negroes they want having nothing for them to subsist upon[.] A great number of Guiney Men in this Harbour at Present not sold." He listed seventeen ships in the harbor with 5,452 Africans. Unsurprisingly it took from 25 May until 24 July for "all but four" of his cargo to sell.[53]

By contrast, the *Zong* seems to have sold its slaves in one day, 9 January 1782.[54] And knowledge of the large subvention that Britain was planning to give to Jamaica for hurricane relief was very well known by the middle of 1781 when the *Zong* left Anomabu. The campaigns to help the victims of the 1780 hurricane were "the largest and most significant relief effort of the eighteenth century." The newspapers carried appeals on behalf of Jamaicans and Barbadians for week after week in early 1781. Parliament also debated the question of relief extensively in

January 1781.[55] Arriving at Black River just as money from the hurricane relief fund became generally available may have just been dumb luck for a ship plagued by misfortune. But there is at least a plausible argument that the men making decisions on the *Zong* were manufacturing their own luck.

The importance of the *Zong* lies beyond the simple facts of the case. It became the cause célèbre that galvanized the antislavery movement from being a minor campaign by a set of marginal figures infected by evangelical enthusiasm to becoming the most significant moral campaign in British history. That remarkable change happened in a few short years in the mid-1780s. The *Zong*, as the most infamous atrocity in the Atlantic slave trade, played a pivotal role in the transformation of the antislavery movement. Opponents of the slave trade never failed to express disgust at what had happened off the seas of Jamaica in late 1781 and seldom missed a chance to fulminate against the inhumanity of a legal system that saw the murder and forced suicide of 132 Africans solely as an interesting example of marine insurance law. Fifty-eight years later, it was rehearsed once more in the grand celebratory narrative of the triumph of the antislavery movement written in 1839 by Thomas Clarkson, the great old man of the campaign whose long career in abolitionism was in part stimulated by his outrage about what had happened on the *Zong*.[56]

The subsequent history of the *Zong*—it struck a chord with the public and resonates in memory until the present day—might have given the Gregsons and Thomas Gilbert and his underwriting team pause. From the very long perspective, Gilbert's decision not to accept the Gregsons' claim and the Gregsons' determination to fight that rejection in a court of law was seriously misjudged. From that decision, in part, came the beginnings of the destruction of their lucrative business. That happened in 1807. In retrospect, the underwriters should have accepted their loss on the premium. They probably would have done so if there had been only two sets of killings on the *Zong*. The claim then would have been for the replacement value of ninety-two slaves at £30 each, making a claim of £2,880, or a loss of £1,200 for the underwriters once the cost of the premium was deducted. That loss was worth bearing. It is unlikely, given the verdict of the March 5 trial, that the underwriters would have been successful in their case at court if only the first two killings had been under discussion. The rules of marine insurance law in respect of killing to prevent insurrection were sufficiently clear to make the Gilberts' case hard to win. But one can understand why the Gilberts refused to pay. They became suspicious when they realized that the killings had gone on after the "perilous necessity" of preventing insurrection had diminished. So too did Mansfield become suspicious. It was the murder of twenty-six slaves and suicide of a further ten Africans sometime between 1 and 22 December that cast doubt on the credibility of the two witnesses and on the judgment of the deceased captain everyone wanted to blame for the murders.

The reverberations from the *Zong* case also had a major effect on the plantation economy of Jamaica, an island that went, in half a century, from being the jewel in Britain's imperial crown to an especially disgraceful part of Britain's empire in the aftermath of the Morant Bay killings in 1865. No Jamaican was called to give witness in the case of the *Zong*. Jamaica, however, was crucial in the case. We cannot understand the decisions that were made in late November and December onboard the *Zong* without placing it in the context of a Jamaican commercial and political world facing severe disruption because of the turmoil of the American Revolution and accompanying natural disasters. Because this was an unusual time, it led to unusual slaving voyages. What was most unusual about the *Zong* was not that slavers threw Africans overboard to claim on insurance. That crime happened with regularity. It was an unusual voyage because of the way that the ship was acquired as a prize and unusual because the crew on the ship needed to be particularly attuned to some very specific commercial and political rhythms in Jamaica in 1781. Unusual cases seldom make for lasting legal precedents. This unusual case, however, led to a lasting change in the relationship between Britain's leading slave colony and the imperial state.

Chapter 8

Loyalism and Rebellion
in Plantation Societies

In rural southeastern Jamaica, up a dirt track leading to nowhere, lies a monument (a gravestone) erected around 1814 that reminds one of Percy Bysshe Shelley's great poem of January 1818 "Ozymandias." The monument, like the poem, is a poignant reminder of the fleeting nature of fame and the tendency of all empires to crumble into dust. The gravestone in rural St. Thomas, now the poorest parish in a poverty-stricken nation but in the early nineteenth century a place full of highly productive sugar estates, lies just beyond a rusted car and past humble rural homes and an electricity substation. It honors the life of Simon Taylor, noted as "a loyal Subject, a firm Friend and a honest Man." Taylor was born in Kingston on 3 January 1740, and died on 14 April 1813, choosing to be buried on Golden Grove, his most profitable sugar estate. When he died he was the richest man in Jamaica and possibly in the British Empire. His wealth has now all gone, and so too has his fame as a proud Jamaican patriot. No one in Jamaica now remembers his name. If he is remembered at all, it is because his Kingston house, Prospect Pen, first purchased in 1785 in the garden suburbs of Liguanea, is the current residence of the prime minister of Jamaica. It was from this gracious home that Taylor wrote most of an extensive correspondence with friends and business associates, mostly in Britain, making him one of the more richly documented Jamaicans of his period.[1]

Taylor is not that well known today. Certainly, his eighteenth- and early nineteenth-century fame has not passed down the ages as it has for his near contemporary and fellow planter Thomas Jefferson (1743–1826). Both men were born as British subjects in far-flung reaches of an expanding British Empire. But whereas Jefferson decided to join in the rebellion of thirteen of Britain's twenty-six American or West Indian colonies in 1776 and then became a citizen and leader of the newly formed United States of America, Taylor never lost his allegiance to his monarch and to his country—a country he thought was Britain as well as being Jamaica. There are reasons why Taylor has been eclipsed in the annals of posterity by Jefferson besides the fact that Taylor was a resident of an island whose global importance has rapidly declined since Taylor's death and leaving aside Jefferson's intellectual and political accomplishments. Taylor was never a revolutionary but, like most white Jamaicans, was a fervent patriot and "loyal Subject," as his gravestone spells out. He never questioned his loyalty to the British Empire and to George III (1738–1820), also a near contemporary, even in 1775 when he described the coming conflict between Britain and the thirteen colonies as "truly alarming."[2]

He feared for the continuing prosperity of Jamaica when its plantation system was so dependent on supplies of food and provisions from North America. Nevertheless, he remained optimistic about Jamaica's future even as war between America and Britain became irreversible. He thought that Jamaica would "do tolerable well without America" and that the war itself might have some beneficial consequences, lancing a boil between mother country and colonies so that "good will arise from evil." Whatever sympathy he might once have had toward North Americans, however, had disappeared. In 1774, he wrote to a friend that he fully supported Britain's harsh measures against a recalcitrant Massachusetts, stating that "after what the Americans have done Britain cannot give up the point [as] it would only be making them more arrogant than they are at present and I look upon them as dogs that will bark but dare not stand when opposed [and are] loud in mouth but slow to action."[3]

Taylor's confidence was misplaced. Although Jamaica suffered relatively little from the depredations of war before France and Spain entered the war in 1778 and 1779 respectively, the following years, especially the period from October 1780, when Jamaica suffered a devastating hurricane, through to April 1782, when Rodney's victory over the French at the Battle of the Saintes prevented a Spanish-French invasion of the island, were desperate times. Taylor recognized how terrible times were, though his main concern was what he considered an avaricious government. He grumbled that he contemplated moving off the island, fulminating in 1781 that he might leave Jamaica "to go to some other Government where we might be able to make a shift to live, and not be held in Egyptian bondage. The irony of a slaveholder making use of a discourse of slavery to describe his own discomfort was probably lost on him.

But Taylor had nowhere to turn to, either during or after the American Revolution. The trauma of a civil war with fellow planters had ended only a few years before what he considered the idiocy of British abolitionism commenced, led by "that Madman Wilberforce." From 1788, his hopes for his island and his ever-growing personal affluence were at odds with what he considered a world gone crazy, with the French Revolution and what he saw as the ultimate disaster of slave revolt in Saint-Domingue compounding his woes.

His self-image as a determined patriot with a highly conservative political outlook made it impossible for him to view revolutionary actions with anything other than great disdain. He was very disdainful of what he considered the excesses of liberty in the United States of America and could not think of moving there, even though the United States was a place where slavery was not just protected but was becoming ever more important. His loyalty to Britain, however, was tested. He felt that the abolition of the slave trade—a trade essential to the continued success of the plantation system in Jamaica—was a colossal act of folly that could have happened only through the corruption of "a systemick plan . . . to ruin and destroy the West Indian colonies, by breaking every compact and agreement with them." The abolitionists, especially that "hell-born imp," Wilberforce, were nothing but "traitors, democrats and men who were using every act to bring in a revolution." The very concept of revolution, he thought, was politically and morally wrong. The revolution in France was especially outrageous, but America's devotion to republican principles was equally reprehensible. He could never live in the United States, owing to its contamination by an overly democratic spirit of liberty. It was full of the "cant of philosophy, liberty and equality," ideas that he considered "the most pernicious vermin that were ever created." American ideology was designed, he thought, "to overcome all established government and in order to establish in their room murder, anarchy and confusion."

Taylor thus had no place to go except St. Thomas in eastern Jamaica. Britain had betrayed him. Going to the United States would mean that he would have "to learn the new philosophy of rights of Men as they are called." Moreover, his beloved Jamaica was by the early nineteenth century a place over which "some evil genius seem[s] to preside." His massive wealth (he died a millionaire) could not protect him from the ravages of a revolutionary world and from the betrayal of his reflexive loyalty to monarch, church, the Tory party, and a hierarchical order in which slavery sustained an agreeable culture of deference. He felt bereft. It was perhaps fitting, therefore, that his gravestone was humble rather than grand, and that the encomium on it less than fulsome.[4]

—

Starting an chapter on the American Revolution with a vignette on Simon Taylor tells us something about how the field of early American history is being changed by what we might call a spatial turn. In the last twenty years, early British American history has been transformed by the rise of Atlantic history, which connects the history of Europe, Africa, and the Americas, putting a strong emphasis on transnational and trans-imperial relations between these four continents.[5] The reconceptualization of the spatial dimensions of early America has taken a while to affect the writing of the American Revolution. Given that it is the foundational event in American history, in most accounts the crisis starts quite suddenly in 1763 and ends just as abruptly with the creation of the United States of America in 1788. It is often depicted solely in American terms—as a historical watershed with exclusively American origins and resulting in changes that pertained mainly to the historical trajectory of the American nation.[6] These works include some of the most distinguished works in the field.[7]

But as Kathleen Duval notes, as a scholar attuned to seeing the American Revolution in a wider geographical compass, the American continent in the eighteenth century was a place of multiple sovereignties. It might seem today, when the borders of the United States so easily extend from east to west, that this expansion was part of what nineteenth-century Americans called America's "manifest destiny" to stretch from the Atlantic to the Pacific. The first leaders of the United States made no such an assumption. The European-driven resolution of the American War for Independence presumed that North America and the Caribbean would always be a place of multiple and varying sovereignties, including European empires, sovereign nations, Native American confederations, and even, after 1804, a pariah black republic.[8]

The American Revolution was a stage in a larger process of global change brought on by revolution. It lasted from as early as the start of the Seven Years' War in 1756 to as late as the War of 1812 and perhaps the subsequent wars for Spanish American independence. Possibly it continued even up to the end of the American Civil War in 1861.[9] This was the world in which men like Simon Taylor and Thomas Jefferson lived and died. It was also the world that men like Frederick North, Lord North (1732–1792), prime minister of Britain between 1770 and 1782, lived in. North was a well-meaning aristocrat who struggled to understand a strange overseas empire that Britain seemed to have acquired almost by accident. We need to pay attention to the position of North within the British ruling class—a ruling class where it mattered that a prime minister might be the mere eldest son of an earl rather than a territorial magnate like the Dukes of Bedford and Devonshire or the Marquess of Rockingham.[10]

The genial and dutiful Lord North and the high-minded evangelical responsible for the colonies in the early 1770s, William Legge, Lord Dartmouth (1731–1801), were not such wealthy and cosmopolitan men of the world but were instead the very model of the best kind of parochial English aristocrat. Their experience of life, however—rooted in the villages, churches, and stately homes of the Home Counties and in the imperial grandeur of London with precious little knowledge of Europe, let alone the wilds of America—ill fitted them for dealing with the revolutionary ideas and radical notions of the men who propelled Massachusetts into rebellion. As Nick Bunker writes of the system that produced North and his ministers, "the system set mental boundaries that they could not transcend, raised as they were in a culture where the landscape and the parish church bore everywhere the signs of privilege. They found the rebels unthinkable. Nothing in rural Oxfordshire could prepare Lord North for an encounter" with them and their ideas.[11]

These well-meaning but bumbling and narrow-sighted men failed to understand a growing and increasingly complicated empire. Their attempts to make that empire conform to the values they held dear cost George III (a man of similar type to Lord North) much of his American possessions.[12] Simon Taylor was a similar kind of conservative—of which there were many more in America and the West Indies than we are exposed to in accounts that focus on radicals and revolutionaries—even if his world would have been one utterly unfamiliar to most British statesmen. One feels that George III would have liked him, although he would have disapproved of Taylor's unwillingness to get married and proclivity for cohabiting with colored domestics. One guesses also that he would not have taken the sort of instant dislike to him that seems to have been the case when he met Thomas Jefferson.[13] When the British ruling elite encountered imperial subjects at school, at university, in the merchant houses of the City of London, and at Westminster, it tended to be West Indians, rather than North Americans, with whom they had contact.[14]

It was men like Rose Fuller and his nephew Stephen, in turn agents for Jamaica, and above all William Beckford—Lord Mayor of London, best friend of William Pitt, Earl of Chatham, and Taylor's predecessor as the richest man and largest slaveholder in Jamaica—who informed British rulers about American attitudes. Beckford's biographer Perry Gauci rightly calls him "the first prime minister of the London Empire." He worked hard during the imperial crises in the Americas in the 1760s to "act as a bridge between colonial and metropolitan societies . . . endeavour[ing] to highlight the common calamities facing Britain and its empire." We cannot argue that his death in 1770 eliminated one of the few men who might have been able to steer British politicians away from the abyss of conflict with the thirteen colonies from 1773 but losing one of America's great champions at a crucial time was indeed a loss for the many Britons, Americans, and West Indians seeking compromise over constitutional and imperial issues.[15]

The "global turn" has encouraged scholars of the American Revolution to take a more expansive view of the ways in which the American Revolution had an impact on all parts of the world, not just the thirteen colonies. In some ways, of course, it was the younger brother of the French Revolution, an event whose origins partly lie in a financial crisis caused by its overexpenditure in the War for American Independence and also out of disappointment that its ambition to weaken Britain's hold in the Caribbean had not succeeded.[16] Nevertheless, some areas and some topics in the vast literature on the American Revolution are better covered than others. The American Revolution and its effects in the British and French Caribbean, for example, are understudied. There are a few essays on what happened in the Caribbean during the American Revolution but only one, albeit magnificent, full-length study, from 2000, by Andrew O'Shaughnessy.[17]

A few historians have taken up the challenge of incorporating the Caribbean into accounts of the American Revolution. For example, in a encyclopedic collection of essays on the American Revolution, in which the Caribbean is mostly dealt with tangentially, Christopher Brown, in an essay that deals with the problem of slavery for revolutionaries, argues that the Caribbean islands were the "economic engine of the Atlantic empire," and that by the final years of the war the defense of the West Indies took priority over retaining the thirteen colonies. Following O'Shaughnessy, Brown notes that Britain's defeat at Yorktown was largely caused by the British navy withdrawing to the Caribbean to stop the French fleet leading an invasion force to conquer Jamaica. It was Britain's victory at the Battle of the Saintes that was the end of the War of American of Independence, not Yorktown, meaning that Britain entered the Peace of Paris not as a loser but high on a major triumph against its principal European rival. For Britain, the end of the War for American Independence was one of muted triumph. "If the thirteen settler colonies were lost," Brown concludes, "the Caribbean empire and its hundreds of thousands of slave laborers and millions of pounds in revenue remained intact."[18]

The preservation of Jamaica allowed British merchants to intensify their efforts to establish commercial dominance in the Caribbean and Latin America, free from American competition. The British decision to ban American shipping after 1783 from the West Indies was bitterly resented by West Indian planters but aided British merchants, especially those involved in the slave trade. The decade following the end of the American War, from 1783 to 1792, saw the greatest ever influx of African enslaved people into the West Indies, an influx that more than compensated for a substantial decline in Britain's share of the slave trade into North America, where the Atlantic slave trade was becoming increasingly marginal compared to a booming internal market.[19]

—

The limited attention given to the West Indies in accounts of the American Revolution (even if, as Edward Gray and Jane Kamensky comment, "losing the colonies from New Hampshire to Georgia was, in part, the cost of defending the Greater and Lesser Antilles—a cost the British government was willing to bear" and, as Stephen Conway makes clear, "the war in America . . . was to be continued [after 1778] mainly for West Indian ends")[20] suggests that looking at the American Revolution from the vantage point of the British Caribbean—a view from Kingston, in short, rather than the customary view from Boston, London, or Williamsburg—is worth doing. Of course, this view of the American Revolution from Kingston—perhaps as seen from the veranda of Simon Taylor's Prospect Pen great house—lacks some of the glamour and excitement of perspectives drawn from places that went to war, but it does allow us to get a different slant on well-known events.

The American Revolution looks different when seen from Kingston. For example, the paucity of Native Americans in the British West Indies takes away one of the major themes shaping the revolution on the North American mainland.[21] Similarly, the Stamp Act, the Townshend duties, and the Coercive Acts are also less important in a Caribbean context. They were not unimportant, however. West Indians objected to the Stamp Act but less vociferously than in North America. Colonists in Antigua, Jamaica, and Barbados complied with the act, with Jamaica paying more taxes than the other British American colonies combined.[22] West Indians also objected to aspects of the legislation enacted from 1774 against Boston. They implored the king to resolve what they saw as an escalating and dangerous crisis. The most fervent expression of their support for embattled Americans came from the Jamaican Assembly in 1774 in an address to the Continental Congress in which they gave their support to the American cause and directly referred to the links between West Indians and Americans as being "a part of the *English* people, in every respect equal to them, and possessed of every right and privilege . . . which the people of *England* were possessed of."[23]

Many West Indians had considerable sympathy for the American position in the years preceding the Declaration of Independence, even if that sympathy did not encourage them to join the Continental Congress. Moreover, it was a sympathy that quickly dissipated once rebellion had been announced. Some of the most prominent opponents of Lord North and his policies toward North America at Westminster in debates over the coercive acts in 1774 were Jamaicans, such as William Beckford before his death in 1770 and Rose Fuller, MP for Rye. Fuller urged caution to a house roused to a fever pitch of outrage against the "flagitiousness of the offence in the Americans." Fuller explained that Britain could implement the Coercive Acts only through force and that it did not have the troops necessary to make Bostonians comply with British demands. If they tried to enforce unjust legislation, Fuller argued, with the soldiers they had in 1774 in Boston, "the Boston militia would immediately cut them to pieces." If they sent an army to Boston to make Bostonians comply, the results would be counterproductive: "the Americans would debauch them, and that by these means we should only hurt ourselves." Unlike most British statesmen, Fuller had a good sense both of American opinion and of how settlers aspiring to be thought of as British had to be cajoled rather than forced into accepting parliamentary authority. He also understood the realities of trying to use an army to make people obey authority in a continent that was vaster than Europe.[24] Nevertheless, there was never any serious push in the British West Indies for independence. The West Indian position regarding the revolutionary crisis was one more of anxiety about the economic consequences of war between Britain and the thirteen colonies than outrage over British acts and American opposition.[25]

The reason for West Indian indifference to the effects of imperial legislation was that they remained instinctively loyal through the revolutionary crisis. That Barbados was loyal was probably unsurprising—despite its recalcitrance in the revolutionary war to support British defense measures, forcing the British to move its troops and their headquarters to St. Lucia: it was an island that conspicuously prided itself on its fervent devotion to Britain and all its values.[26] That Jamaica stayed loyal needs a bit more explanation. Jamaica had a well-deserved reputation for refusing to obey imperial edicts when it felt they violated settler rights. It was well known for its trenchant expressions of its disdain for kowtowing to governors, the Board of Trade, or even the British Parliament.[27] Indeed, if experience had been a guide to which colony was likely to cause Britain most trouble in the 1770s, Jamaica would have been at the top of the list of potential imperial troublemakers. It remained, for example, the only colony to be formally censured by the British Parliament, an event that happened in 1757 after years of Jamaican opposition to firm imperial edicts issued from London.[28]

Nevertheless, Jamaican opposition to imperial authority was less engrained within the white Jamaican psyche than its many complaints and assertions of independence seemed to show. As Sarah Yeh perceptively argues, Jamaicans adopted a rhetorical strategy from the second quarter of the eighteenth century that stressed that their complaints against imperial actions arose from an acute appreciation of their vulnerability as slave-dominated islands in a sea full of Britain's imperial enemies. To an extent they adopted this posture to overcome their well-deserved seventeenth-century reputation as fractious troublemakers. But it became a reflexive response to adversity that in the long run undermined their ability to confront imperial rulers on the grounds of colonial rights. Jamaicans made arguments for equal rights as Englishmen, but they did so within a discourse, Yeh argues, that emphasized strongly Caribbean weakness. West Indians' habit of making out that they were hapless dependents in "low and languishing circumstances," continued into the revolutionary crisis. Caribbean colonists presented themselves as "weak, loyal, and devoted to their merciful British brethren." West Indians, culturally models of excess and inveterate risk takers in their social and economic lives, pretended that they were models of moderation and restraint in their political worlds.[29]

As Jack P. Greene notes about the quiet Barbadian response to the Stamp Act, by "turning the defences of Barbadian behavior . . . from an apology for their own timidity to a celebration of their own prudence and realism, [they] sharpen[ed] and reinforce[d] the image of themselves as a people among whom moderation was 'the Prevailing Principle.'"[30] North Americans easily saw through this disingenuous posturing. When the Jamaican Assembly in their 1774 petition to the Continental Congress stated that they wanted to act as neutral mediators between America and Britain but lamented, using their customary rhetoric of political helplessness, that their colony was so "weak and feeble" that they could not "now intend, or ever could have intended, resistance to Great-Britain," the Continental Congress replied with withering scorn. They sent to Jamaica "the warmest gratitude for your pathetic mediation on our behalf with the Crown."[31]

Americans had taken a long time to agree to rebellion. It took years of political agitation and the formation of radical opposition movements, all done within mainland colonies whose common antagonism to imperial actions had led, between 1765 and 1775, to an unprecedented degree of intercolonial contact and interaction.[32] West Indians had not shared such political consciousness-raising. They were never invited to join the Continental Congress and thus never got to share in the sense that they were involved in a common cause with colonists living elsewhere in the Americas. West Indians did not believe, in any case, that such common cause existed. Their loyalism was natural in a white population dominated by recently arrived British migrants. It was reinforced by a strong belief that West Indians could defend their most vital imperial interests through the representations of well-connected West Indians living in Britain. The West Indian presence in London was much stronger than that of North Americans, and experience told West Indians that their great wealth and considerable social and political presence in the metropolis enabled them to get their own way over most pieces of imperial legislation that affected the Caribbean.[33]

In the case of Jamaica, it stayed in the empire in 1776 from choice. Jamaica was loyalist, not revolutionary, in outlook, as seen in the worldview of Simon Taylor. Thomas Iredell, uncle of James Iredell from North Carolina, an early U.S. Supreme Court justice, was similarly ill disposed to ideas of revolution. He disinherited his nephew in January 1775 for "having taken an oath of Allegiance to Congress in violation of your first to this Country."[34] Jamaica was also largely happy with its place in the empire, especially immediately before the American Revolution, which were years of great prosperity. And Britain tended to reward it because they were convinced that the West Indies were central to imperial geopolitics and because their economic contributions to Britain were too great to ignore. They retained, for example, their preferential tariff on sugar and rum over American opposition because such tariffs conformed to prevailing mercantilist theories. Jamaica got a great deal from the imperial state and, unlike colonies in North America, wanted more rather than less imperial intervention. Even Edward Long, as detailed in Chapter 2, was able to reconcile his Whiggish principles about corrupt colonial governors in order to urge Britain to copy absolutist France and send more soldiers to Jamaica, to interfere more often and more vigorously in supporting agricultural development, and to allow for more local autonomy within a system of government where everyone recognized the sovereignty of the British Parliament in all places ruled by the British Crown.[35]

Moreover, the customary reason advanced as to why Jamaica stayed loyal—that they were afraid of declaring independence in a society in which whites were outnumbered by blacks by over 9 to 1—is not convincing. Certainly, Jamaican planters were apprehensive of what their enslaved people might do to them if given their chance. They thought it foolish to become too devoted to proclaiming their support for American notions of liberty when potential black rebels might hear them. A slave conspiracy discovered in Hanover in 1776 seemed to confirm to some Jamaicans that colonial revolution and slave rebellion went together. Reverend John Lindsay commented how "the topic of the American Revolution has been by the disaffected among us, dwelt upon and blandished off with strains of Virtuous Heroism" at dinner tables "where by the Bye every Person has his own waiting man behind him." It turned, he thought, "even the creole Negroes who were the savers of their Masters and Mistresses in the Rebellion of 1760" into potential insurrectionaries.[36] Lindsay's mention of the rebellion of 1760 is significant, however, because, ironically, it was this event that reassured Jamaicans that they could resist British authority when they wanted to without having enslaved people rise up against them. Tacky's Revolt had been put down so successfully and so severely that Jamaicans were convinced that they had the means and the will to overcome whatever enslaved people could throw at them. As far as they were concerned, violent repression worked. Before events in Saint-Domingue from 1791 changed slave owners' views about slave rebellion irrevocably, it was Tacky's Revolt—dangerous but subdued reasonably effectively—that provided the model of slave rebellion in the Caribbean, showing Jamaican planters that they need not be paralyzed by fear of slave revolt.[37]

—

West Indian planters' resilience in the face of possible slave revolt did not mean that they were indifferent to imperial politics. They were especially sensitive to policies that were a relaxation in the absolute control planters demanded that they should have over their enslaved property. The politics of slavery was essential not only to West Indians' relationship with the Crown but to their very identity.[38] If the constitutional issues separating America from Britain had revolved around issues of slavery rather than around taxation, then there may have been more cooperation between West Indian and North American colonists. The most significant attack on settler rights in the West Indies, therefore, was not the Stamp Act or the Coercive Acts but was the decision by Lord Mansfield in the *Somerset* case of 1772 to stop a planter resident in Britain from sending his slave to Jamaica. Chapter 6 deals at length with this case and about how it outraged West Indian opinion. The case provided the one possibility of connection between the West Indies and other plantation regions in British America, especially Virginia.[39]

Somerset was a galvanizing event that convinced southern and West Indian planters that Britain was prepared to compromise colonial liberties. It caused, as Alan Taylor notes, "a sensation in the colonies," especially in Virginia, where some enslaved people tried to take advantage of what they thought was a loophole giving them freedom by imperial authority. In Virginia, the news of *Somerset* coincided with concerns among legislators (Jefferson being a notable example: his anxieties about British imperial policy about the slave trade later found their way into the first draft of the Declaration of Independence) that Britain was telling them what to do with respect to their enslaved people. Mansfield's ruling was linked with Virginians' anger over an imperial veto over the colony's latest attempt to discourage further slave imports into a region where the slave population was increasingly naturally. It is hard to know how much importance can be attached to the *Somerset* decision in propelling Virginia into rebellion, but it probably intensified antagonism to Britain at a crucial time, when Virginia was suffering economically. As Taylor argues, Virginians interpreted imperial taxes through the ominous prism of the *Somerset* ruling, the imperial vetoes of Virginia's legislature over the slave trade, and news (erroneous as it turned out) of slave unrest in the West Indies. It made many Virginians sympathetic to the plight of Boston in 1774–75.[40]

Why, then, did shared outrage against *Somerset* not serve to unite Virginians, South Carolinians, Georgians, and West Indians? Why did the responses to it in the colonies not transform the debate over British constitutional pretensions in British America from an arcane battle over taxation and representation to one that focused on the important issue of the future of slavery in British America? After all, it was this issue that was to divide Americans from 1776 to 1861. In part, the danger that *Somerset* posed to planters was never realized. The significance of Mansfield's ruling was apparent only in the 1780s when abolitionism exploded into British politics following the scandal of the *Zong* (as noted in Chapter 7 of this book). *Somerset* was a great triumph for the early abolitionist campaigner Granville Sharp, but he was unable to build on that triumph, in large part because the start of the American Revolution diverted British attention from the horrors of the slave trade. In the perennial dispute about whether the American Revolution retarded or advanced the cause of black liberty, an argument could be made that at least in Britain the start of the conflict hindered rather than helped abolition. It is interesting to speculate on whether, if war had not intervened, the abolitionist campaign that erupted in 1787–88 might have emerged a decade earlier.[41]

In addition, *Somerset* was an abstraction in British North American plantation societies, rather than an issue of practical concern. It was not followed by any legislation that curtailed planters' rights in the colonies. Its main effect was to limit the power of slaveholders in Britain to sell enslaved people as they pleased (though, in fact, sales of enslaved people in London continued for many years after Mansfield's decision). Virtually no slave owners from Virginia or Maryland and few from South Carolina or Georgia spent any time in Britain before the American Revolution, meaning that a ruling that stopped them from sending enslaved people they had taken to Britain somewhere else in the colonies without a slave's consent was mostly a theoretical impediment rather than a real hindrance.[42]

Moreover, it was hard to unite planters in a common cause when they had so little contact with each other. That was especially true for Virginian planters. George Washington's only trip outside the thirteen colonies was to Barbados, and neither he nor other Virginian planters had many connections, either personally or economically, with the West Indies. South Carolinians, owing to their trade links with the Caribbean, were more connected, but even their knowledge of the West Indies during the period of the American Revolution was limited and fragmentary.[43] West Indians tended to be more closely connected to merchants in the American North than to southern planters. Their often-hostile attitudes to North America were developed out of antagonistic relationships in trade and warfare from the 1740s onward.[44]

The West Indies was more affected by *Somerset* than was British North America. As outlined in chapter 6, the leading West Indian thinkers and proslavery advocates quickly realized how dangerous Mansfield's principles were to the long-term interests of the British West Indies and wrote tracts denouncing it. But no protests about *Somerset* from the West Indies upset the imperial balance. West Indians were too comfortable in 1773–74 to protest something that did not affect them directly They were also mollified by Lord Mansfield's decision in another case, *Campbell v. Hall*, which supported West Indian arguments that their constitutional rights were not subject to the whims of the royal prerogative. As always before the American Revolution, West Indians managed to get their way over imperial issues that worried them in ways that Americans never achieved.[45]

This satisfaction with a resolution of a major constitutional issue may explain why West Indians did not get outraged in 1775 when Lord Dunmore—following the suggestion made in Parliament by the former governor of South Carolina and Jamaica, William Henry Lyttleton—made an infamous (to Virginian planters, anyway) proclamation offering liberty to servants and enslaved people willing to desert American patriots unwilling to submit to imperial rule. Their shared commitment to slavery was not enough to overcome their horror at Virginians' contempt for royal authority. Moreover, as explained in Chapter 4 in this volume, arming enslaved people to fight in imperial disputes had been a common practice in the West Indies since before the Seven Years' War—which was one reason why Lyttleton suggested it as an imperial strategy. Virginians, and later historians, who invariably treat Dunmore's proclamation as a singular innovation, were unaware of the many Caribbean precedents for what Dunmore did.[46]

Finally, it was less the politics of slavery than the economics of the institution that best explain why planters and merchants in the West Indies were highly unlikely to ever join the thirteen colonies in rebellion. The West Indies was reliant on the imperial state for defense and, more important, for supporting the Atlantic slave trade. The American South was not so reliant on the British state. Slave populations started to increase by natural reproduction rather than by additional inputs of labor from Africa first in Virginia by the 1720s and then in the low country of South Carolina and Georgia by the 1770s. By the early 1770s, wealthy planters in Virginia (though not, significantly, ordinary white men) believed that the colony had more than enough enslaved laborers and that they could subsist better without the Atlantic slave trade than with it.[47]

South Carolinians were less willing to see the slave trade halt, but they too could join in rebellion with some expectation that natural population increase among enslaved people would mean that their plantations could survive whatever Britain threw at them. That expectation was overly optimistic. The region experienced massive loss of enslaved people running away to the British during a vicious military campaign after 1780 designed to crush the South. But it was an expectation based on demographic changes in the enslaved population as the revolution approached that was impossible to contemplate in the West Indies. There, they were utterly dependent on an active slave trade to get enough enslaved people to replace the 3 to 4 percent of the enslaved population who died from malnutrition and overwork every year. Jamaica was in a "weak and feeble" condition and unable to resist the imperial state not so much because they feared slave revolt, or even because of their insecurity living on an island adjacent to powerful outposts of the French and Spanish empires, but because they could not function without the continuing presence of the Atlantic slave trade.[48]

Without the slave trade, demographic decline would have been catastrophic —as was shown during the American Revolution when the slave trade, especially after France entered the war in 1778, slowed to a trickle. It was the rapid decline in the supply of African enslaved people to Jamaica after 1778 that was mainly responsible for what Richard Sheridan identified many years ago as a crisis in slave subsistence. White Jamaicans attributed massive population loss among enslaved people during the late 1770s and early 1780s to a combination of the British blockade of shipping from North America, thus stopping provisions coming to the island, and a series of devastating hurricanes between 1780 and 1786. These were important. But the rate of decline in the enslaved population from 1776 to 1783 can be explained almost solely with reference to a declining supply of Africans arriving on the island to replace those enslaved people who had died in an enslaved population not enjoying natural population increase.[49] The Atlantic slave trade was thus central to determining revolutionary allegiances.

—

One explanation for the different courses that the West Indies and British North America took in the 1770s is that they were quite different places, with different demographics and different sets of problems, imperial and domestic. It is possible that these differences were sufficiently great to outweigh the similarities, most obviously a similar experience as people mostly of British heritage living in colonial settings but demanding the rights and privileges of British subjects living in Britain. Certainly, as early as the 1760s some American colonists had come to see that there was a chasm opening between the two regions of British America. Possibly, as Richard Dunn argues, this gap had started to develop as early as the 1690s.[50] But by the 1760s, differences between the West Indies and North America were noticeable enough as to excite comment—and disdain. James Otis, for example, the John the Baptist of the revolutionary movement in Boston, fulminated as early as 1764 that the Caribbean colonies were a "compound mongrel mixture of *English, Indian,* and *Negro,*" as opposed to New Englanders who were "freeborn *British white* subjects, whose loyalty has never yet been suspected." Otis continued that planters, used to being slave owners, would "for a little present gain" make other Americans "worse enslaved people if possible" than the Africans they mistreated so terribly.[51]

Otis made explicit what Benjamin Franklin had kept implicit in his pioneering work on political economy published a few years previously, *Observations on the Increase of Mankind.* Despite his son marrying the daughter of a prosperous Barbadian planter and thus assuring his fortune, Franklin was highly contemptuous of West Indian planters and even more so their enslaved people. Franklin's text was the first major work written by an American that suggested that the two regions were not unified. It was a sustained attack on West Indian pretensions and made a powerful argument that slavery diverted resources from worthwhile and sustainable long-term projects. His encomium to the rapid increase of white people in North America was a rebuke to the demographically challenged West Indies. It was also a suggestion to imperial authorities that they needed to make a choice: did they support a society like Pennsylvania, where free-born white men devoted themselves to useful improvements, or a society like Jamaica, which was based on slavery and where moral degradation accompanied great, but temporary, wealth making?[52]

By the 1770s, some Britons also began to make the association that Otis and Franklin encouraged: Jamaicans were natural tyrants. Josiah Tucker argued in 1781 that if they set up "an independent Government," headed by a king, they "would tie up his Majesty's hands as much as possible, and make him little more than a Cypher," leaving them free "to whip and scourge the poor Negroes, according to their own brutal Will and Pleasure."[53] It is interesting to speculate on what might have happened if the American Revolution been fought around the preservation of slavery rather than around imperial taxation. Possibly another split, as in 1861, might have occurred, and over slavery, but with the American north and Canada siding with abolitionist Britain against a united plantation south and the West Indies.[54]

What West Indians thought about North Americans is not so clear, but what little information we have suggests that there was a degree of mutuality in each region's disdain for the other. Isaac de Pinto, a Jamaican Jew resident in Paris, lambasted New Englanders as "fanatics and barbarians," claiming that if New Englanders came to dominate America it would be a disaster because "by reason of their great population and natural hardiness," they would seek to "invade and subjugate Mexico and Peru."[55] Jamaican views on the American North were echoed in Britain, where the people of Massachusetts were generally disliked, as being religious zealots with a tendency to violence. When the British cabinet after the Boston Tea Party of 1773 considered the people of Massachusetts, they "beheld a mirage, in which the mobs of Boston or Rhode Island stood for forces of sin and disorder." New Englanders, they thought, "were nothing but criminals led by fanatics."[56] British antagonism to New Englanders was extended to that famous son of Boston, once resident in Philadelphia and now famous in London, Benjamin Franklin. He was snubbed by Lord Hillsborough, the colonial secretary, in 1773; was distrusted by Hillsborough's successor, the Earl of Dartmouth; and was publicly humiliated in a room called the Cockpit on the western side of Whitehall when summoned in winter 1774 to answer charges that he had passed on secret letters of Thomas Hutchinson, the governor of Massachusetts, about the Boston Tea Party. Franklin was subjected to the full scorn of a British elite baying for blood before being fired from his post as deputy postmaster for the colonies.[57]

How Franklin was treated was disrespectful. It reflected as much opposition for the place he came from and what that place represented as it did personal antipathies to a man believed to have behaved in a deceitful manner. Franklin was a booster for the northern provinces of British America because he thought their demographic success was of massive long-term benefit to a growing British Empire. His key finding was his calculation that the abundance of land in places like Pennsylvania made people marry earlier and produce more children, all of whom added to the national strength of Britain just by needing to be fed, housed, and entertained. Franklin gloried in how the mainland colonies of America, by his estimation, would have more people than England in a century. "What an Accession of Power," he gloated, that this growing population would bring "to the *British Empire* by Sea as well as Land!"[58]

By the eve of the American Revolution there were three different demographic regimes in British America. The first, celebrated by Franklin, was in the American North, where population was increasingly rapidly, where the number of blacks was comparatively small, and where the economy and wealth patterns resembled more those of Britain than of plantation America. The second region was centered on the Chesapeake in the eighteenth century but moved during the early nineteenth century to encompass most of the American South. It had growing white and black populations and a mixed economy in which slave-based plantation agriculture combined with other forms of economic activity. The final region was the West Indies, where there was a very small white population and large black populations, and which had a flourishing economy based on plantation agriculture and slavery in which ownership and control was increasingly in the nineteenth century concentrated outside the Americas.[59]

The subsequent history of the United States, and our tendency to share eighteenth-century European assumptions that the test of a country's vitality was how many citizens it had, encourages us to accept as sensible Franklin's views about which sort of colonial societies Britain ought to have fostered in the mid-eighteenth century, though we probably would demur from Franklin's explicit racism about how the best societies were those full of white people. But that is not necessarily how British politicians thought of their imperial possessions. Their aim, in a mercantilist age, was to have colonies that made lots of money and that contributed geopolitically to enhancing Britain's security, power, and influence in Europe.[60]

The settler colonies of the nonplantation areas of North America neither made Britain much money from goods sent to it (at least when compared to those in the West Indies, though British North America was an increasingly important market for British exports) nor, after the Seven Years' War removed the threat of France from most of the North American continent, were especially helpful in countering France's quest for dominance in Europe. As P. J. Marshall notes, there was a strong tradition in Britain of believing that the purpose of colonies was so that the country could avoid costly continental European engagements. The aim of British foreign policy was that Britain would use their naval dominance to disrupt French trade and then seize French colonies as a way of weakening their rival. That such a strategy worked was the main lesson Britain drew from their successes in the Seven Years' War. In that conflict, British ability to destroy the French navy, capture French colonies, and hinder French trade had severely affected France's ability to carry on a continental war, leading eventually to a magnificent victory.[61]

The concerns of North American colonists, especially their zeal for taking Native American lands and thus disturbing the fragile balance of power in the interior, were not those of British policy makers. Moreover, their insistence on colonial rights, their assaults on the sovereignty of Parliament, their unwillingness to pay their rightful share of colonial defense costs, and their propensity for conflict and dissent were profoundly irritating to aristocratic leaders who expected subjects to obey or at least stay silent.[62] Even if by the middle of the eighteenth century the North American colonies ranked slightly ahead of the West Indies and well ahead of India in Britain's assessment of its global interests, they were very problematic parts of a growing British empire. Britain had an American problem, and that problem was that the American colonies were largely outside its effective authority. When it tried to make imperial authority more coercive and efficient, the result was American rebellion.[63]

Britain acted as it did not just because it was tone-deaf to the feelings of Americans in the northern provinces, though that was certainly the case, but because it had models of empire drawn from other parts of the British Empire that impressed it more than the chaotic organization of the thirteen colonies. By the end of the Seven Years' War, Britain had another model of how colonies should work that proved increasingly influential and that seemed to reflect how the empire was developing in the West Indies. This model served better and fitted in with foreign affairs strategies and with mercantilist policy more than did Franklin's vision of a settler empire in British North America. That model was how the East India Company operated in Bengal.[64] Britain did not take Franklin's long view of how colonies of settlement were superior to colonies of extraction, even if on balance that was a sensible view to take. Their concerns were more mundane and immediate: how did colonies bring Britain wealth and global bargaining power?

The East India Company, despite its tendency toward massive corruption and its arcane business structure, seemed a better model of how colonies should run than the anarchic and troublesome models favored by British American colonists.[65] The East India Company shaped its political engagement in Bengal around commercial advantage with little concern about insisting on its rights and privileges. British statesmen could work with such an organization. Indeed, it was their concern to fix the economic problems of the company in 1772–73 that led them to the disastrous policies forcing tea on America that resulted in the calamity of the Boston Tea Party.[66] What the East India Company's success in Bengal showed was that it was possible to have a profitable empire without a large and troublesome population of white settlers.[67] That this was the desired model for British colonization in the second half of the eighteenth century can be seen in how Britain envisioned empire in the parts of the West Indies that it settled after 1763. The Seven Years' War and the shock of Tacky's Revolt in Jamaica in 1760 had stopped white West Indian and imperial plans for increasing white settler numbers in the West Indies.[68]

Instead, British officials and investors focused their attentions on places in the southern Caribbean and, during the Napoleonic Wars, in northeast South America where whites were few, and mostly were supervisors rather than owners of enterprises. In these places most of the population were African enslaved people. They were colonies that were devoted to plantation agriculture without troublesome assemblies and irritating settler populations fixated on their "rights."

British colonization efforts in the late eighteenth and early nineteenth centuries avoided schemes of white settlement for models of colonization that were like what had been established in India through the East India Company. The ideal British colony was somewhere like Demerara, where hardly any whites went but where fertile and abundant land made plantation slavery extremely profitable. Even Australia, the settler colony par excellence in the Victorian period, was founded, in the last act of the British response to the American Revolution, not to be a place where white settlement was to be encouraged but as a far distant open-air prison where Britain cast off its undesired criminals.[69]

Including the West Indies into frameworks of the American Revolution helps us to understand therefore one of the underlying mysteries of the American Revolution: why would Britain go out of its way to provoke conflict with the northern sections of British America when it should have been clear that if Britain just left that region alone and allowed it develop and grow without imperial interference, then the natural bonds of affection that connected Britons in Britain and Britons living in New England, New York, and Pennsylvania would have led to a stronger, richer, and more connected British Empire?[70] In short, if Britain had recognized that authority in their eighteenth-century imperial system was "negotiated," as Jack P. Greene insists, rather than imposed, then Franklin's fond dream of a British Empire in which the majority of white people lived in America, enjoying a form of independency within, rather than outside, the British Empire, would have been realized. America could have been another (and the largest) part of what historians of the nineteenth-century settler colonies call a British world. It would have been inside rather than outside the empire, occupying a place analogous to that of Canada, Australia, and New Zealand in the imperial imaginary.[71]

One way of conceiving empire after 1763 was to see it as an empire favorable to the expansion of slavery through the development of new settlements in the West Indies like Grenada, Trinidad, Demerara, and Berbice. Such a model was contested, of course, with several influential theorists making a pitch for establishing colonies where slavery was excluded.[72] The founding of Botany Bay in 1788 was the culmination of such a viewpoint, though as a colony that was also a prison it was not exactly a homeland of British liberty.[73] Moreover, the popular agitation for the abolition of the slave trade after 1787 and the outcry over British abuses in India in the mid-1780s suggest that a substantial body of British opinion in the late eighteenth century believed that empire was a form of contamination of British culture and morals. Abolitionists and imperial officials unimpressed by slavery did their best, as Simon Taylor lamented, to hobble the West Indies by cutting off their essential supply of labor from Africa.[74] But if we look at what the British did, rather than what they said they might do, the dominant official line was to preserve and extend plantation slavery while reducing or ignoring the claims of planters such as those in

Jamaica that they were entitled to the sorts of rights that Britons had refused to Americans in the previous decade.

Abolitionism was only one trend, and not the dominant one in late eighteenth-century British politics, at least until the "Ministry of all the Talents" took over from the younger Pitt's administration in 1806.[75] This ministry, headed by Pitt's cousin, William Grenville, son of the prime minister who was most responsible for starting the American Revolution, was full of abolitionists and was buoyed up by Nelson's famous victory over the French at Trafalgar in 1805. As Roger Anstey and David Ryden have explained, Grenville and William Wilberforce seized the unique opportunity of 1806 to pursue new tactics that appealed to Britain's patriotic tendencies by first stopping slave trading to foreign nations before turning their attention to the British slave trade to the West Indies. These patriotic tendencies were very strong in 1806. The abolition of the slave trade occurred when the West Indian economy faltered and when French defeat and the creation of a black republic in Haiti encouraged Britons to think they could afford to show their moral superiority over Napoleonic France and thus confirm the conservative political order that had saved Britain from the horrors of the French Revolution.[76]

For most of the period between 1763 and 1807, however, Britain was strongly committed to the extension of plantation slavery in the Americas. Their commitment to slavery is often overlooked in recent scholarship, which emphasizes the extent to which enslaved Americans tried to get their own freedom by running away to the British during the revolution.[77] But unlike the British army in the War of 1812, who skillfully used slave runaways to undermine the American war effort in Virginia, and the United States of America army during the American Civil War of 1861 to 1865, who actively encouraged southern enslaved people to run away from their masters, the British army in the American Revolution (the largest single slaveholder in the South by 1781) had no desire to see slavery end. As Christopher Brown comments, "slavery in the British Empire survived the American Revolution because the British government wanted it to."[78]

British commitment to slavery, if not to West Indian planters, who were increasingly traduced in Briton as cruel and sex-crazed tyrants, can be seen in several ways. The slave trade, for example, was at its height in the last twenty-five years of its existence, delivering hundreds of thousands of Africans to the West Indies, including foreign locations. Britain kept on acquiring land suitable for slavery, especially during the Seven Years' War and the Napoleonic Wars, meaning that its West Indian empire increased by six colonies in 1763 and by a further five colonies between 1797 and 1803. It might have increased even more if the British invasion of Saint-Domingue/Haiti during the Haitian Revolution had been successful.[79] Moreover, as revealed in analyses of where compensation money for freed enslaved people went in Britain in 1834, the tentacles of slave ownership reached deep into the commercial and cultural life of Britain.[80]

—

Viewing the American Revolution from Kingston shows that the revolution was a more ambiguous event than it is often depicted. It is not correct, for example, to declare that America "won" the revolution with the help of France, while Britain "lost." Britain gave little indication during the 1780s—one of its best ever decades, when the economy boomed, when international affairs worked in its favor, when it had in William Pitt the Younger one of the very greatest of Britain's leaders and when even the weather was so good that harvests were bountiful— that it had suffered a major setback at Yorktown in 1781.[81] As J. G. A. Pocock quips, "The Revolution was less of a traumatic shock to the British than a display of their capacity for losing an empire without caring too deeply."[82]

Such a statement may be a touch too flippant, but it does point to a remarkable fact, which is that the loss of the Americas caused less of a rethinking of imperialism than might have been thought normal given a catastrophic military defeat.[83] Perhaps the British did not feel that concerned about the loss of America because they did not feel they had really lost. When France entered the war in 1778 and the War for American Independence became a global war with the British at odds with France and Spain, the chances of British victory in North America became very slim. British statesmen concentrated after 1778 and especially after 1781 on saving the West Indies. In stopping the invasion of Jamaica, they achieved what they wanted to get from the second phase of the war. They may have lost America, but overall Britain did not do too badly out of the war. They kept Canada, India, and Jamaica and gained Gibraltar. Indeed, the end of the war saw the start of Britain's real push for global empire.[84]

And their "loss" of America was more theoretical than real. Britain quickly regained its commercial predominance in America and established amicable relations with a government struggling to find its feet. It was for this reason that it was happy for the United States to be awarded most of the land in the west that it wanted at the Peace of Paris in 1782 and was prepared to accept the loss of sixteen colonies from what they had in 1776 (the thirteen colonies that became the United States plus the Floridas and Tobago). They knew that their victories in the Caribbean and at Gibraltar had greatly enhanced their bargaining power in Europe. Moreover, their principal rival in Europe—France—had been tamed. The result of the American Revolution for France was financial disaster, eventually leading to the French Revolution. France was happy that Britain had lost control over North America, but it had hoped to make gains in the Caribbean. At Paris, it was reduced to cutting its losses from what was a ruinous war in which it had made few concrete territorial or political gains. Spain did not do much better than France. It lost Gibraltar, much to its horror, but regained Florida and kept control of navigation on the lower Mississippi. And all sides abandoned Native Americans, failing to uphold any of the promises they had made to Indian sovereign nations.[85]

It is also not clear that southern planters "won" the war. A Virginia planter became president, followed by three others between 1800 and 1824, with six other slave owners or descendants of slave owners (Jackson, Harrison, Tyler, Polk, Taylor, and Andrew Johnson) holding office between 1828 and 1869. Slavery became entrenched in the Constitution and expanded rapidly and successfully into the fertile lands of the Deep South and the Mississippi valley.[86] Yet the older part of the South was hit hard by the concerted campaigns of the British against it during the War for American Independence. The tobacco-growing regions of the Chesapeake went into rapid economic decline, facing a fully-fledged subsistence crisis by 1816–17, with large planters struggling and ordinary people cast into dire poverty. The rice planters of South Carolina recovered better than their counterparts further north, but ordinary white southerners never recovered fully from the perfect storm that hit the South between 1775 and 1782, with the number of ordinary white laborers shrinking by as much as 25 percent, much due to wartime privation.[87]

Among the biggest losers of the American Revolution were West Indian planters. Their loyalty, in the end, counted for relatively little to a British establishment rethinking its ideas about empire. The empire after 1783 was different from the one that had existed in 1776 insofar as it was dominated by people of color rather than by white Protestants claiming all the rights of Englishmen. For the first time, West Indians were unable to get their way in imperial counsels. They were forced to accept, for example, British restrictions on U.S. trade with British colonies. Simon Taylor was aghast. He thought that this was a ministerial plot to ruin planters and accused Britain of treating West Indian planters as Gibionites (biblical hewers of wood and fetchers of water).[88]

The problems white West Indians faced were mostly cultural and political, not economic. The plantations recovered remarkably quickly from the travails of war and revolution. But British public opinion had turned against West Indian planters. The American Revolution had shown Britons various deficiencies in the West Indian planter character. Britons' poor estimation of the West Indian planter as a social and cultural type was amplified by the sudden birth of abolitionism in 1787–88. West Indian planters came to realize that the real threat to their prosperity and position came from an assertive, self-confident imperial state with moralizing attitudes and centralizing tendencies, rather than from American patriots, French invaders, or potential slave rebels. It made some planters regret staying loyal to such a perfidious mother country. Simon Taylor may have been one of those men who regretted his loyalty to a state that he believed had betrayed him.[89]

And the enslaved men and women who made up a considerable proportion of the populations of the plantation regions of British America and up to 90 percent of the population of the West Indies? How did they fare in the American Revolution? Unlike the American Civil War, the causes of which are inextricably linked to slavery, its retention, and its expansion, the American Revolution was not fought primarily over slavery. Its legacy is as much about other issues—liberty, equality, and democracy—as about the preservation of slavery. But for many planters the preservation of slavery was linked strongly with what they believed they were fighting for. Thus, for many African American and Afro-Caribbean enslaved people, American victory in the American Revolution was no victory for them. At its most basic level, the result of the American Revolution condemned most African American enslaved people to a further generation of captivity compared to their counterparts in Jamaica, Barbados, and Demerara.

The American Revolution showed, however, that African Americans were far from the passive and contented creatures that were imagined by nineteenth- and early twentieth-century historians. Thousands fled their masters to fight for the British, where they had a mixed experience in an army more committed to maintaining slavery than to destroying it.[90] Thousands more left America as loyalists, including going to places like the Bahamas, Jamaica, and Sierra Leone.[91] In the northern states, many enslaved people gained their freedom as gradual abolitionism took hold.[92] Yet for many African Americans in the United States, the results of the American Revolution were intensely disappointing. Slavery was growing rather than disappearing from those areas of the country where it was strongest. And even in the abolitionist North, freedom was accompanied by an often-crippling racism.

The idea that America stood for universal liberty, an idea spouted forth at Fourth of July celebrations, seemed to most blacks to be a sham. For this reason, many blacks in cities like New York and Philadelphia shunned what they considered the false claims of the Fourth of July and instead had marches and speeches on 5 July. From 1834, the principal day of celebration was 1 August, the anniversary of the emancipation of enslaved people in the British Empire. African Americans at these parades read out Britain's Act of Emancipation and pointedly quoted Jefferson's phrase in the Declaration of Independence that "all men were created equal." This celebration of a British event, rather than an American one, reflected the increasing tendency of African Americans after 1815 (when the Royal Navy became the principal enforcer of bans against the international slave trade) to see Britain as playing a crucial role as a trustee of black rights.[93]

Such open participation by African Americans in ceremonial occasions where Britain was praised for its efforts in ending the slave trade and slavery would have appalled Simon Taylor. It may have made him reflect on whether he and fellow planters made a mistake when staying loyal to Britain rather than joining the United States. At least America was run by planters who understood slavery and who saw, as Thomas Jefferson did, Britain's antislavery position as a cynical vehicle for advancing Britain's global interests.[94] John Adams argued after the Peace of Paris that if the American Revolution started again, white West Indians "would now declare for us if they dared." There was some evidence to back up Adams's view. Bryan Edwards, Jamaica's leading intellectual in the 1780s, made an impassioned argument that denied William Blackstone's contention that the British Parliament had absolute sovereignty in the British Empire.[95] It was very reminiscent of arguments made against parliamentary sovereignty by British North Americans in the 1760s.

It was too little too late. The American Revolution had shown that Britain would not budge on this question. But Taylor's time—and the time of Jamaicans arguing for settlers' rights—had passed. By 1813, when Taylor died, the issues that had led to the creation of the United States, the consolidation of Canada, the reconfiguration of the British Empire, and the diminishment of the West Indies after the American Revolution were long over. For Taylor, as for other men of his age in Britain, North America, and the West Indies, the American Revolution, and the split in plantation British America that resulted from this conflict, remained a defining event. He needs—as do other West Indians, white and black—to be part of the stories we tell about the American Revolution.

Chapter 9

Slavery and Industrialization

The "New History of Capitalism" and Williams Redux

Historians know that slavery was an important part of the eighteenth-century British Empire. They have long wondered about its influence on the economic development of Britain, especially on Britain's transition, the first in the world, to a highly developed industrial economy in the late eighteenth century. The historian who gave the most compelling answer to this question was the Trinidad scholar, and later prime minister, Eric Williams. Williams wrote a highly influential book in 1944, *Capitalism and Slavery*, that suggested that the link between West Indian slavery and British industrialization was real and substantial.[1] Although he never made an argument that slavery "caused" the Industrial Revolution, he suggested, as Barbara Solow was later to note, that it "played an active role in its pattern and timing."[2]

Anyone interested in the relationship between the growth of slavery in the early modern Americas, especially the British Caribbean, and the origins of the Industrial Revolution, is aware of the Williams's thesis and its galvanizing effect on scholarship. The Williams thesis (or, rather, his several overlapping theses about West Indian growth and decline) attracted an avalanche of scholarship among historians of the British West Indies, especially in the 1970s and 1980s. That outpouring of scholarship settled into a consensus that Williams's arguments were provocative rather than persuasive.[3] David Eltis and Stanley Engerman produced an important article in 2000 that provided a lucid refutation of the more dramatic claims made by Williams, notably that Britain would not have been able to industrialize without slavery. They also cast doubt on Williams's assertion that slave-produced sugar was more important than any other British product in the build-up to industrialization.[4]

Recently, however, the Williams thesis has had a fresh lease on life because of a historiographical movement among U.S. historians of the early Republic and antebellum periods who write under the self-penned title of the "New History of

Capitalism." One sign of how prominent the movement has become is that it recently was the subject of a lengthy conversation in one of the most important journals in American history.[5] The proponents of this movement take as apodictic the central role slavery played in creating modern capitalism.[6] Indeed, slavery is capitalism's "beating heart." They trumpet the novelty of their view—one claim being that "for too long historians saw no problem in the opposition between capitalism and slavery." They argue that historians have depicted the history of American capitalism as developing without slavery—although the more careful among this New History of Capitalism movement admit that discussions of slavery and capitalism are not new.[7]

They set their face against the many interpretations of the origins of the Industrial Revolution that see it as a product of bourgeois virtues, unexplained Enlightenment, or the happy coincidences of a Glorious Revolution in 1688, which guaranteed rights in property. This reading of their work is implied rather than stated, as they largely ignore the extensive literature on bourgeois dignity and on the institutional consequences of the Glorious Revolution. It seems clear, however, that their approach is opposed to culturally based interpretations explaining Britain's precocious industrialization and that they support interpretations of capitalism in Britain and in the United States that emphasize how economic growth was fostered by the formation, integration, and enhancement of the financial capacities of states geared for commercial activities.[8]

The proponents of the New History of Capitalism are unabashed fans of the Williams thesis and wish to extend its insights backward and especially forward in time. Seth Rockman, for example, argues that while the Williams thesis can be criticized as an explanation for British economic growth, Williams's "famous juxtaposition of slavery and capitalism" still warrants consideration for the United States.[9] Unlike Williams, however, the advocates of this new movement concentrate more on the nineteenth rather than the eighteenth century. They argue that in that century slavery "was at the core of the American economy" and was inextricable from the development of modern industry in the United States. They emphasize cotton as key to establishing the United States in the global economy, creating markets not just for southern but also for northern agriculture. But, like Williams, they argue that capital accumulation from slavery had a lasting impact on a range of industries, "re-animating profits in other sectors of the global economy." And, again like Williams, its proponents make extravagant claims for the global impact of New World plantations based on enslaved persons' labor. Because of slavery, it is claimed, Europe "was able to escape the constraints on its own resources," developing "innovations in long distance trade, investment of capital over long distances and the institutions in which this new form of capitalist globalization were embedded." These were all derived, it is suggested, "from a global trade dominated by slave labour and colonial expansion."[10]

That money derived from West Indian slavery, and even more so that profits accumulated from slave-based labor in the antebellum South, influenced how factories developed and thrived in the early stages of the Industrial Revolution is a fundamental theme in this literature. This chapter will look once again at the relationship between slavery and capitalism, especially in the development of British industry in the second half of the eighteenth century, in the light of new assertions of a revised and enhanced Williams thesis. It will concentrate on Harvard historian Sven Beckert's wide-ranging and influential book *Empire of Cotton: A New History of Global Capitalism* (2014), which is to date the major achievement of this new historical movement.

Beckert extends Williams's argument backward in time through his invention of a new form of economic organization, which he calls "war capitalism" (defined as "slavery, the expropriation of indigenous peoples, imperial expansion, armed trade, and the assertion of sovereignty over people and land by entrepreneurs at its core").[11] Williams's concern was mostly with the British West Indies, not the United States of America. He thought the American Revolution an event that marked the end of a mercantile capitalism favorable to planters and the start of an industrial capitalism where planters, at least in the British West Indies, became expendable. By contrast, Beckert largely ignores the American Revolution. He contends, moreover, that war capitalism was not supplanted by industrial capitalism but that several forms of capitalism coexisted until well after the end of the American Civil War. He argues that there was a seamless shift from the dominance of West Indian planters in the production of cotton to the rapid growth of cotton planting in the United States in the nineteenth century. The United States brought war capitalism and industrial capitalism together, but such coexistence did not work elsewhere in the world in the late nineteenth century. In an echo of Williams's argument that slavery started mercantile capitalism, but that industrial capitalism killed it, Beckert argues that after war capitalism had created the modern world, including industrial capitalism, European colonialism in the second half of the nineteenth century prevented places outside western Europe and the United States from making the next step toward industrialization. Thus, "colonialism allowed industrial capitalism in some parts of the world while making it less likely elsewhere."[12] Does this argument breathe new life into the Williams thesis?

—

Eric Williams based his arguments on the relationship between the development of industrial capitalism and British West Indian slavery and the slave trade from his understanding that the slave trade was highly profitable and that the British West Indies was central to Britain's mercantile economy between the Restoration

of 1660 and the American Revolution of 1776–83. The British West Indian plantation system reached its eighteenth-century height around the time of the conclusion of the Seven Years' War. It did so just at the time that Britain was in
the early stages of an explosive growth in industrial innovation, including the
development of a distinctive factory system. The British American plantation
system involved badly exploited workers, overwhelmingly enslaved persons of
African descent, producing tropical goods, like sugar, cotton, and tobacco, for
European markets. The simplest explanation of West Indian prosperity is that
planters made money through systematically stealing the rightful rewards
enslaved people deserved for their labor. Slaves working in sugar in Jamaica circa
1774 produced around £16 per annum for their employer while receiving income,
mostly in kind, of less than £6.[13] The number of slaves involved in plantation
agriculture was considerable—around 555,000 in 1750, of whom 295,000 lived
in the British Caribbean and 247,000 in British North America. Plantation profits
reached an all-time peak during the Seven Years' War, averaging 13.5 percent
return on capital. Profits still averaged around 9 percent between the Peace of
Paris in 1763 and the start of the American Revolution in 1776. Profits from the
slave trade were less, below 10 percent, but they were still healthy, given that
returns on government bonds were 3–3.5 percent and returns from agricultural
land were between 4 and 6 percent.[14] This impressive economic performance
encouraged Williams to argue that slavery contributed to the remarkable economic transformation that propelled Britain to the industrial leadership of the
world beginning in the second half of the eighteenth century.

Indeed, his argument was stronger than just that historians had not realized
that the British West Indies was a vital part of empire in the eighteenth century
before the start of the American Revolution. He argued, first, that slavery was
key to the Industrial Revolution insofar as "the profits obtained [from slavery]
provided one of the main streams of that accumulation of capital which financed
the Industrial Revolution."[15] In addition, he argued that wealth derived from
slavery was important to the social, cultural, and political fabric of eighteenth-
century Britain.[16] Third, he insisted that the West Indian slave economy went
into decline from 1783, possibly as early as 1763. Here, he was repeating an
argument made by the pioneering American scholar of the British West Indian
economy Lowell Ragatz, the man to whom Williams dedicated his book.[17] Finally,
he argued that West Indian planters changed from being progressive forces within
mercantilism to becoming a reactionary and backward-looking group, opposed to
industrial capitalism and increasingly abandoned by industrialists as protection-
ists and economic misfits with no place in modern Britain. In a famous formulation, he contended that West Indian planters provided the material basis that
allowed industrialization to occur but were then cruelly abandoned by an industrializing British state, which no longer had any need for them. In short, "the

capitalists had first encouraged West Indian slavery and then helped to destroy it." Indeed, he argued that "when British capitalism found the West Indian monopoly a nuisance, they destroyed West Indian slavery," despite having "ignored or defended" slavery when "British capitalism depended upon the West Indies."[18]

One of his subsidiary aims was to attack the imperial school of British historians who had taught him at Oxford in the 1930s. He disliked them for their ethnocentric celebration of abolitionists as altruistic humanitarians. He argued instead that abolitionism was founded on baser economic motivations. Another aim was to contribute to a developing "Third World scholarship" of nationalistic anticolonialism, with a West Indian audience more in mind than a European or American one.[19] A more pressing imperative, however, was to counter the assumptions made by Adam Smith that the West Indies was less a source of wealth than a drain on British resources with the large capital outlays of Britain put into plantation agriculture in the country being a major misallocation of funds. Smith argued that the "overflowing" of capital from Britain to the colonies showed that the empire concentrated in the West Indies was "a project which has cost, which continues to cost, and which, if it is pursued in the same way as it has been hitherto, is likely to cost immense expence, without being likely to bring in any profit."[20]

For S. D. Smith, revisiting Adam Smith, the problem that Adam Smith addressed was one of monopoly profits and the excessive protectionism demanded by West Indian planters for their products within the British market.[21] Williams denied that money flowed mainly from Britain to the colonies, as Smith suggested, but he accepted that Smith was an important figure in pushing British politicians, influenced by a new class of British industrialists, away from a protectionist mercantilist regime toward free trade and away from slavery in favor of wage labor. Williams argued that Smith was "the intellectual champion of the industrial middle-class and the major proponent of the idea that slavery was more expensive than free labor," thus treating, Williams believed, what "is a specific question of time, place, labour and soil" as "an abstract proposition."[22] Williams's book was thus as much an argument about political economy as about economic history: he wanted to connect the decline of the West Indies after 1783 to new ideas about free trade and industrial capitalism and to the discarding of policies of mercantilism in which West Indian planters had heavily invested.[23]

Williams overestimated Adam Smith's influence in policy making in the late eighteenth century. He saw the move from mercantilism to industrial capitalism as a compressed process when in fact free trade did not become a serious feature of British political life until after the abolition of slavery in 1834. But Williams made an important point that West Indian prosperity rested very much on vested interests in Britain, including colonial merchants, slave traders, and the unreformed House of Commons. Those vested interests, he argued, could not maintain West Indians' privileged economic position after the end of the American Revolution. British consumers were increasingly unwilling to support special interests advocating for protection when new industries created by Britain's technological prowess looked less for protection from competition than for access to new export markets.[24]

The part of the Williams thesis that has stood up least well to empirical investigation is his assertion that the British West Indies were in economic decline after the end of the American Revolution, leading to a diminishment of planters' political power in Britain at the same time as the power of industrial capitalists was growing and as British imperial policy turned eastward to focus on India. Historians no longer see such a "move to the east" in British imperial policy after 1783. The British Empire in the Americas remained central to imperial policy and to the imperial imagination until at least the abolition of the slave trade in 1807 and probably for some time after.[25] It was vital to imperial geopolitics, especially during the long Napoleonic Wars and during the conflagration of the Haitian Revolution in the 1790s and 1800s.[26] In addition, the British made several new acquisitions in the southern Caribbean and South America as a result of war between it and France and the Netherlands. Moreover, the profits from the plantation complex remained very high, including both the West Indies and the United States of America as British trading partners, totaling perhaps 11 percent of British GDP per annum in the period 1800–1810.[27]

Britain had good reason to covet Caribbean territory. The West Indies, despite a small economic dip during the American Revolution, remained a highly profitable part of the empire at least until the abolition of the slave trade in 1807.[28] Indeed, some of Britain's new acquisitions, notably Trinidad and what became British Guiana, formed a new frontier of high plantation profits for the first thirty years of the nineteenth century.[29] The profitability of the plantations did not decline, despite Williams's predictions, after the end of the American Revolution. The abolition of the slave trade, as Seymour Drescher argued in a famous intervention against a major plank of the Williams thesis, was not in the economic interests of Britain but was a variation of "econocide."[30] This abolition occurred in a period when planters were usually making good money from slavery. The one exception to this general tale of plantation prosperity from around 1790 is a short-term period of economic difficulty around 1805–7, which had, David Ryden tells us, a significant effect on the politics leading directly to the abolition of the slave trade.[31]

⁓

Sven Beckert's prize-winning global history of cotton presents a version of the Williams thesis that is stronger than anything Eric Williams considered. It has been a highly praised book, garnering the sort of laudatory reviews most authors can only dream about. The argument evades easy synopsis as it is a grand undertaking that covers a large range of topics and many countries, although it is decidedly centered on the United States. Even critics acknowledge that his account of the global history of cotton is a highly readable, engaging, and comprehensive book. Most commentators have been more generous than this in their judgments. For Eric Foner, it is "global history as it should be written." For Thomas Bender it is "masterly" and an "astonishing achievement."[32] At the very least, it is the book that is setting the pace in the New History of Capitalism movement. It also makes the Caribbean pivotal to the making of the modern world and reinvigorates the Williams thesis in a remarkable way.

The Williams thesis is central to how Beckert conceptualizes the development of war capitalism and its eventual morphing, as he sees it, into industrial capitalism. Cotton grown by slaves in the Caribbean "motivated and financed the unprecedented incorporation of newly depopulated territories into the world economy," Beckert asserts. The results of such cotton production "created the expansive, and elastic, global cotton supply network necessary for the Industrial Revolution, and with it the mechanisms through which the needs and rhythms of industrial life in Europe could be transferred to the global countryside."[33] Beckert is drawing here on Marx's famous dictum that "slavery is just as much the pivot of bourgeois industry as machinery, credits etc. Without slavery you have no cotton; without cotton you have no modern industry."[34]

One problem with this argument about the importance of the Caribbean to British industrialization is that Beckert makes mistakes about Caribbean history. For example, he argues that owing to an ant invasion and hurricanes Barbados's traditional crop, sugar, had been "decimated" by the 1770s. The result, he argues, is that Barbados was "transformed essentially into a huge cotton plantation." Caribbean historians will raise their eyebrows at this suggestion. Certainly, cotton production increased greatly in Barbados during the 1780s, peaking in the years after the 1780 hurricane and after peace with America had been declared. Nevertheless, sugar always remained king in Barbados during the late eighteenth and early nineteenth centuries, employing 49.5 of all slaves in 1834 as compared to 5.8 percent in "cotton and other agriculture."[35]

A more serious error is Beckert's treatment of Samuel Greg, an early cotton industrialist and the owner of Quarry Bank Mill, a pioneering cotton factory established in 1784 with a "few newfangled spinning machines, so-called water frames, a collection of orphaned children, putting-out workers from surrounding villages, and a supply of Caribbean cotton." Beckert emphasizes that what Greg did in developing a machine to process cotton was because he was a slave owner, in Dominica, who got cotton from merchant relatives in Liverpool and used the finished product to clothe his own slaves. "From this local spark," Beckert claims grandly, "industrial capitalism would emerge and eventually spread its wings across the globe. From this local spark, the world as most of us know it emerged."[36] The one problem with this soaring rhetoric is that Samuel Greg was not, in fact, a Caribbean planter when he established his cotton-manufacturing mill. It was his uncles who owned a slave-run plantation in Dominica called Hillsborough, which Samuel took control of as sole proprietor only in 1819, following the death of his aunt, who had a life interest in the property from 1795.[37] Thus, Greg was not, as Beckert claims, sending his manufactured cotton to dress his own slaves in Dominica. Given the crucial role Beckert assigns Greg as a revolutionary shaping the future, this otherwise small error takes on large significance.

Where the Caribbean is important to Beckert is that it was where Christopher Columbus landed in 1492, thus setting off a chain of unfortunate developments that led to war capitalism and the rise of Europe and the United States to a global dominance they have yet to relinquish. Beckert's use of "war capitalism" to describe what was behind a long process of change in the cotton industry is vague, with little attention paid to its theoretical roots. In a book that is replete with footnotes and is generous about other scholars' influence, it is noticeable that in the theoretical sections of the book dealing with war capitalism Beckert does not footnote the works to which he is intellectually indebted.[38]

War capitalism is a form of economic organization intimately tied to the series of transformations of the Americas unleashed in the aftermath of the Columbian Encounter and what Beckert calls "the recreation of the world."[39] His central tenet is that after Columbus "Europeans united the power of capital and the power of the state to forge, often violently, a global production context." Europeans, who were the active agents in the process of the recreation of the world, developed "new ways of organizing production, trade and consumption" through slavery, land expropriation, and imperial expansion. These processes were rooted in violence and "flourished not in the factory but in the field," with the victims Native Americans and Africans. He argues that war capitalism lasted from the sixteenth century until well into the nineteenth century and was based "not on free labor but on slavery." It involved a bifurcation of the world into an "inside" Europe, in which rules of law operated and where "state enforced order ruled." By contrast, in "outside" and colonized parts of the world, "frontier capitalists" could act with impunity, even when behaving criminally, such as when they engaged in large-scale theft of land and resources, when they involved themselves in the decimation of indigenous peoples, and when they employed the untrammeled use of violence to exert dominance over less fortunate people.

Slavery was the emblematic institution of war capitalism. It remained so, even as war capitalism mutated into "a new different form of integration of labor, raw materials, markets and capital in huge swaths of the world." The "true importance of Caribbean planters," Beckert argues, was not the cotton (or sugar) that they grew "but the institutional innovation that the Caribbean experiment produced." That innovation involved "the recreation of the countryside through bodily coercion, something only possible under war capitalism."[40] The main problem with this definition is due to its origin in world systems theory, which assumes that modern capitalism is the work of active European agents over passive non-Europeans in ever-widening spaces. Europe and neo-Europes, such as the United States, are core regions while other parts are peripheries.[41] The Western European economy from the fifteenth century onward is dynamic and ever changing and was dominated by a powerful merchant class, supported by European military power. Beckert's formulation gives a great deal of weight to the ability of Europeans to shape and direct the rest of the world through their manipulation of various institutions and sometimes, it seems, by willpower alone. Europe acts; the rest of the world reacts. It is only the West, for example, that develops any form of war capitalism and only the West that engages in slavery, land expropriation, and imperial expansion. Thus, war capitalism, as Sanjay Subrahmanyam notes, "represents the apogee of an unapologetically Eurocentric world history, which is dismissive of the dynamic potential of most non-European societies, whose fate seems to be to await more or less their formal conquest or informal 'incorporation' by European agents." These non-European societies seem to be assimilated into "a sort of historic slumber of homeostasis from which only contact with Europe will awaken them."[42]

Was European overseas expansion crucial to European economic supremacy in the early modern period? It is doubtful, though overseas expansion was not insignificant. The numbers are too small to make the conclusions that the proponents of the New History of Capitalism movement want to make, at least for the early modern period. Atlantic trade made up only a small percentage of European gross national product, even in Britain, where the Atlantic trade was largest and most dynamic. The relatively small size of overseas trade before the late eighteenth century means that we cannot argue that plantation agriculture was all that decisive in driving economic growth.[43] That proposition remains correct, even if it is acknowledged that Atlantic trade was growing faster than other sectors of the economy, that it was becoming more and more important within the overseas trade sector of the British and possibly the French economy after the Seven Years' War, and that Atlantic trade encouraged considerable feedback effects, linkages, and "invisibles" like shipbuilding, insurance, and other international services. Until the late eighteenth century, Atlantic trade was always subsidiary, often very subsidiary, to inter-European trade for both exports and imports. In addition, overseas trade was continually dwarfed by the domestic economy. Foreign commerce was important for the Industrial Revolution, but it was nowhere near as important as the New History of Capitalism historians claim, or as Williams suggested.[44]

Moreover, what money came from colonial trade only sometimes went to the state. The early modern European state—which Beckert describes as central to building war capitalism in the early modern period—was not the powerful state of the nineteenth century, let alone the twentieth century. Until at least 1815, almost all states were concerned less with profit than with power and with securing the authority of rulers over the ruled when there were few coercive possibilities available to enforce that authority. Their other ambition was to use what money they could raise to pay armed forces to keep their territory safe. One reason why early modern states were so weak and comparatively inefficient is that few had a decent fiscal base to provide the economic capacity to make and enforce political decisions. What fiscal base they had seldom came from overseas trade. Patrick O'Brien—a notable advocate nowadays for the importance of colonial trade for the development of the British economy—comments that the "total flow" of "colonial" tribute into state coffers "cannot be depicted as important for the construction of productive and viable fiscal systems for the long term growth of metropolitan economies."[45] Only Portugal and Spain "succeeded in sustaining notable increases to the flow of fiscal resources to support centralizing states by way of conquest, annexations, and colonization" though Spain's imports of expropriated American silver was expended on religious warfare and debilitating European conflicts that reduced rather than increased its economic potential. In general, it seems that European states received but minimal tax flows from imperialism before the nineteenth century.[46]

More profits went to individuals, of course, and some of those profits ended up in the hands of the state through excise taxes on colonial "luxuries." But, as O'Brien argues, states became fiscally powerful not though colonial expropriations but through constructing "fiscal and financial regimes with sufficient powers and organizational capacities to penetrate deeply into local economies for purposes of taxation, and to obtain access through loans and credits to the incomes, wealth, and expenditures of the populations over which they claimed sovereignty."[47] War capitalism and the growth of powerful European states went together, but not in the way that Beckert suggests. It was not external wars in the Americas but internal wars in Europe that pushed the state, at least in Britain, into greatly enhancing its revenues from taxation.

O'Brien connects the rise of the fiscal state in England (later Britain) with the furnace of conflict in the English Civil War in the mid-seventeenth century. In part the fiscal state arose because England had been expelled from Europe since 1453 and thus had become a semi-independent island realm relatively uninvolved until the eighteenth century in European power politics. The main factor, O'Brien insists, in why Britain came to be a high-taxing, fiscally powerful state was that wealthy elites were so devastated by the destruction of the Civil War that they were prepared after the restoration of the monarchy in 1660 to support enhanced taxation. They were willing to make a government fiscally powerful in order that this government would pass legislation that secured individual property rights. Thus, England was able to develop a fiscal state in which revenue from taxation was much higher than anywhere else in Europe and in which the burdens on the populace were correspondingly large.[48]

In addition, as Barbara Hahn notes in a critical review of Beckert's book, one question that the practitioners of the New History of Capitalism find difficult to answer, given their theoretical assumptions and adherence to world systems teleology, is whether there is any alternative to this vague form of capitalism. If war capitalism and industrial capitalism are the same, only with a more powerful state determined to protect capitalist interests, then after industrialization was established what can ever change? As Hahn asks, "if capitalism and the modern nation-state developed hand in glove, how did that connection emerge from or oppose the older relationship between guilds and local governments?" She believes that "'capitalism' is not an entirely satisfactory answer," especially regarding explaining how slavery embodies capitalism even as it draws on old hierarchies and structures (a question, of course, crucial to Marxist analyses). Is the alternative socialism? If so, how would this make things better, given what we know of the patchy history of socialist experiments? There is a decided feel of old-fashioned social history about Beckert's approach: before the arrival of the Europeans, everything was better, and people lived in happy communities where they controlled their own labor and persons.[49]

Moreover, the process Beckert describes seems inevitable, bound to happen in the way that it in fact happened. Williams does much the same thing, not stopping to engage in counterfactuals that might complicate a relatively simple story. It was never inevitable either that Europe (which knew little about cotton before it began to manufacture it) would become dominant in this product's production and consumption or else that only cotton would have led to industrialization in its British manifestation. The result would have been different, and perhaps less satisfying for Britain, but neither sugar nor cotton needed to be among the raw materials that Britain needed for industrialization. Britain could have industrialized using woolens (exploiting Australia's great possibilities for sheep) or through linen (if Russia produced enough flax to make the process worthwhile).[50] We need to be careful not to assume that the patterns of plantation development and British industrialization that developed in the eighteenth century had to develop that way. Giorgio Riello is right to emphasize that there was no one factor that explains Europe's comparative economic advantage over other parts of the world by the nineteenth century. Rather, he suggests that Europe's economic path after circa 1750 was the result of a "layering" of different factors and circumstances, some of which were peculiar to Europe, some of which came from the Americas, and some of which arose from trial and error over many years.

These factors produced synergies and catalyzed change. But these changes did not lead to predetermined results. The Industrial Revolution did not have to happen in the way that it did. Certainly, one can imagine ways that it could have developed without slavery being essential.[51]

—

What is especially curious about the agenda of the New History of Capitalism movement is that it depends on war in general for its interpretative position but ignores wars when describing historical causality.[52] This criticism is the sort of criticism that a "splitter" gives to a "lumper" but is worth indulging in anyway. If war capitalism is to have any real interpretative power, then it should be connected to particular wars as a means of showing that violence and war making were indeed fundamental elements in what connected Europe, specifically Britain, to slavery and the plantation system in British America and the United States (the French Empire is neglected in Beckert's book, except for the cataclysm of the Haitian Revolution).

The omission of individual wars is unfortunate because war is central to how slavery in British America and in the United States evolved.[53] The transition to African slavery occurred first in Barbados during the British Civil Wars in the mid-seventeenth century. The Nine Years' War and the War of the Spanish Succession around the turn of the eighteenth century solidified support for slavery and consolidated the essential features of mercantilism, which lasted until the American Revolution. Wars between Britain, Spain, and France from 1739 to 1763, including the Seven Years' War, which was the first major war between European powers fought largely in the Americas, arose out of imperial competition for an Atlantic trade in which slavery was essential. All these wars are important in respect to the Williams thesis, especially the Nine Years' War between 1689 and 1697. This war supports a moderate version of the Williams thesis in demonstrating that Williams was right in pinpointing overseas trade based on plantation slavery as essential to the development of the financial revolution. The Glorious Revolution of 1688, in turn, played an important role in consolidating the major features of the emerging fiscal-military state that underpinned the remarkable growth of plantation agriculture and the British economy in the half century before the beginnings of the Industrial Revolution.[54]

State support for planters, the plantation complex, slavery, and the slave trade in Britain and British America was not constant. There was just one period—from the Glorious Revolution in 1688 until the end of the Seven Years' War in 1763—in which Britain supported the planter interest almost without reservation. Williams got right how much support West Indian planters received from the state in this period. Planters in the West Indies and the American South enjoyed healthy profits, an increasingly effective and efficient slave trade, favorable imperial legislation, and minimal public opposition to slavery.[55] It is important to note, however, that this support was due mainly to the power of the West Indies, which was more to the forefront of imperial attention than was the American South, especially given the strength of the West Indian interest in Parliament.[56] We need to remember, also, that the West Indies and the American South belonged to the same polity—the British Empire in the Americas. The plantation interest was thus much more powerful than it was to become in the aftermath of the American Revolution when planters divided between a section that stayed loyal to Britain (a country that from the 1780s had a substantial abolitionist movement) and another section that joined the American North (a region increasingly hostile to slavery).[57]

State support for the plantation complex was not immediate. In the first half of the seventeenth century, state involvement in the establishment of slavery in English America was minimal. The colonies were a long way away and were economically marginal, and the most significant changes, notably in Barbados, occurred during the British Civil Wars in the 1640s and 1650s, when the English and Scottish states imploded and when colonies were largely left to their own devices.[58] The Western Design of 1655, in which Jamaica was conquered from the Spanish, meant that more attention was focused on the value of the plantations to imperial growth.[59] The implementation of the Navigation Acts from 1651 and the creation of a new Royal African Company and a new Committee of Trade and Plantations in 1672 showed that the Crown was intent on making the colonies conform to metropolitan wishes and pay their own way. The spectacular growth of the West Indian economies and those also of the Chesapeake and Carolina low country from 1600 to 1700 made that wish more of an imperative.[60]

The British state began to provide unwavering support for the plantation system only after William and Mary took power in 1689. Nuala Zahedieh has provided the political economy arguments to support this claim, while Richard Dunn has explained the partisan politics. Zahedieh shows that while by 1700 England's transoceanic trade was not overwhelmingly large, it had a significance greater than its ostensible value. Not only was Atlantic trade growing rapidly, thus making it key to developing prosperity, but it was essential for new industries, like copper, and for sustaining industries like shipbuilding that were also vital for British defense. Moreover, it was central in encouraging the financial innovations that we consider essential to what historians have termed a commercial "revolution" in the period. In short, Zahedieh argues, the endogenous responses to the market opportunities created by imperial expansion led to advances in London's commercial leadership in Europe, better transport networks, improvements in early manufacturing capacity, and an increase in "useful knowledge" as people acquired mathematical and mechanical skills necessary for complicated trade such as was common in Atlantic commerce. Many of the advances were hindered by vested interests diverting capital and enterprise into rent-seeking activities, but what the state realized from around 1700 was that the success of the American plantations, especially in the West Indies, showed that mercantilism worked. As Zahedieh concludes, the highly performing plantation trade not only outperformed other sectors; it stimulated "adaptive innovations which took the country to a new plateau of possibilities from which Industrial Revolution was not only possible but increasingly likely."[61]

Unsurprisingly, such an important trade, in the absence of antislavery sentiment, attracted government support, as Williams insisted. Planters had chosen the right side in the Glorious Revolution, supporting protests against James II. They proved highly effective in persuading the new government of William and Mary to modify Crown colonial policy in their favor. West Indian planters, the richest men in the colonies, benefited most of all and got many items of legislation that they wanted during the difficult years of the 1690s, as well as greater military aid. Along with London merchants, they convinced Parliament to allow private traders to supplant the Royal African Company, meaning that the volume of the slave trade to the islands immediately doubled. Sugar planters, whose authority had been challenged between 1675 and 1688, were firmly in charge from 1689. They asked William III for reduced Crown taxes, expanded slave imports, better military support, and full protection against foreign slave competition and got most of these requests granted. As Dunn argues, "the revolutionary settlement gave them these things, crystallizing their dependent status." That dependent status, as Williams rightly discerned, was perfectly satisfactory if the parent government let them do as they pleased in the colonies and if it protected them within the imperial system.[62]

The result of these multiple changes was that planters had an influence in imperial counsels and the support of the British state in ways that they were never to receive again. In the period between the Treaty of Utrecht in 1714 and the Peace of Paris in 1763, slave colonies in both the West Indies and British North America were nurtured within an empire that gave them ample support through generous land grants, state-sponsored negotiations with Native Americans that provided temporary peace, massive incentives for private trading in the slave trade, and protected markets for slave-produced products. Britain used its growing naval power to defend colonial slave societies that were especially vulnerable to invasion or to slave rebellion and legitimized hierarchies of power by accepting local political assemblies. It supported the claims by a wealthy planter ruling class that they had the authority and ability to legislate on most things that they wanted to do that did not conflict markedly with imperial policies. That right to legislate included, importantly, the right to make laws on colonial slavery. Britain also facilitated colonial leaders' access to imperial power brokers and metropolitan merchants, and it cultivated a political system that systematically favored colonial commerce. Moreover, it used its state power to validate slavery in the courts at a time when the vast majority of Britons were either comfortable with or indifferent to the system of racial domination that sustained planter rule and enabled plantation wealth.[63] We do need to do lots more on this period, especially for the West Indies, where the early eighteenth century was a period of transformative change but is historically a statistical dark age.

The Seven Years' War marked, in retrospect, the peak of planter power within the empire. It was not fought over plantations, but protecting the plantation sector, especially in the British West Indies, was a major factor in the outcome. And in the Peace of Paris, in part at the behest of the West Indian planter elite, the British gave back Guadeloupe to the French in return for getting Canada. Williams was right to say that the Peace of Paris was another victory for the West Indian interest.[64] He was also correct in thinking this a pyrrhic victory. The Seven Years' War marked a turning point for the plantation colonies. They did not decline economically, contrary to Lowell Ragatz's arguments from 1928. West Indian planters continued to make great profits in the West Indies at least until the 1820s and perhaps beyond, though many planters in the American South, especially the Old South, as discussed below, never really recovered from the American Revolution.[65]

Where West Indian planters started to lose out was when Britons came to realize that West Indian wealth was based on cruelty toward Africans. The image of the West Indian planter went into decline just as the first stirrings of abolitionism began in the 1760s. And the American Revolution led to a split in the planter class, with West Indians staying loyal and many southern planters opting for rebellion. Those Americans who left the empire and who wanted the federal state they helped to create to be a proslavery state made the right choice. In contrast, and in part owing to the defection of the greatest number of slaveholding whites who had belonged to the mid-eighteenth-century British Empire, by the late eighteenth century and certainly into the nineteenth century Britain was defining itself as an antislavery nation.[66] But southern planters' continued ability to persuade the American state to protect their interest in slaves and to foster the westward expansion of slavery came at considerable cost. Economic ascendancy increasingly shifted to the North and, as in Britain, a strong abolitionist movement began to develop in that part of America that was increasingly the most economically dynamic, the most culturally powerful, and eventually the most politically dominant part of the Union.[67]

—

Thus, although Williams was very insightful about the extent to which American and West Indian planters were supported before the end of the Seven Years' War, his arguments about later historical periods are less convincing. It was in looking at the American Revolution where Williams went astray. The American Revolution was a major short-term correction in West Indian finances but it did not lead to more than a temporary decline in the profitability of West Indian plantations. But the American Revolution was bad for plantation profitability in another part of the British American plantation empire. That region was the American South. Historians such as Allan Kulikoff understand that the American Revolution was an economic catastrophe in the American South, especially in low-country South Carolina and Georgia.[68] Peter Lindert and Jeffrey Williamson suggest that real income per capita in British North America generally and in the American South dropped precipitously between 1774 and 1790. They argue that the American Revolution in the thirteen colonies saw "America's greatest income slump ever, in percentage terms." The revolutionary war hit the American South especially hard. Its commodity exports fell in real per capita terms by 39.1 percent in the Upper South and 49.7 percent in the Lower South. They conclude that "the South Atlantic underwent a reversal of fortune between 1774 and 1840, dropping from the richest American region to the poorest."[69]

As Lindert, Williamson, and Kulikoff have shown, the predominance of the North in the American economy predated the beginnings of American industrialization and arose from the conflict that Boston initiated but in which the South suffered. In the colonial period, it was the plantation colonies of the American South and even more so the British West Indies that were the centers of wealth in America. In 1774, the richest British American regions were plantation areas. Their economies were based on slavery, and their white residents treated their enslaved property with enormous amounts of violence and crass callousness, especially in the West Indies.[70] But these colonies were wealthy places. Not only were southern and West Indian whites the richest people on average in the British Empire, but they lived in societies marked by considerable equality within white populations and limited white poverty. Lindert and Williamson, working on data provided by Peter Mancall, Joshua Rosenbloom, and Thomas Weiss, claim that virtually no white male household heads in the southern colonies in 1774 were very poor or destitute.[71] They show, however, that in the years between 1774 and 1800 the American South endured a prolonged depression, with gross personal income in 1840 plummeting from $91.77 in 1774 to $64.46 in 1800. They conclude that the South Atlantic suffered what Daron Acemoglu, Simon Johnson, and James Robinson have termed a "reversal of fortune," where they went from being the richest to the poorest region in the United States.[72]

The effects of the revolutionary war are probably greater for the South than Lindert and Williamson suggest as their dates do not separate out the decade of the 1790s, when the American economy everywhere in the United States picked up, with the South benefitting from the explosion in cotton production allowed by the invention of the cotton gin in 1794. But between 1776 and 1790 the South suffered huge infrastructural damage. It was also punished by the British in trade policy. Commodity exports fell by a catastrophic 49.7 percent in the Lower South. Per capita income dropped in the United States by 18 percent overall, but in the South it probably dropped much further. Kulikoff confirms such speculations and adds more empirical information on how different sectors of the white population of the American South fared from the revolution. He estimates that the number of white laborers in the South shrank by nearly 25 percent between 1776 and 1780 because of wartime privation. Enslaved people ran away in large numbers, and to the financial detriment of the planters who owned them. Even if they did not run away, enslaved people proved harder to manage and more unwilling to obey orders. The overall result was that the region's per capita wealth, exclusive of slaves, declined from 14.5 percent above the national average in 1774 to 36 percent below it in 1799. The South's share of national wealth dropped in this period from over half to less than a third.[73]

What Williams got right therefore was that the American Revolution caused a crisis in planter prosperity. He just picked the wrong region. It was the American South not the British West Indies that never quite recovered, at least in relative terms to the North. Of course, the South remained wealthy into the nineteenth century. As late as 1860, two-thirds of the wealthiest Americans lived in the South, and the nation's gross national product was only 20 percent above the value of southern-owned slave property.[74] Yet the relative decline of the South after the American Revolution, especially if the West Indies is included as part of plantation British America, is palpable. The New History of Capitalism historians tend to focus on the booming cotton frontier of the southwest after 1820, where profits were especially high and planters particularly rich, not just in cotton but also in sugar.[75] But the core tobacco-growing region of Tidewater Virginia and Maryland (where many more southerners lived than in the southwest) suffered enormously after the American Revolution, with endemic poverty and declining plantation profits.[76] Lindert and Williamson suggest that the share of gross total national income held in the South Atlantic dropped from 58 percent in 1774 to 48 percent in 1800, a drop that mirrored a similar drop after the Civil War.[77]

Southerners knew, moreover, who was responsible for their relative decline. It was the British. P. J. Marshall tells us that hostility to Britain after 1783 was intense in South Carolina, which had been invaded and had its economy wrecked, and strong in Virginia. In New England, by contrast, opposition to Britain soon declined after the Peace of Paris, despite the pivotal role of Boston in starting the Revolution. And New York, steeped in Anglophilia, became a bridgehead for British influence.[78] Slaveholders may have held the office of U.S. president for fifty of the first seventy years of the nation, but the balance of economic power had shifted to the North—and to Britain. Slavery remained profitable in the West

Indies and very profitable in the American South, planters continued to make lots of money, and there was always a sizable body of people, especially in the United States, who were favorable to slaveholder concerns. But, as Williams intuited, the real money after the American Revolution was being made elsewhere, and especially in industrial capitalism in places like Lancashire and the American Northeast.

And in both Britain and the American North the principal ideological orientation was away from slavery, not in support of it. Before the American Revolution, not only were the South and the West Indies easily the richest parts of British America; they also faced virtually no opposition to their commitment to slavery. After the American Revolution, that was no longer the case.[79] The paradox is that, contrary to what Williams thought, the effect on the West Indies of the American Revolution was not economic (the West Indies stayed rich) but was cultural and political (the image of the West Indian planter crashed in the 1780s and never recovered while the political influence of the West Indian lobby slowly declined). The effect of the American Revolution on the American South, especially in its major region of the Chesapeake, was the opposite—economic decline but continued political and cultural power.[80] Eventually, however, the relative economic decline of the South made a difference, leading in 1861 to the election of Abraham Lincoln as president and the ascendancy of the Republican Party, a sectional party of the North tending toward antislavery.[81] Southerners always overestimated the extent of their political power. Where Williams remains influential to the changing fortunes of planters is in his recognition that countervailing forces to planters emerged from time to time. Williams got the motivation of these opponents wrong because as a material determinist he was dubious that altruism and religious conviction played any role in imperial politics. But he did acknowledge that planters faced opposition.

One of the ironies of the New History of Capitalism movement is that in their eagerness to reassert the importance of planter wealth and influence globally, they rely uncritically on statements from planters themselves trumpeting their power and overestimating their capacity to determine politics. The opponents of planters—the people who eventually overcame planter pretensions and forced them to accept the end of slavery—are not given a voice in these histories. Beckert, for example, ignores abolitionism as a growing political movement and discounts, any explanations for abolitionism that suggest it was a campaign of moral reform, unlike Williams, who always acknowledged the importance of the abolitionist movement. Thus, Beckert argues that Britain "bowed to a century of abolitionist pressure" when it abolished slavery in 1834 and did so largely because manufacturers could still use slave-produced cotton coming from America to obviate any short-term difficulty caused by emancipation. Moreover, any opposition in Britain to "the lords of the lash" from "the lords of the loom" was not based on distaste for how southern planters used their political muscle to degrade enslaved people. Instead, it was derived from a fear that cotton planters were growing too powerful: "raw material producers had to be politically subordinate to the will and direction of industrial capital."[82]

There is never any hint in the works of the New History of Capitalism movement that ordinary Britons and Americans, black and white, male and especially female, signed petitions against slavery based on their belief that slavery was a sin and a national disgrace. Many white abolitionists were inspired by African American abolitionists who successfully established a "moral cordon" around the American South. This "antislavery wall" was intended, as Frederick Douglass argued, so that "wherever a slaveholder went, he might be looked down upon as a man-stealing, cradle-robbing, and woman-stripping monster."[83] Douglass's message worked. An essential part of British self-definition in the nineteenth century was that it was an antislavery nation.[84] The material advantages brought by cotton outweighed the moral disgust Britons from all walks of life felt against slaveholders.[85]

Southern planters thought that what they did was so important that it was indispensable to Europe's astonishing material advances in the nineteenth century. South Carolina senator James Henry Hammond, on the floor of the Senate in 1858, famously boasted that "England would topple headlong" if slave-produced cotton stopped, meaning that "no power on earth dares to make war upon it. Cotton *is* king." Yet the American North waged war on slave-produced cotton, and it won that war, winning moreover without stopping cotton from being produced in the South, except during the worst years of northern occupation of southern land. The South produced more cotton after the Civil War than before it, and without the benefit of slavery.[86] And Britain never stood by the Confederacy, despite Hammond's prediction. Bumper cotton crops from just before the war might have lessened the need on American cotton (though the thousands of men left destitute in Manchester because of cotton slowdowns in Britain during the Civil War suggest that the pain of ceasing cotton imports from America was not inconsiderable).[87]

But Britain was never going to support a war to defend slavery, having based its national prestige on having abolished slavery in 1834, and having expended large amounts of money on antislaving naval activities between 1834 and 1861.[88] Moreover, the signs of what Britain would do if the South went to war against the North had been signaled in another war, that entered by Britain against the United States in 1812, and won decisively two years later. As with the American Revolution, the War of 1812 was not started over the American commitment to plantation slavery. Many of its most important consequences were in the North, not the South. But it originated in a catastrophic miscalculation by Jeffersonian Republicans in the period 1807–9 that Britain was so reliant on American raw materials like cotton that an embargo on such goods being shipped to Britain would force the greatest imperial power on earth to its knees. Britain merely moved to other markets, causing a major crisis in the American economy. Britain tended to favor America in its trading relationships, partly because it was economically beneficial for it to do so, and partly because, as Lord Palmerston argued, "commercially, no doubt we should gain by having the whole American continent occupied by an active enterprising race like the Anglo-Saxons instead of sleepy Spaniards."[89] But it was prepared to act against America—as in 1812 and again in 1861—when national interest, including its self-definition as an antislavery nation, was threatened.

This symbiotic commercial connection between Britain and the United States formed a "single, integrated Atlantic economy" in the nineteenth century, one in which the West Indies, as Williams noted, played an increasingly minor role.[90] But the United States was always the subordinate partner in this relationship. The relationship between Britain and the United States in the early nineteenth century was almost the inverse of today's "special relationship," as Britain imagines it. The American republic remained an economic vassalage of the Britain. As Henry Clay lamented, the United States was in danger of remaining a "sort of set of independent colonies of England—politically free, commercially slaves."[91] And Britain was never worried about pushing its weight around. It was especially unconcerned about protecting the rights of American slaveholders over their enslaved property. Virginians received a huge shock in the War of 1812 when British military officers followed through on the threat by Lord Dunmore in 1775 and encouraged slaves to run away from their plantations and adopted several policies intended to harm planters and the institution of slavery.[92]

How the British acted in Virginia in the War of 1812 showed the truth of Williams's argument that slaveholding interests had been supplanted in the early nineteenth century by more powerful forces who were prepared to cast slaveholders overboard, even if this caused, as it did in Lancashire in the mid-1860s, considerable localized economic difficulty. It is interesting to speculate here on what Britain might have done in the War of 1812 regarding protecting the interests of slaveholders if, as in the American Revolution, it had to consider how actions against slavery in North America would alarm slaveholders in the West Indies. It was a sign of West Indian slaveholders' declining influence after the American Revolution that made Britain unconstrained in how it dealt with slave owners it conflicted with. Such an interpretation is in line with Williams's general arguments about West Indian political decline. It contrasts, however, with assumptions that planters in the American South had unparalleled political power in the 1810s through to the 1850s.

—

So where are we left with the Williams thesis after seventy-five years of debate? It remains a provocative thesis because it connects major economic change (the Industrial Revolution) to the Age of Revolution. A new generation of scholars, notably those ascribing to the New History of Capitalism movement, have given the Williams thesis a new lease on life, even if their arguments extending Williams's thesis out of the West Indies and the late eighteenth and early nineteenth centuries into the United States and until at least the middle of the nineteenth century are not convincing. Certainly, their "strong" version of Williams—that slavery was principally responsible for British industrialization and that the growth of capitalism in Europe and America is inexplicable without reference to the role that colonial exploitation played in its rise—is overstated.

Nevertheless, it is a good thing that we are returning to look at Williams's insistence that slavery, and the Caribbean, mattered to Britain at a critical time in its long history. It reminds us of the trauma of slavery and the open wounds that remain, especially for people of African American and African Caribbean descent. We do not want to return to the times when slavery was invisible in British and American history and where West Indian history (where slavery was always considered important) was relegated to a cul-de-sac of historical inquiry. Even the great Marxist historians celebrated by the New History of Capitalism movement shared this indifference to race, slavery, and the Caribbean. Eric Hobsbawm, for example, showed little interest in the work of Eric Williams.[93] Such lack of interest was unfortunate and is no longer defensible today.

But we want to be careful not to redress the absence of slavery in accounts of the past by now overstating its importance. Slavery was important, and so too were slave owners. But their importance needs to be kept in perspective. We do not want to adopt the attitude of proslavery spokesmen for the eighteenth-century West Indian interest or James Henry Hammond for antebellum cotton planters, who made insistent claims for themselves and their value to Britain and America. It is not just New History of Capitalism historians and Eric Williams who trumpet how important slavery and slave owning was to industrial Britain. Edward Long, the historian of Jamaica and a pro-planter voice, argued in 1774 that the sugar colonies were a source of immense wealth and power. After listing how many Britons relied on wealth from the West Indies, he argued that "we may from thence form a competent idea of the prodigious value of our sugar colonies, and a just conception of their immense importance to the grandeur and prosperity of their mother country."[94] While we should be prepared to rethink our histories of the beginnings of industrial capitalism so that we find more space for enslaved people, we need to remember that our efforts may end up with us sharing space with some uncomfortable bedfellows.

Epilogue

Jamaica and the State in the Age
of the American Revolution,
1760–88

Eric Williams's *Capitalism and Slavery* (1944) remains one of the fundamental works in West Indian history. Even though Williams was subsequently prime minister of Trinidad and Tobago, his analysis of the political economy of Caribbean slavery and freedom in the seventeenth through the nineteenth centuries was heavily dependent on the histories of the "older" West Indian colonies, notably Jamaica and Barbados.[1] For Williams, the American Revolution was the beginning of the end, "the first stage in the decline of the sugar colonies." The American Revolution itself was a period of economic difficulty, but, Williams argued, the consequences of the revolution were much worse. It separated, he noted, the West Indies from its vital American markets—the demand for American products remained, but the supply of such goods was made more difficult. American traders turned increasingly to trading with the French islands, with Saint-Domingue becoming a colonial superpower. By 1789, Williams argued, the "sugar colonies had become vastly more essential to France than they were to England"; the Caribbean "ceased to be a British lake"; and the "center of gravity in the British Empire shifted from the Caribbean Sea to the Indian Ocean, from the West Indies to India." The decline in importance of the West Indies heralded by the American Revolution, in this account, allowed the abolition of the slave trade to develop and put a final nail in the coffin of proud West Indian planters.[2]

Williams's views on the economic importance of the American Revolution in initiating West Indian economic decline no longer hold up very well, although as historians point out, Williams was really arguing for a shift in the balance of forces, so that the West Indies was in relative decline, compared to an industrializing metropole.[3] Moreover, as I have written elsewhere, the cultural decline of West Indian planters within British popular culture was notable during the 1780s, in part because of British perceptions of planters during the American Revolution that they were fundamentally un-British.[4] But Williams was wrong on the issue on which he placed greatest importance. The British West Indies' economic difficulties during the American Revolution did not mean long-term economic decline. The problems of the American Revolution in Jamaica—much less serious than in the Lower South of the United States—were overcome by the late 1780s. Indeed, the implosion of Saint-Domingue in 1791 gave British sugar planters wealth they had never enjoyed previously. The end of the American Revolution was emphatically not the end of Jamaican wealth and geopolitical importance.[5] Its importance was never higher than in the 1790s, when Britain devoted massive amounts of money to war and colonial expansion in the Caribbean .[6] It was not the American but the French Revolution that marked a decisive caesura in the history of the British West Indies, even if the consequences of that major world event in the Caribbean were especially ambivalent.[7]

Yet something was different in Jamaica after the end of the American Revolution. In 1782, when news of Admiral George Rodney's great victory at the Battle of the Saintes reached Jamaica in late April, the island (or at least its white inhabitants) erupted in joy. Simon Taylor, Jamaica's wealthiest planter, thought the worst of times were over. He gushed that "Rodney's victory was the only thing that could have saved us." He joined with other white Jamaicans in gawking at Rodney's battered fleet bringing into Kingston harbor the French flagship, *Ville de Paris*. He helped host a huge celebratory dinner for three hundred and more people in Kingston; was involved in organizing a grand ball in Spanish Town; and contributed to paying for a massive statue of Rodney, still extant, in the center of Spanish Town. It seemed a triumph for those planters who had remained loyal to the British Crown, unlike white settlers in British North America. But the triumph was pyrrhic. As Christer Petley concludes, "the planters of the sugar islands also shared something in common with their fellow loyalists from the American mainland: they were soon to find that they were among the principal losers of the American Revolution." They did not lose their homes or their property, but Jamaican planters suffered as "the Revolution and its consequences disrupted the old British Atlantic system, weakened long-established ties between the Caribbean and North America, and gave people in Britain fresh reasons to look at transatlantic colonial practices (and not least slavery) through increasingly critical eyes."[8]

If white Jamaicans were losers in the long term from the consequences of the American Revolution, that does not mean that enslaved Jamaicans were beneficiaries of the conflict. Historians have difficulty in studying slavery as a temporal institution and establishing when slavery was better or worse for slaves—a somewhat nebulous proposition at any rate, given that no form of enslavement was good in the Atlantic World between the seventeenth and nineteenth centuries. Slavery has a timeless quality about it, meaning that analyses of this institution— easily the most important socioeconomic institution in Jamaica during the period of plantation agriculture and over time—seldom consider how slavery was different when, as in late seventeenth-century Jamaica, most enslaved people lived in slaveholdings that contained fewer than one hundred slaves and when almost all enslaved people were born in Africa, from slavery in the middle of the eighteenth century when the large integrated plantation with two hundred or more slaves working producing sugar was predominant in Jamaica and when ideas of amelioration had no purchase on the island. It was different again in the early nineteenth century, when the end of the slave trade meant that an ever-increasing percentage of the slave population had been born in Jamaica, when abolitionist pressure forced planters to adopt some limited measures of amelioration that made slavery slightly less brutal, and, perhaps most important, when the success of Haitian rebels in overthrowing slavery in Saint-Domingue had made slaves everywhere realize that planter power was not as overwhelming as it had seemed when Jamaicans crushed slave rebels in Tacky's Revolt in 1760.[9]

But for enslaved people alive when the American Revolution ended, any improvements in their condition and especially in how they were worked were relatively insignificant. A child born as Rodney defeated the French at the Battle of the Saintes would have been fifty-two, if she survived, when the abolition of slavery occurred and fifty-six when "full freedom" happened in 1838. As Richard Dunn has recently shown in his marvelous demographic history of the Mesopotamia estate in the late eighteenth century, through to emancipation Jamaican slavery remained notably harsh and life sapping. It might have been materially better because of ameliorationist policies from the 1790s than in the hard-driving 1770s, but the differences probably seemed marginal to enslaved people, and Jamaican slavery remained far more brutal and unhealthy than slavery in other parts of the Americas, notably North America, into the nineteenth century.[10]

—

This work on Jamaica in the age of the American Revolution operates in a historiographical world where the advent of Atlantic perspectives on British American history and a new movement of "Britain in the World" has meant that historians increasingly connect Jamaica to the history of America and to Britain in ways that were unusual even a few decades ago.[11] The most important survey of Jamaica during the age of the American Revolution—Edward "Kamau" Braithwaite's 1970 survey of Jamaica as a Creole society—hardly dealt with the American Revolution as a global event with worldwide consequences separate from its impact on Jamaica.[12] Similarly, until recently syntheses of the American Revolution seldom mentioned Jamaica. That situation is changing. Of the four most important recent general accounts of the American Revolution, all except one pay some attention to Jamaica. The island is both prominent in the authoritative collection of essays in a volume edited by Ed Gray and Jane Kamensky and notably present in the short but incisive synthesis by Stephen Conway and the more continentally focused and larger synthesis by Alan Taylor.[13]

Here, I provide empirical evidence and historiographical information that will encourage more scholars to consider what happened to Jamaica during the period of the American Revolution. While my introduction to this volume summarizes most of the themes covered in the book, broadly conceived, this epilogue looks at one aspect of long-term change in Jamaica that has been only touched on in the previous chapters and that is connected to new scholarship on eighteenth-century Jamaica.[14] If we see this period of Jamaican history through the prism of state making, then the period that elsewhere might be called the age of the revolution starts early, in 1739, at the end of the First Maroon War, which opened up the interior of Jamaica to white settlement and resulted in a treaty that resolved Jamaica's major security problem through making Maroons the allies of whites against slaves. The next date of significance from a Jamaican context is 1760, Tacky's Revolt, and the end of ideas that Jamaica might become a white settler colony. The end of the revolution might be thought to be 1788, the year that white Jamaicans realized that abolitionism was not a minor irritation but a massive threat to their economic and social interests. Their society was based upon a highly destructive slave system in which annual declines in slave numbers were matched by large numbers of arrivals through the Atlantic slave trade. One might also make an argument for 1795 being crucial, as the outbreak of the Second Maroon War in the middle of Caribbean-wide conflict between revolutionary France and the British Empire initiated the end of a comfortable period when white Jamaicans enjoyed both political autonomy and economic prosperity on an island that was relatively secure from internal disruptions.[15]

Certainly, the dates that span the American Revolutionary period in Jamaica are a little different than in British North America. For the thirteen colonies, the usual dates historians think important are 1748, when a new imperial regime in British North America first became manifest; 1765, when the Stamp Act was met with universal resistance on the mainland colonies; 1773, when the Boston Tea Party inflamed an already perilous imperial relationship; 1776, when the Declaration of Independence signaled war and a break with Britain; 1783, when the American Revolutionary War ended; and 1787–88, when the U.S. Constitution was put in place. In Jamaica, the age of the American Revolution started earlier and ended later. The dates of importance were Tacky's Revolt in 1760; the *Somerset* case in 1772; France and Spain's entry into the revolutionary war in 1779; Rodney's defeat of the French at the Battle of the Saintes in 1782; and the shock of parliamentary action to abolish the slave trade, which began in earnest in 1788. I would also argue that in the long run the events off southwest Jamaica in late 1781, when the sailors on the *Zong* murdered captive Africans for insurance purposes, was another date of real significance for Jamaica during the revolution, even if its importance became clear only in the first abolitionist campaign of 1787–88. One aim of this volume is to suggest that a different chronology of the American Revolution might be profitable for scholars to contemplate.

What follows, then, is some consideration of Jamaica as a colonial state during this period of global change, starting with the First Maroon War in 1739. The conclusion of this war allowed white Jamaicans to enter their period of greatest prosperity and most imperial influence—prosperity and influence that stayed relatively constant for much of the rest of the eighteenth century. The end of the First Maroon War was mostly important for establishing internal order on the island, thus allowing for greater settlement in some of the most fertile and productive regions of Jamaica. Between 1739 and 1807, the plantation economy dominated Jamaica more than at any other time in its history. These years thus constitute a distinct period in Jamaican history. Examining Jamaica and the evolution of its colonial state in the second and third quarters of the eighteenth century thus helps us place the chapters in this book, on various aspects of Jamaica in the period of the American Revolution, in a larger context.

—

The colonial Jamaican state was the major beneficiary of the settlement of the First Maroon War, even more than Maroons themselves, though Maroon success in not being defeated in the 1730s allowed them a remarkable amount of autonomy within the British Empire.[16] The signing of the treaty between Britain and the Maroons in 1739 signaled the beginnings of the transformation of state power in Jamaica through the same kinds of fiscal-military tactics that were revolutionizing state formation in Britain and Ireland. Maroons fell comfortably into a new imperial role as policemen and auxiliaries in wartime. As Michael Craton notes, "slaves came to treat them with a mixture of envy, fear and enmity, while the planters found their presence reassuring as well as picturesque."[17]

Writing under the pseudonym Veridicus, it was probably Thomas Fearon, a contemporary of the Jamaican governor responsible for the treaty, Edward Trelawney, and the chief justice and father-in-law of Nicholas Bourke, who praised Trelawney's "mild Government" of the Maroons for transforming white Jamaicans' greatest internal foe into a considerable supporter of the plantation regime. Trelawney, the anonymous correspondent argued, had "quite delivered" the Maroons into "good Subjects, and the most useful tractable Servants in the whole Island." Among their many services, Maroons "opened the Highways, [and] prevented or suppressed all Insurrections of the Negroe and other Slaves." In addition, they "never failed to bring back all runaway Slaves to the Service of their respective Masters" and "kept the Peace and found Ease and Prosperity under the

Protection of the Laws." The result was a dramatic increase in "new Settlements
. . . in all the extreme Parts of the Country." The writer argued that this "multipli-
cation of small Settlements" was welcomed everywhere on the island, including
in Kingston, as it contributed "very greatly towards the Security of the Island"
and was "a most immediate Safeguard against the Depredations of foreign or
Domestick Foes."[18]

The results of a more secure supply of slaves from Africa and the pacification
of the Maroons and subsequent settlement in western and northeastern Jamaica
became apparent a decade later. By 1750, the potential of Jamaica was turning
into reality, with the number of sugar plantations increasing and the economic
prospects of the island being realized through a boom in exports and an increase
in imports. An anonymous tract written in 1750 but published in 1757 confirmed
earlier reports by Robert Dinwiddie, surveyor general of customs for the Southern Department in the 1740s, that Jamaica was easily Britain's most valuable
American possession based on its contribution to imperial wealth. It generated
substantial public revenues to Britain through the importation of sugar into Britain, a commodity that attracted high taxes. The island in 1750 had capital
resources of £7,602,165 and exported £579,610 of goods to Britain. It imported
£505,662 of goods from Britain and a further £67,260 from North America. The
author calculated that Jamaica's annual contribution to the British economy
between 1749 and 1752 was £579,649. Its wealth also meant it had substantial
public revenues on the island, even if these revenues were low compared to what
was required in the form of taxes from the 1760s through the 1780s. From public
revenues of £28,234, Jamaica could pay its governor £3,571 per annum, making
him easily the best remunerated governor in the British Empire, save perhaps
Ireland.[19]

Such wealth did not go unnoticed in Britain. After many years of quiescence,
the Board of Trade under the energetic leadership of George Dunk, Earl of Hali-
fax, decided to promote a number of measures designed to strengthen imperial
authority in the colonies, many of which had been proposed years earlier but had
been set aside by ministers during the career of the Duke of Newcastle, secretary
of state for the Southern Department. Jamaica was the colony on which this newly
active Board of Trade focused much attention. It harassed Trelawney so much in
order that he might reform politics in the colony and promote an enhanced
royal prerogative that Trelawney eventually resigned in 1752. What especially
disturbed the Board of Trade was a belief that great planters had illegally monopo-
lized large tracts of land on which they paid little to no revenue. The board was
anxious to take advantage of Jamaica's growing wealth to get more money to take
on more imperial activities. Trelawney's replacement, Admiral Charles Knowles,
was more subservient to the board's aims and tried to inquire into landholding
and conduct surveys to ascertain just how much wealth Jamaicans had in order

that taxation could be increased, Jamaicans would pay more for the cost of imperial services, and the authority of the imperial state could be enhanced.

Knowles met with massive resistance from powerful planters and their even more powerful advocates in Britain. Despite this knockback and despite planter opposition to imperial inquisitions into the capacity of the Jamaican economy to pay more to imperial coffers, imperial surveillance increased during the 1750s, as might have been suspected after pronouncements made about parliamentary prerogatives concerning Jamaica by the House of Commons in 1757. One instance of increased imperial interest in ordering Jamaican affairs was the production of three large maps of each of Jamaica's three counties by Thomas Craskell and James Simpson in 1763, based on surveys conducted between 1756 and 1761.[20] The result of the controversies of the 1750s between the Jamaican legislature, the governor, and the British Parliament was a sizable increase in government revenue in this decade that more than matched the economic growth of the island. Revenues rose from £10,000 sterling per annum in the early 1720s to £35,000 in 1749 and £53,000 in 1761. The total population increased over this period, almost entirely within the enslaved population, though this was also the first period when free people of color became sizeable, meaning that average taxation per head stayed relatively constant. The number of wealth holders, however, either did not increase or else declined from 1710 to 1760, so that the burden on the well-off became greater, especially as a large proportion of internal revenue was gained from the deficiency tax levied on slaveholders who did not have the required number of white people on their estates in proportion to the slaves they owned.[21]

White Jamaicans learned many lessons from their near-death experience in 1760–61, when their slaves seemed likely to overwhelm the island.[22] The major lesson learned was the need for constant vigilance over what slaves were doing and ready resort to maximum application of force without concern for its lawfulness or morality. The events of 1760 also altered irrevocably white Jamaicans' understanding of what kind of society they were creating in the Greater Antilles. Jamaica was not going to be a settler society on the model of colonies in British North America. Abruptly from 1760, Jamaica stopped the many settlement schemes on which it had expended vast sums to no effect in the 1750s.[23] One of the major complaints from governors between 1739 and 1760 was that the comparatively large sums raised for government revenue were wasted on poorly thought through schemes intended to increase white population levels and to gratify Jamaicans' superior sense of themselves through the building of grand public edifices. Governor Trelawney despaired in private about how irresponsible Jamaican legislators were in their parochial disbursement of government money and in their reckless purchasing of Africans while doing little to improve internal security. His successor, Charles Knowles, lamented that while Jamaican defenses were neglected Jamaicans were happy to spend massive sums on "useless public buildings" at Spanish Town and a "ridiculous" spa at Bath while engaging in corrupt "gratifications and donations" to political friends.[24] Jamaicans stopped such expenditures after 1760. The year 1760 was more important in changing imperial relations in Jamaica than 1763, when the Seven Years' War ended and when Britain started to reorganize its empire.

⸺

The third major event of consequence in the period between 1739 and 1807 was an event in which Jamaica was not directly involved and that white Jamaicans had hoped to avert at all costs.[25] Traditionally, the start of the American Revolution has been seen as the beginning of the long-term decline of the Jamaican planter class. There is no doubt that the American Revolution was a difficult period for many white Jamaicans and almost all black Jamaicans, especially after 1779 when the French and Spanish entered the war, turning it into global conflict as well as a civil war. The disruption in trade in plantation supplies from North America caused hardship to planters and famine for enslaved people. The island experienced lengthy periods of martial law and the great threat of a possible invasion by the French and Spanish in 1782. The imperial government did not help when it launched a foolish and costly expedition against the Spanish in British Honduras. The slave trade floundered owing to wartime exigencies, leading to what in retrospect was the most disastrous event of the American Revolution for Jamaicans, the murder of captives on the slave ship the *Zong* for insurance money, an event that catalyzed the burgeoning abolitionist movement in the mid-1780s. To political difficulties can be added the most devastating hurricane in Jamaican history in October 1780, which flattened most of western Jamaica, forcing planters and their slaves into ruin and, for slaves, starvation.[26]

Imperial policies in the 1780s, especially the British government's determination to treat the United States as a foreign power and restrict American access to West Indian ports, thus disrupting a trade vital to the plantation economies, only increased West Indian alienation from their government. Historian-planter Bryan Edwards bitterly commented that Britain made "war under the name of peace against the most valuable of her plantations" while Jamaica's richest man, Simon Taylor, lamented that Britain saw Jamaica and other West Indian colonies as "Objects of taxation," exclaiming that "if we are the most favored subjects, then God help the rest."[27]

Nevertheless, the travails of the American Revolution were temporary rather than permanent. Seymour Drescher showed in his 1981 book *Econocide* that the arguments of Lowell Ragatz in the 1920s and Eric Williams in the 1940s that the American Revolution started an irrevocable decline in the Jamaican economy were wrong. Ahmed Reid sums up the current state of play on this much-debated historiographical controversy by arguing that the Jamaican economy retained its dynamism and efficiency until the end of the slave trade in 1807. The Jamaican economy recovered remarkably quickly from the American Revolution in the 1780s, and once the Atlantic slave trade began to flourish again profits returned to what they had been in the 1760s and 1770s.[28]

Moreover, there was enough encouragement for West Indian planters coming from Britain to convince planters that the abolitionist movement, which erupted suddenly in 1787–88, could be easily confronted.[29] Jamaicans responded to the shock of hearing about the 1788 debate in the House of Commons, led by William Wilberforce, on the evils of the slave trade, in their traditionally aggressive manner in regard to protecting their rights. They did not, moreover, couch their aggression as they often did within a language that emphasized Jamaican "weakness" and "dependency" on British troops nor with reference to how loyal and dutiful Jamaicans had always been.[30] The Jamaican Assembly addressed the House of Commons in 1789 (its petition arriving just after the French Revolution started) in robust language. It stated that "the rights of the British colonists are as inviolable as those of their fellow-citizens within any part of the British dominions" and warned Parliament that the constitution of the empire did not "give omnipotence to a British parliament." It defended the slave trade as the cornerstone of the Jamaican economy and argued it had to be defended both for reasons of economic common sense and lest Jamaicans' affections be alienated from the "parent state."[31]

Jamaica's resolute opposition to the beginnings of abolitionist agitation in the late 1780s showed, as Christer Petley argues, that "British statesmen might have encountered severe, even violent opposition from slaveholders had Parliament chosen to end the slave trade at the beginning of the 1790s."[32] The strong response of Jamaicans to abolitionist petitions, combined with concern at the start of the French Revolution that any concessions to reform in any area might lead to similar disturbances as in France, meant that the only concrete action Parliament took against the slave trade in this period was the Dolben Act in 1788, a year before the fall of the Bastille foreclosed any further parliamentary action.[33] White Jamaicans were happy with the response to their strongly worded statement about supporting the slave trade. At least until 1793, they felt confident that most of Britain's ruling class was on their side and that their dependency on British military protection was matched by British dependence on the West Indies to feed their addiction to sugar and their love of rum.

What really changed the position of Jamaican planters were the twin challenges of slave revolt in Saint-Domingue, especially after the destruction of Cap Français in 1793, and the outbreak of war between Britain and revolutionary France. The key year was 1793, when the radical turn in the French Revolution horrified white Jamaicans and made them realize how defenseless they were as a slave society in a hostile sea of revolutionary tumult. The slave uprising in Saint-Domingue brought short term benefits to Jamaica, as planters benefited from the destruction of Saint-Domingue's sugar economy through increasing sugar output while sugar prices soared to a peak in 1795 that had not been reached since 1755 and 1756.[34] The long term prognosis, however, was much less positive. War with France in the Caribbean was enormously expensive. Between 1793 and 1798 the war cost a massive £20 million, and between 1792 and 1802 over forty-five thousand British soldiers lost their lives in Caribbean conflicts.[35]

More important, the shock of slave revolt and war in a revolutionary world after 1793 forced slaveholders into a political and economic dependence on Britain that was different than before insofar it was not voluntary but arose from force of circumstances. In short, white Jamaicans were forced to give up most of their local autonomy, and their confidence that they had control over how their taxes were spent, in favor of relying on the imperial state to secure their lives and properties from slave insurrection and French invasion. Planters fell in line with imperial dictates. Thus, while it appeared on the surface that planters were able to survive the 1790s with their slave system intact and abolitionism in retreat, the balance between the imperial government and local legislatures had altered permanently.

Not only was the level of taxation raised to much higher levels than before 1793; the imperial government was much firmer in how they asserted political control over the empire in the West Indies in the 1790s. Britain refused to allow any new West Indian colony a legislative assembly and enforced new legislation that forced planters to modify their slave management practices.[36] Thus, the political influence of Jamaican planters within the British Empire was radically reduced after 1793. When short-term economic problems were added to their growing political impotence after 1805, abolitionists had the chance to use the imperial government to finally outmaneuver and defeat Jamaican and other West Indian planters.[37] Planters were helpless in the face of metropolitan onslaughts on them and their devotion to slavery. Their fear about slave revolt from below in the wake of the Saint-Domingue revolt forced planters to accept any policies that Britain chose to impose on them. The abolition of the slave trade in 1807 heralded a quick decline in real estate prices, a gradual deterioration in the Jamaican economy, and a determined but defensive and ultimately unsuccessful resistance to the long struggle to end slavery.[38]

—

The years between the end of the First Maroon War in 1739 and the end of the Second Maroon War in 1796 marked one caesura in Jamaican history—the period when planters were dominant, the plantation economy was flourishing, and security from internal and external enemies was manageable. Within this nearly sixty-year period, the age of the American Revolution as defined in this book—the years between 1760 and 1788—has enough common characteristics to make it a distinct subperiod, one that both marked out Jamaica from what went before and what went after, and which made Jamaica both similar to and somewhat different from other plantation societies in the Greater Antilles and British America. I have treated the second point in my two previous books.[39] I would like to end this book, however, with a reflection on what features of Jamaican history were most significant between 1760—Tacky's Revolt—and 1788—the beginning of the end of Jamaican planter and merchant power to do largely as they pleased within an imperial system that indulged them.

This period is like, but not contiguous to, the customary dates of the American Revolution when studied from North America: 1763–88. It was a period in which white Jamaicans were widely disliked by residents of nonplantation societies like Massachusetts. New Englanders and Pennsylvanians resented white Jamaicans' pretensions, their wealth, and their lack of attention to religion or social decorum. James Otis spoke for many when he denounced West Indians as a "compound mongrel mixture of English, Indian and Negro" who did not deserve to be thought better than "free born British white subjects."[40] They were criticized as men, as an opponent described them in the 1760s , who were "habituated by Precept and Example, to Sensuality and Despotism." They were unrestrained, godless narcissists who allowed their desire for pleasure and excessive self-expression to overwhelm any sense of decency and order.[41]

Yet for all their hedonistic gaucheness and crudity, white Jamaicans in the age of revolution had developed a common culture—"as modern as it was repugnant," in the words of Barry Higman—that worked in a fashion.[42] Having given up after 1760 on the idea that Jamaica could become a settler society on the North American model with lots of white settlers, mostly owning black slaves, Jamaican planters and merchants perfected a dynamic economic system that was based on a plantation form of agriculture in which slaves were driven so hard that their lives were truncated. It was also a colony that contained a vibrant urban sector and opportunities for smaller planters outside the sugar plantation sector. Planters dominated politics and society, but Kingston merchants, though subservient to planters, were also rich, powerful, and influential. The shock of the near success of the slave rebellions of 1760 meant that white Jamaicans realized they were dependent on the support of an imperial state, but their ability to largely control that state especially in its colonial manifestation in the 1760s and 1770s meant a very large degree of comfort with being dependent.

Confident that they had the support of the people who counted in Britain for the maintenance of their social system, and especially their unconstrained control of their slaves, Jamaicans luxuriated in being "the jewel in the imperial crown." The late 1770s and early 1780s were difficult economically and politically, but by 1787, just before the onset of abolitionism changed their world, white Jamaicans believed that they would continue to prosper in an empire seemingly set up to work for their interests. Their happiness, however, was predicated on the misery of the more than two hundred thousand enslaved men, women, and children who did the work that made white Jamaicans rich. The question that began to be asked after this period of Jamaican history ended was whether this happiness, self-satisfaction, and undeserved wealth was sustainable. The answer was that it was not. White Jamaicans eventually learned that the halcyon years of the age of the American Revolution were, in retrospect, the last period in which they exercised real power and autonomy. But the power they possessed in the period when planters had most power both over their slaves and within imperial counsel—roughly between 1739 and 1796—means that a study of events and themes in eighteenth-century Jamaican history has relevance not just for students of Jamaican history but for analysts of the Atlantic World in its fullest context—encompassing Africa, Europe, and the Americas.

Notes

Introduction

1. For the wealth of Jamaica, see Trevor Burnard, "'A Prodigious Mine': The Wealth of Jamaica Before the American Revolution Once Again," *Economic History Review* 54 (2001): 505–23; and Trevor Burnard, Laura Panza, and Jeffrey Williamson, "Living Costs, Real Incomes and Inequality in Colonial Jamaica," *Explorations in Economic History* 71 (2019): 55–71. For Jamaica's place within the British Empire in the second half of the eighteenth century, see Trevor Burnard, *Planters, Merchants, and Slaves: Plantation Societies in British America, 1650–1820* (Chicago: University of Chicago Press, 2015), ch. 4.

2. P. J. Marshall, *The Making and Unmaking of Empires: Britain, India and America c. 1760–1783* (Oxford: Oxford University Press, 2005), 363–64.

3. Andrew Jackson O'Shaughnessy, *An Empire Divided: The American Revolution and the British Caribbean* (Philadelphia: University of Pennsylvania Press, 2000), 210, 230–32.

4. Richard S. Dunn, *Sugar and Slaves: The Rise of the Planter Class in the English West Indies* (Chapel Hill: University of North Carolina Press, 1972); Carl Bridenbaugh and Roberta Bridenbaugh, *No Peace Beyond the Line: The English in the Caribbean, 1624–1690* (New York: Oxford University Press, 1972).

5. For the most recent work, see, inter alia, Carla Gardana Pestana, *The English Conquest of Jamaica: Oliver Cromwell's Bid for Empire* (Cambridge, Mass.: Harvard University Press, 2017); and Edward Rugemer, *Slave Law and the Politics of Resistance in the Early Atlantic World* (Cambridge, Mass.: Harvard University Press, 2018).

6. Vincent Brown, *The Reaper's Garden: Death and Power in the World of Atlantic Slavery* (Cambridge, Mass.: Harvard University Press, 2008).

7. Jack P. Greene, *Settler Jamaica in the 1750s: A Social Portrait* (Charlottesville: University of Virginia Press, 2016).

8. Richard S. Dunn, *A Tale of Two Plantations: Slave Life and Labor in Jamaica and Virginia* (Cambridge, Mass.: Harvard University Press, 2014).

9. Emily Senior, *The Caribbean and the Medical Imagination, 1764–1834: Slavery, Disease and Colonial Modernity* (Cambridge: Cambridge University Press, 2018); Sasha Turner, *Contested Bodies: Pregnancy, Childrearing, and Slavery in Jamaica* (Philadelphia: University of Pennsylvania Press, 2017); Katherine Gerbner, *Christian Slavery: Protestant Missions and Slave Conversion in the Atlantic World, 1660–1760* (Philadelphia: University of Pennsylvania Press, 2018); Daniel Livesay, *Children of Uncertain Fortune: Mixed-Race Jamaicans in Britain and the Atlantic Family, 1733–1833* (Chapel Hill: University of North Carolina Press, 2018); Brooke N. Newman, *A Dark Inheritance: Blood, Race, and Sex in Colonial Jamaica* (New Haven, Conn.: Yale University Press, 2018); Christer Petley, *White Fury: A Jamaican Slaveholder and the Age of Revolution* (Oxford: Oxford University Press, 2018); and Aaron Graham, *Slavery, Society and the State in Jamaica, 1770–1840* (Oxford: Oxford University Press, forthcoming).

10. Robert S. DuPlessis, *The Material Atlantic: Clothing, Commerce, and Colonization in the Atlantic World, 1650–1800* (Cambridge: Cambridge University Press, 2016).

11. Among his many works, see the following that are directly connected to Jamaica: B. W. Higman, *Slave Population and Economy in Jamaica, 1807–1834* (Cambridge: Cambridge University Press, 1976); B. W. Higman, *Montpelier, Jamaica: A Plantation Community in Slavery and Freedom 1739–1912* (Kingston: University of the West Indies Press, 1998); B. W. Higman, *Plantation Jamaica 1750–1850: Capital and Control in a Colonial Economy* (Kingston: University of the West Indies Press, 2005); B. W. Higman, *Proslavery Priest: The Atlantic World of John Lindsay, 1729–1788* (Kingston: University of the West Indies Press, 2011).

12. See also Orlando Patterson, *The Sociology of Slavery: An Analysis of the Origins, Development, and Structure of Negro Slave Society in Jamaica* (Cranbury, N.J.: Fairleigh Dickinson University Press, 1969); Michael Craton and James Walvin, *A Jamaican Plantation: A History of Worthy Park, 1670–1790* (London: W. H. Allen, 1970); and Michael Craton, *Searching for the Invisible Man: Slaves and Plantation Life in Jamaica* (Cambridge, Mass.: Harvard University Press, 1978).

13. Edward Braithwaite, *The Development of Creole Society in Jamaica, 1770–1820* (Oxford: Oxford University Press, 1971); Trevor Burnard, *Creole Gentlemen: The Maryland Elite, 1691–1776* (New York: Routledge, 2002). For creolization among the enslaved, see Richard D. E. Burton, *Afro-Creole: Power, Opposition, and Play in the Caribbean* (Ithaca, N.Y.: Cornell University Press, 1997).

14. One aim of this work, and other works I have written on seventeenth- and eighteenth-century Jamaica, is to help provide that foundational empirical work on early Jamaica that allows other scholars to move beyond static pictures of Jamaica's past to more temporally discriminative studies and provide analytical frameworks that can help inform more detailed social, cultural, and political history of this place and time. For an appreciation of what I have done that summarizes my work better than I can do, see Greene, *Settler Jamaica*, 5–8.

15. James Robertson, "Jamaican Archival Resources for Seventeenth- and Eighteenth-Century Atlantic History," *Slavery and Abolition* 22 (2001): 109–40.

16. For other important work on eighteenth-century Jamaica, see, inter alia, Kathleen Wilson, "Rethinking the Colonial State: Family, Gender, and Governmentality in Eighteenth-Century British Frontiers," *American Historical Review* 116 (2011): 1294–322; Louis P. Nelson, *Architecture and Empire in Jamaica* (New Haven, Conn.: Yale University Press, 2016); James Robertson, *Gone Is the Ancient Glory: Spanish Town, Jamaica, 1534–2000* (Kingston: Ian Randle, 2005); J. R. Ward, *British West Indian Slavery, 1750–1834: The Process of Amelioration* (Oxford: Clarendon, 1988); and J. R. McNeill, *Mosquito Empires: Ecology and War in the Greater Caribbean, 1620–1914* (New York: Cambridge University Press, 2010).

17. Thomas C. Holt, *The Problem of Freedom: Race, Labour, and Politics in Jamaica and Britain, 1832–1938* (Baltimore: Johns Hopkins University Press, 1992); and Gad Heuman, *The Killing Time: The Morant Bay Rebellion in Jamaica* (London: Macmillan, 1994).

18. For travelogues that describe, often in negative terms, modern Jamaica, see Ian Thompson, *The Dead Yard: Tales of Modern Jamaica* (London: Faber and Faber, 2009); and Joshua Jelly-Schapiro, *Island People: The Caribbean and the World* (New York: Knopf, 2017). For a balanced academic account, see Patrick Bryan, *Edward Seaga and the Challenges of Modern Jamaica* (Kingston: University of West Indies Press, 2010).

19. Marlon James has an acute appreciation for how the history of Jamaica has an impact on the present. Marlon James, *Book of Night Women* (New York: Riverhead, 2009); and Marlon James, *A Brief History of Seven Killings* (New York: Riverhead, 2014).

20. Richard S. Dunn, "The Glorious Revolution and America," in *The Oxford History of the British Empire*, vol. 1, *The Origins of Empire*, ed. Nicolas Canny (Oxford: Oxford University Press, 1988), 463–65.

21. Allan Kulikoff, "'Such Things Ought Not to Be': The American Revolution and the First National Great Depression," in *The World of the Revolutionary American Republic: Expansion, Conflict, and the Struggle for a Continent*, ed. Andrew Shankman (New York: Routledge, 2014), 134–64.

22. Alan Taylor, *American Revolutions: A Continental History, 1750–1804* (New York: Norton, 2016), 408–9; Max Edling, *A Hercules in the Cradle: War, Money, and the American State, 1783–1867* (Chicago: University of Chicago Press, 2014), chs. 2–3.

23. Peter H. Lindert and Jeffrey G. Williamson, "American Incomes Before and After the Revolution," *Journal of Economic History* 73 (2016): 725–65.

24. This debate was started in Eric Williams, *Capitalism and Slavery* (Chapel Hill: University of North Carolina Press, 1944). For the American Revolution as economic disaster, see Selwyn Carrington, *The Sugar Industry and the Abolition of the Slave Trade, 1775–1810* (Gainesville: University Press of Florida, 2002). For the 1780s as difficult but not terminal, see John J. McCusker, "The Economy of the British West Indies, 1763–1790: Growth, Stagnation, or Decline?" in *Essays on the Economic History of the Atlantic World* (New York: Routledge, 1997), 330. For the American Revolution signaling a change in British racial attitudes, see Christopher Leslie Brown, *Moral Capital: The Foundations of British Abolitionism* (Chapel Hill: University of North Carolina Press, 2006).

25. Christer Petley, ed., "Rethinking the Fall of the Planter Class," special issue, *Atlantic Studies* 9 (2012); and Christer Petley, "Slaveholders and Revolution: The Jamaican Planter Class, British Imperial Politics, and the Ending of the Slave Trade, 1775–1807," *Slavery and Abolition* 39 (2018): 53–79.

26. This theme has been explored more fully for Saint-Domingue than for Jamaica. See Doris Garraway, *The Libertine Colony: Creolization in the Early French Caribbean* (Durham, N.C.: Duke University Press, 2005), 194–292; and Joan Dayan, *Haiti, History, and the Gods* (Berkeley: University of California Press, 1995).

27. Madeleine Dobie, *Trading Places: Colonization and Slavery in Eighteenth-Century French Culture* (Ithaca, N.Y.: Cornell University Press, 2010), 256. For Haiti in wide context, see Robin Blackburn, "Haiti, Slavery, and the Age of Democratic Revolution," *William and Mary Quarterly*, 3rd ser., 63 (2006): 643–74.

28. Laurent Dubois, "An Enslaved Enlightenment: Rethinking the Intellectual History of the French Atlantic," *Social History* 31 (2006): 1–14.

29. Trevor Burnard, *Mastery, Tyranny, and Desire: Thomas Thistlewood and His Slaves in the Anglo-Jamaican World* (Chapel Hill: University of North Carolina Press, 2004); Burnard, *Planters, Merchants, and Slaves* and Trevor Burnard and John Garrigus, *The Plantation Machine: Atlantic Capitalism in French Saint-Domingue and British Jamaica, 1748–1788* (Philadelphia: University of Pennsylvania Press, 2016).

30. See Trevor Burnard, "Slavery and the Enlightenment in Jamaica, 1760–1772: The Afterlife of Tackey's Rebellion," in *Enlightened Colonialism: Imperial Agents, Narratives of Progress and Civilizing Policies in the Eighteenth Century*, ed. Damien Tricoire (Basingstoke: Palgrave Macmillan, 2018); Trevor Burnard, "Plantations and the Great Divergence," in *Economic Change in Global History*, ed. Giorgio Riello and Tirthankur Roy (London: Bloomsbury, 2018), 227–46; Trevor Burnard, "Harvest Years: Reconfigurations of Empire in Jamaica, 1756–1807," *Journal of Commonwealth and Imperial History* 40 (2012): 533–55; Trevor Burnard, "Ireland, Jamaica and the Fate of White Protestants in the British Empire in the 1780s," in *Ireland in the World: Comparative, Transnational, and Personal Perspectives*, ed. Angela McCarthy (London: Routledge, 2015), 15–33; and Trevor Burnard, "The Founding Fathers in Early American Historiography: A View from Abroad," *William and Mary Quarterly*, 3rd ser., 62 (2005): 745–64.

31. Robert G. Parkinson, *The Common Cause: Creating Race and Nation in the American Revolution* (Chapel Hill: University of North Carolina Press, 2016).

32. Richard K. Macmaster, "Arthur Lee's 'Address on Slavery': An Aspect of Virginia's Struggle to End the Slave Trade, 1764–1774," *Virginia Magazine of History and Biography* 80 (1972): 141–57; Katherine Paugh, *The Politics of Reproduction: Race, Medicine, and Fertility in the Age of Abolition* (Oxford: Oxford University Press, 2017), 26–29; and Woody Holton, *Forced Founders: Indians, Debtors, Slaves, and the Making of the American Revolution in Virginia* (Chapel Hill: University of North Carolina Press, 1999), 66–69.

33. Burnard, *Planters, Merchants, and Slaves*, ch. 5.

34. For the wider context of colonialism and theatre, see Elizabeth Maddock Dillon, *New World Drama: The Performative Commons in the Atlantic World, 1649–1849* (Durham: Duke University Press, 2014).

35. Higman, *Plantation Jamaica*, 5.

36. Richard B. Sheridan, "The Crisis of Slave Subsistence in the British West Indies During and After the American Revolution," *William and Mary Quarterly*, 3rd ser., 33 (1976): 615–41.

37. Justin Roberts, "Uncertain Business: A Case Study of Barbadian Plantation Management, 1770–93," *Slavery and Abolition* 32 (2011): 253–55; and Heather Cateau, "The New 'Negro' Business: Slave Hiring in the British West Indies," in *In the Shadow of the Plantation: Caribbean History and Legacy*, ed. Alvin O. Thompson (Kingston: University of the West Indies, 2002), 100–120; Robert William Fogel, *Without Consent or Coercion: The Rise and Fall of American Slavery* (New York: W. W. Norton, 1989), 124.

38. Cited in Betty Wood, ed. *The Letters of Simon Taylor of Jamaica to Chaloner Arcedeckne, 1765–1775* (London: Royal Historical Society, Camden Misc., 2002), 87.

39. Simon Taylor to Chaloner Arcedeckne, 30 January, 11 June, 29 October 1782, Bundle 2/10, Vanneck Mss., Cambridge University Library.

40. Trevor Burnard, "Powerless Masters: The Curious Decline of Jamaican Sugar Planters in the Foundational Period of British Abolition," *Slavery and Abolition* 32 (2011): 185–98.

41. Resistance, however, was less important than survival for most enslaved people. They had to cope under the terrible circumstances the planter class imposed on them where "the work regime made the slaves into interchangeable units of labor—alienated, expendable, interchangeable—as if they lacked individuality or any personal power." The conditions of life were so dreadful—slaves were "commonly undernourished, deprived of sleep, and adequate food, often beaten"—that survival in the face of overwhelming power was usually the best that slaves could do. Sidney W. Mintz, *Three Ancient Cultures: Caribbean Themes and Variations* (Cambridge, Mass.: Harvard University Press, 2010), 11. For survival, see Randy Browne, *Surviving Slavery in the British Caribbean* (Philadelphia: University of Pennsylvania Press, 2017). My theoretical position on the complicated relationship between slave agency, resistance, and survival is informed by the work on strategies of resistance and tactics of opposition outlined by Michel de Certeau, who argues that resistance is possible only when the dominated group or dominated individuals act outside the system of domination that encloses them. De Certeau, "On the Oppositional Practices of Everyday Life," *Social Text* 3 (1980): 3–43.

42. Burnard, *Planters, Merchants, and Slaves*; Steven Pincus, *The Heart of the Declaration: The Founders' Case for an Activist Government* (New Haven, Conn.: Yale University Press, 2016), 30–31; and Abigail L. Swingen, *Competing Visions of Empire: Labor, Slavery, and the Origins of the British Atlantic Empire* (New Haven, Conn.: Yale University Press, 2015), ch. 7.

43. David Barry Gaspar and David Patrick Geggus, eds., *A Turbulent Time: The French Revolution and the Greater Caribbean* (Bloomington: , 1997); Petley, "Slaveholders and Revolution."

44. Catherine Hall et al., *Legacies of British Slave-Ownership: Colonial Slavery and the Formation of Victorian Britain* (Cambridge: Cambridge University Press, 2014); Sherylynne Haggerty and Susanna Seymour, "Imperial Careering and Enslavement in the Long Eighteenth Century: The Bentinck Family, 1710–1830s," *Slavery and Abolition*, 39 (2018), 642–62; Madge Dresser et al., *Slavery and the British Country House* (London: English Heritage, 2013); S. D. Smith, *Slavery, Family and Gentry Capitalism in the British Atlantic: The World of the Lascelles, 1648–1834* (Cambridge: Cambridge University Press, 2006), 2–6.

45. For that earlier efflorescence of scholarship, see B. W. Higman, *Writing West Indies History* (London: Macmillan, 1999).

46. Philip D. Morgan, "Slavery in the British Caribbean," in *The Cambridge World History of Slavery*, vol. 3, *A.D. 1420–A.D. 1804*, ed. David Eltis and Stanley L. Engerman (Cambridge: Cambridge University Press, 2011), 378–406; and Trevor Burnard, "British West Indies and Bermuda," in *Oxford Handbook of Slavery*, ed. Mark M. Smith and Robert L. Paquette (New York: Oxford University Press, 2010), 135–53.

47. See the thoughtful discussion in David Abernethy, *Dynamics of Global Dominance: European Overseas Empires, 1415–1980* (New Haven, Conn.: Yale University Press, 2000), 387–407.

48. See especially Justin Roberts, *Slavery and the Enlightenment in the British Atlantic, 1750–1807* (New York: Cambridge University Press, 2013), 2–3.

49. Trevor Burnard, "The Planter Class," in *The Routledge History of Slavery*, ed. Gad Heuman and Trevor Burnard (London: Routledge, 2010), 187–203.

50. Marisa Fuentes, *Dispossessed Lives: Enslaved Women, Violence, and the Archive* (Philadelphia: University of Pennsylvania Press, 2016), 1–2, 145–47; Marisa Fuentes, "Power and Historical Figuring: Rachel Pringle Polgreen's Troubled Archive," *Gender and History* 22 (2010): 564–84. For a meditation on power and narrative in Caribbean historiography, see David Scott, *Conscripts of Modernity: The Tragedy of Colonial Enlightenment* (Durham, N.C.: Duke University Press, 2004). For my own thoughts on the "problem of the archive," see Trevor Burnard and Sophie White, eds., *Hearing Slaves' Voices: African and Indian Slave Testimony in British and French America, 1700–1848* (New York: Routledge, 2020).

51. Ira Berlin, "Time, Space, and the Evolution of Afro-American Society on British Mainland North America," *American Historical Review* 85 (1980): 44–78.

52. David Scott, "The Paradox of Freedom: An Interview with Orlando Patterson," *Small Axe* 17 (2013): 96–242 (quotations 222 and 236).

53. Burnard, Panza, and Williamson, "Living Costs, Real Incomes and Inequality in Colonial Jamaica."

54. Vincent Brown, "Spiritual Terror and Sacred Authority in Jamaican Slave Society," *Slavery and Abolition* 24 (2003): 27–29; Amanda Thornton, "Coerced Care: Thomas Thistlewood's Account of Medical Practice on Enslaved Populations in Colonial Jamaica, 1751–1786," *Slavery and Abolition* 32 (2011): 535–59.

55. Vincent Brown, "Social Death and Political Life in the Study of Slavery," *American Historical Review* 114 (2009): 1236.

56. Ibid., 1242–46.

57. Katherine Gerbner, *Christian Slavery: Conversion and Race in the Protestant Atlantic World* (Philadelphia: University of Pennsylvania Press, 2018); Livesay, *Children of Uncertain Fortune*. See also Christine Walker, "Pursuing Her Profits: Women in Jamaica, Atlantic Slavery and a Globalising Market, 1700–60," *Gender and History* 26 (2014): 478–501; and Christine Walker, *Jamaican Ladies: Gender, Authority and Atlantic Slavery* (Chapel Hill: University of North Carolina Press, forthcoming).

58. Petley, *White Fury*; J. R. Ward, "The Amelioration of British West Indian Slavery: Anthropological Evidence," *Economic History Review* 71 (2018): 1199–1226; Graham, *Slavery, Society and the State*; Justin Roberts, "The 'Better Sort' and the 'Poorer Sort': Wealth Inequalities, Family Formation and the Economy of Energy on British Caribbean Sugar Plantations," *Slavery and Abolition* 35 (2014): 458–73.

59. For a study of neighboring Saint-Domingue that focuses on the planter class and the plantation system, but which sees that system as bound to fail under its own contradictions, see Paul Cheney, *Cul de Sac: Patrimony, Capitalism, and Slavery in French Saint-Domingue* (Chicago: University of Chicago Press, 2017).

60. Jennifer L. Morgan, *Laboring Women: Reproduction and Gender in New World Slavery* (Philadelphia: University of Pennsylvania Press, 2004); Diana Paton, "Punishment, Crime, and the Bodies of Slaves in Eighteenth-Century Jamaica," *Journal of Social History* 34 (2001): 923–54; and Dawn P. Harris, *Punishing the Black Body: Marking Social and Racial Structures in Barbados and Jamaica* (Athens: University of Georgia Press, 2017).

61. Turner, *Contested Bodies*; Paugh, *Politics of Reproduction*; Audra Diptee, *From Africa to Jamaica: The Making of an Atlantic Slave Society, 1776–1807* (Gainesville: University Press of Florida, 2010); and Colleen Vasconcellos, *Slavery, Childhood and Abolition in Jamaica, 1788–1834* (Athens: University of Georgia Press, 2015). See also Daina Berry et al., *Sexuality and Slavery: Reclaiming Intimate Histories in the Americas* (Athens: University of Georgia Press, 2018). A useful survey of current historiographical trends is Diana Paton, "Maternal Struggles and the Politics of Childlessness Under Pro-Natalist Caribbean Slavery," *Slavery and Abolition* 38 (2017): 251–68.

62. Greene, *Settler Jamaica*.

63. James Delbourgo, *Collecting the World: Hans Sloane and the Origins of the British Museum* (Cambridge, Mass.: Harvard University Press, 2017); Deirdre Coleman, *Henry Smeathman, the Fly-catcher: Natural History, Slavery and Empire in the late Eighteenth Century* (Liverpool: Liverpool University Press, 2018), 7. See also Beth Fowkes Tobin, *The Duchess's Shells: Natural History Collecting in the*

Age of Cook's Voyages (New Haven, Conn.: Yale University Press, 2014). A related field, to which my work in this book makes little contribution but which is a vibrant part of recent scholarship on the Caribbean, is the history of science and medicine. See Londa Schiebinger, *Secret Cures of Slaves: People, Plants and Medicine in the Eighteenth-Century Atlantic World* (Stanford, Calif.: Stanford University Press, 2017); and Senior, *Caribbean and the Medical Imagination*.

64. Charmaine A. Nelson, *Slavery, Geography and Empire in Nineteenth-Century Marine Landscapes of Montreal and Jamaica* (London: Routledge, 2016); Simon P. Newman, "Hidden in Plain Sight: Escaped Slaves in Late Eighteenth- and Early Nineteenth-Century Jamaica," *William and Mary Quarterly* (OI Reader app), (June 2018): 1–53 http://doi.org/willmaryquar.newman.

65. Elizabeth A. Bohls, *Slavery and the Politics of Place: The Colonial Caribbean, 1770–1833* (Cambridge: Cambridge University Press, 2014), 3, 9.

66. L. P. Nelson, *Architecture and Empire in Jamaica*.

67. Mintz, *Three Ancient Cultures*, 189–212.

68. Ibid., 10–11; C. L. R. James, *The Black Jacobins: Toussaint L'Ouverture and the San Domingo Revolution*, 2nd ed. (New York: Vintage Books, 1963), 392. See also David Scott, "Modernity that Predated the Modern: Sidney Mintz's Caribbean," *History Workshop Journal* 58 (2004): 191–210; Alan L. Karras, "The Caribbean Region: Crucible for Modern World History," in *The Cambridge World History*, vol. 6, part 1, ed. Jerry H. Bentley et al. (Cambridge: Cambridge University Press, 2015), 395; and Sarah Yeh, "Colonial Identity and Revolutionary Loyalty: The Case of the West Indies," in *The Oxford History of the British Empire: British North America in the Seventeenth and Eighteenth Centuries*, ed. Stephen Foster (Oxford: Oxford University Press, 2013), 205. My book (coauthored) that most stresses these themes is Burnard and Garrigus, *Plantation Machine*.

Chapter 1

1. Michael Ignatieff, *The Warrior's Honor: Ethnic War and the Modern Conscience* (New York: Henry Holt, 1997), 18–19.

2. In contemporary politics, see David Runciman, *The Politics of Good Intentions: History, Fear and Hypocrisy in the New World Order* (Princeton, N.J.: Princeton University Press, 2006); Corey Robin, *Fear: The History of a Political Idea* (New York: Oxford University Press, 2004); and Cass Sunstein, *The Laws of Fear: Beyond the Precautionary Principle* (New York: Cambridge University Press, 2005). In historiography, see Michael Laffan and Max Weiss, eds., *Facing Fear: The History of an Emotion in Global Perspective* (Princeton, N.J.: Princeton University Press, 2012); Joanna Bourke, *Fear: A Cultural History* (London: Virago, 2006); and Jan Plamper and Benjamin Lazier, eds., *Fear Across the Disciplines* (Pittsburgh: Pittsburgh University Press, 2012). In political theory, see Quentin Skinner, *Hobbes and Republican Liberty* (Cambridge: Cambridge University Press, 2008); Mary Nyquist, "Hobbes, Slavery, and Despotical Rule," *Representations* 106 (2009): 1–33; and Mary Nyquist, *Arbitrary Rule: Slavery, Tyranny, and the Power of Life and Death* (Chicago: University of Chicago Press, 2013).

3. Frank Furedi, *Culture of Fear: Risk-Taking and the Morality of Low Expectations* (London: Cassell, 1997); and Barry Glassner, *The Culture of Fear: Why Americans Are Afraid of the Wrong Things* (New York: Basic Books, 1999).

4. Key works are Barbara H. Roseinwein, "Worrying About Emotions in History," *American Historical Review* 107 (2002): 821–45; and William M. Reddy, *The Navigation of Feeling: A Framework for the History of Emotions* (Cambridge: Cambridge University Press, 2001). See also Ruth Leys, "How Did Fear Become a Scientific Entity and What Kind of Entity Is It?" *Representations* 110 (2010): 66–104; and Jan Plamper, "Fear: Soldiers and Emotion in Early Twentieth-Century Russian Military Psychology," *Slavic Review* 68 (2009): 259–83. "Emotionology" was coined by Peter and Carol Stearns to describe "the attitudes or standards that a society . . . maintains towards basic emotions and their appropriate expression and the ways that institutions reflect and encourage these attitudes in human conduct." Stearns and Stearns, "Emotionology: Clarifying the History of Emotions and Emotional Standards," *American Historical Review* 90 (1985): 813–36.

5. Edmund Burke, *A Philosophical Enquiry into the Origin of Our Ideas of the Sublime and Beautiful*, ed. James T. Boulton (Notre Dame, Ind.: University of Notre Dame Press, 1968), 57. For the West Indies and the gothic, see Srinivas Aravamudan, *Tropicopolitans: Colonialism and Agency, 1688–1804* (Durham, N.C.: Duke University Press, 1999), 214–29; and Keith A. Sandiford, *The Cultural Politics of Sugar: Caribbean Slavery and Narratives of Colonialism* (Cambridge: Cambridge University Press, 2000), 150–76.

6. Emily Senior, *The Caribbean and the Medical Imagination, 1764–1834: Slavery, Disease and Colonial Modernity* (Cambridge: Cambridge University Press, 2018), 79.

7. Lizabeth Paravisini-Gebert, "The Colonial and Post-Colonial Gothic: The Caribbean," in *The Caribbean Companion to Gothic Fiction*, ed. Jerrold E. Hodge (Cambridge: Cambridge University Press, 2002), 231–32, 234.

8. Judith N. Shklar, "The Liberalism of Fear," in *Liberalism and the Moral Life*, ed. Nancy L. Rosenblum (Cambridge, Mass.: Harvard University Press, 1999), 21.

9. Bourke, *Fear*, 188, 190, 353.

10. Lucien Febvre, "La Sensibilité et l'histoire: Comment reconstituer la vie affective d'autrefois," *Annales d'histoire sociale* 3 (1941): 12; Bourke, *Fear*, 7.

11. Jean Delumeau, *La Peur en Occident (XVIe–XVIIIe siècles)* (Paris: Fayard, 1978). His later work, much of which is still untranslated into English, dealt as much with remedies to fear as with its manifestations. Guillaume Cuchet, "Jean Delumeau, historien de la peur et du péché," *Vingtième Siècle: Revue d'histoire* 107 (2010): 145–55. For the continuing depiction of early modern Europe as fear filled, see Mark Greengrass, *Christendom Destroyed: Europe 1517–1648* (London: Penguin, 2014).

12. Peter Lake, "Anti-Popery: The Structure of a Prejudice," in *Conflict in Early Stuart England: Studies in Religion and Politics, 1603–1642*, ed. Richard Cust and Ann Hughes (London: Routledge, 1989), 72–106. For a typical pamphlet, see *A Scheme of Popish Cruelties; or, A Prospect of What Wee Must Expect Under a Popish Successor* (London, 1681). For anti-Catholicism in British America, see Shona Johnston, "Papists in a Protestant World: The Catholic Anglo-Atlantic in the Seventeenth Century" (PhD diss., Georgetown University, 2011); Brendan McConville, *The King's Three Faces: The Rise and Fall of Royal America, 1688–1776* (Chapel Hill: University of North Carolina Press, 2006); and for the Caribbean, Kirsten Block and Jenny Shaw, "Subjects Without an Empire: The Irish in the Early Modern Caribbean," *Past and Present* 210 (2011): 33–60.

13. John M. Murrin, "The Beneficiaries of Catastrophe: The English Colonies in America," in *The New American History*, ed. Eric Foner, rev. ed. (Philadelphia: Temple University Press, 1997), 3–4.

14. For fears of slave revolt after Haiti, see Walter Johnson, *River of Dark Dreams: Slavery and Empire in the Cotton Kingdom* (Cambridge, Mass.: Harvard University Press, 2013), 13, 32, 84; Ashli White, *Encountering Revolution: Haiti and the Making of the Early Republic* (Baltimore: Johns Hopkins University Press, 2010); Matthew Clavin, *Toussaint Louverture and the American Civil War: The Promise and Peril of a Second Haitian Revolution* (Philadelphia: University of Pennsylvania Press, 2012); Julia Gaffield, *Haitian Connections in the Atlantic World: Recognition After Revolution* (Chapel Hill: University of North Carolina Press, 2016).

15. Anthony Parent Jr., *Foul Means: The Formation of a Slave Society in Virginia, 1660–1740* (Chapel Hill: University of North Carolina Press, 2003); Michael A. McDonnell, *The Politics of War: Race, Class and Conflict in Revolutionary Virginia* (Chapel Hill: University of North Carolina Press, 2007); Woody Holton, *Forced Founders: Indians, Debtors, Slaves, and the Making of the American Revolution in Virginia* (Chapel Hill: University of North Carolina Press, 1999); Andrew Jackson O'Shaughnessy, *An Empire Divided: The American Revolution and the British Caribbean* (Philadelphia: University of Pennsylvania Press, 2000); Peter Wood, *Black Majority: Negroes in South Carolina from 1670 to the Stono Rebellion* (New York: W. W. Norton, 1974); Robert Olwell, *Masters, Slaves, and Subjects: The Culture of Power in the South Carolina Low Country, 1740–1790* (Ithaca, N.Y.: Cornell University Press, 1998); James Piecuch, *Three Peoples, One King: Loyalists, Indians, and Slaves in the Revolutionary South, 1775–1782* (Columbia: University of South Carolina Press, 2008); Sylvia Frey, *Water from the Rock: Black Resistance in a Revolutionary Age* (Princeton, N.J.: Princeton University Press, 1991); Gary B. Nash, *The Forgotten Fifth: African Americans in the Age of Revolution* (Cambridge, Mass.: Harvard

University Press, 2006); Ira Berlin, *Generations of Captivity: A History of African-American Slaves* (Cambridge, Mass.: Harvard University Press, 2003), 97–158; Robin Blackburn, *American Crucible: Slavery, Emancipation and Human Rights* (London: Verso, 2011); Douglas R. Egerton, *Death or Liberty: African Americans and Revolutionary America* (New York: Oxford University Press, 2009); Gerald Horne, *The Counter-Revolution of 1776: Slave Resistance and the Origins of the United States* (New York: New York University Press, 2015). See also Alan Gilbert, *Black Patriots and Loyalists: Fighting for Emancipation in the War for American Independence* (Chicago: University of Chicago Press, 2012); and J. William Harris, *The Hanging of Thomas Jeremiah: A Free Black Man's Encounter with Liberty* (New Haven, Conn.: Yale University Press, 2009). For slave revolts in the early nineteenth-century British West Indies, see Michael Mullin, *Africa in America: Slave Acculturation and Resistance in the American South and the British Caribbean, 1736–1831* (Urbana: University of Illinois Press, 1992); Emilia Viotti da Costa, *Crowns of Glory, Tears of Blood: The Demerara Slave Rebellion of 1823* (New York: Oxford University Press, 1994); Claudius Fergus, *Revolutionary Emancipation: Slavery and Abolitionism in the British West Indies* (Baton Rouge: Louisiana State University Press, 2013); Gelien Matthews, *Caribbean Slave Revolts and the British Abolitionist Movement* (Baton Rouge: Louisiana State University Press, 2006); and Edward Rugemer, *The Politics of Atlantic Slavery: Jamaica and South Carolina from the Seventeenth Century to 1838* (Cambridge, Mass.: Harvard University Press, 2018).

16. Alan Taylor, *The Internal Enemy: Slavery and War in Virginia, 1772–1832* (New York: W. W. Norton, 2013).

17. James Madison to William Bradford, 26 November 26 1774, in William T. Hutchinson et al., *The Papers of James Madison* (Chicago: University of Chicago Press, 1962), 1:129–30.

18. Byrd to Egmont, 12 July 1736, in "Colonel William Byrd on Slavery and Indentured Servants, 1736, 1739," *American Historical Review* 1 (1895): 89.

19. Cited in Holton, *Forced Founders*, 70.

20. Cited in Russell R. Menard, *Sweet Negotiations: Sugar, Slavery, and Plantation Agriculture in Early Barbados* (Charlottesville: University of Virginia Press, 2006), 116.

21. [Edward Trelawney], *An Essay Concerning Slavery, and the Danger Jamaica Is Expos'd to from the Too Great Number of Slaves* . . . (London: C. Corbett, 1746), reprinted in *Exploring the Bounds of Liberty: Political Writings of Colonial British America from the Glorious Revolution to the American Revolution* ed. Jack P. Greene and Craig Yirush (Carmel, Ind.: Liberty Fund, 2018), 3 vols., II: 1134, 1144.. For an argument that Trelawney was the author of this work, see George Boulukos, *The Grateful Slave: The Emergence of Race in Eighteenth-Century British and American Culture* (Cambridge: Cambridge University Press, 2008), 4–5.

22. [William Burke], *An Account of the European Settlements in America*, 2 vols. (London, 1757), 2:113.

23. *Journals of the Assembly of Jamaica* . . . *1663–[1826]*, 14 vols. (Kingston, 1811–29), vol. 7, December 31, 1773.

24. O'Shaughnessy, *Empire Divided*, 34, 38, 49, 51, 57.

25. Edmund S. Morgan, "Slavery and Freedom: The American Paradox," *Journal of American History* 59 (1972): 5–29; Edmund S. Morgan, *American Slavery, American Freedom: The Ordeal of Colonial Virginia* (New York: Norton, 1975), 386.

26. Rhys Isaac, *The Transformation of Virginia, 1740–1790* (Chapel Hill: University of North Carolina Press, 1982); Rhys Isaac, *Landon Carter's Uneasy Kingdom: Revolution and Rebellion on a Virginia Plantation* (New York: Oxford University Press, 2004). For a similar portrait of an angst-ridden southern planter, see Kenneth Lockridge, *The Diary, and Life, of William Byrd II of Virginia, 1674–1744* (Chapel Hill: University of North Carolina Press, 1987).

27. For some examples for Virginia, see Parent, *Foul Means*; Holton, *Forced Founders*; McDonnell, *Politics of War*; and T. H. Breen, *Tobacco Culture: The Mentality of the Great Tidewater Planters on the Eve of the Revolution* (Princeton, N.J.: Princeton University Press, 1985). For South Carolina, see Joyce E. Chaplin, *An Anxious Pursuit: Agricultural Innovation and Modernity in the Lower South, 1730–1865* (Chapel Hill: University of North Carolina Press, 1993); and S. Max Edelson, *Plantation Enterprise in Colonial South Carolina* (Cambridge, Mass.: Harvard University Press, 2006). For the

British West Indies, see Richard S. Dunn, *A Tale of Two Plantations: Slave Life and Labor in Jamaica and Virginia* (Cambridge, Mass.: Harvard University Press, 2014); and Richard S. Dunn, *Sugar and Slaves: The Rise of the Planter Class in the English West Indies* (Chapel Hill: University of North Carolina Press, 1972).

28. Kathleen M. Brown, *Good Wives, Nasty Wenches, and Anxious Patriarchs: Gender, Race, and Power in Colonial Virginia* (Chapel Hill: University of North Carolina Press, 2006), 319, 321, 323.

29. Trevor Burnard, *Mastery, Tyranny, and Desire: Thomas Thistlewood and His Slaves in the Anglo-Jamaican World* (University of North Carolina Press: Chapel Hill, 2004), 21.

30. Edward Long, *History of Jamaica . . .*, 3 vols. (London: T. Lowndes , 1774), 2:262–65; Charles Leslie, *A New and Exact Account of Jamaica* (Edinburgh: R. Fleming, ca. 1740), 319.

31. Betty Wood, ed. *The Letters of Simon Taylor of Jamaica to Chaloner Arcedeckne, 1765–1775* (London: Royal Historical Society, Camden Misc., 2002), 29–30; B. W. Higman, *Plantation Jamaica 1750–1850: Capital and Control in a Colonial Economy* (Kingston: University of the West Indies Press, 2005), 197–98. For Taylor, see Christer Petley, *White Fury: A Jamaican Slaveholder and the Age of Revolution* (Oxford: Oxford University Press, 2018).

32. William Wirt, *Sketches of the Life and Character of Patrick Henry* (Philadelphia: James Webster, 1817), 123.

33. François Furstenberg, "Beyond Freedom and Slavery: Autonomy, Virtue and Resistance in Early American Political Discourse," *Journal of American History* 89 (2003): 1295–1330.

34. Daniel Defoe, *Colonel Jack*, ed. Samuel Holt Monk (1722; New York: Oxford University Press, 1989), 128. See Trevor Burnard, *Planters, Merchants, and Slaves: Plantation Societies in British America 1650–1820* (Chicago: University of Chicago Press, 2015).

35. For the languages of slavery, see Francois Furstenberg, "Atlantic Slavery, Atlantic Freedom: George Washington, Slavery, and Transatlantic Abolitionist Networks," *William and Mary Quarterly*, 3rd ser., 68 (2011): 247–86; and Edward Rugemer, "The Development of Mastery and Race in the Comprehensive Slave Codes of the Greater Caribbean During the Seventeenth Century," *William and Mary Quarterly*, 3rd ser., 70 (2013): 429–58.

36. William Bristow, "Enlightenment," in The Stanford Encyclopedia of Philosophy, ed. Edward N. Zalta (Fall 2017 edition), https://plato.stanford.edu/archives/fall2017/entries/enlightenment/.

37. Geoffrey Parker and L. M. Smith, *The General Crisis of the Seventeenth Century*, 2nd. ed. (London: Routledge 1997); and Geoffrey Parker, *Global Crisis: War, Climate Change and Catastrophe in the Seventeenth Century* (New Haven, Conn.: Yale University Press, 2013).

38. Peter Wilson, *Europe's Tragedy: A History of the Thirty Years War* (London: Allen Lane, 2009).

39. Patricia Springborg, ed., *The Cambridge Companion to Hobbes's Leviathan* (Cambridge: Cambridge University Press, 2007).

40. Thomas Hobbes, *Leviathan*, ed. Richard Tuck (Cambridge: Cambridge University Press, 1996), 1.13.62. Subsequent references are to this edition and indicate the part, chapter and section.

41. Richard Tuck, *Hobbes: A Very Short Introduction* (Oxford: Oxford University Press, 2002).

42. Glenn Burgess, *British Political Thought, 1500–1660: The Politics of the Post-Reformation* (Basingstoke: Palgrave Macmillan, 2009), 299.

43. Richard Ashcraft, "*Leviathan* Triumphant: Thomas Hobbes and the Politics of Wild Men," in *The Wild Men Within*, ed. Edward Dudley and Maximilian Novak (Pittsburgh: University of Pittsburgh Press, 1972), 148–54.

44. Nyquist, *Arbitrary Rule*, 269–79.

45. Ibid., 7–10. What Nyquist calls "war slavery" was the product of Roman law. Alan Watson, "Roman Slave Law and Romanist Ideology," *Phoenix* 37 (1983): 53–65; and Alan Watson, *Roman Slave Law* (Baltimore: Johns Hopkins University Press, 1987).

46. Hobbes, *Leviathan*, 2.20.113.

47. Nyquist, *Arbitrary Rule*, 321.

48. The influence of Hobbes in America generally and concerning slave owners specifically is an underresearched topic. Hobbes is absent, for example, from the work of Eugene Genovese and David Brion Davis explicating slaveholder ideology in the first half of the nineteenth century. Trevor Burnard,

"Who Deceived Whom? Eugene Genovese and Planter Self-Deception," *Slavery and Abolition* 34 (2013): 508–14; and David Brion Davis, *The Problem of Slavery in Western Culture* (Ithaca, N.Y.: Cornell University Press, 1969). For discussion of Hobbes in America in a literary context, see Paul Downes, *Hobbes, Sovereignty and Early American Literature* (New York: Cambridge University Press, 2015).

49. Nyquist, *Arbitrary Rule*, 325.

50. Ibid.

51. Michael Craton, "Hobbesian or Panglossian? The Two Extremes of Slave Conditions in the British Caribbean, 1783–1834," *William and Mary Quarterly*, 3rd ser., 35 (1978): 324–56.

52. Thomas Hobbes, *Leviathan*, ed. Richard Tuck (Cambridge: Cambridge University Press, 1996), 465; Runciman, *Politics of Good Intentions*, 123, 188. Corey Robin usefully outlines three implications arising from Hobbes's argument of the wisdom of individuals submitting to the sovereign. First, the state chooses which threats to take seriously (the threat of slave rebellion, for example, in eighteenth-century plantation societies) and which to dismiss (enslaved people's fears of planter violence against them). Second, people will believe themselves obliged to obey the dictates of the state only if they think their security is imperiled or potentially at risk (thus, the state has a strong incentive to claim that potential slave rebellions are always possible). Finally, it is the sovereign that is the judge of our fears and how to respond to them because only the sovereign possesses the suitable unity of will and judgment necessary to deal with a threat (people must not only agree that they are under threat but also agree on how the state might meet this threat). Corey Robin, "The Language of Fear: Security and Modern Politics," in Plamper and Lazier, *Fear Across the Disciplines*, 119–25.

53. For Edwards and his thought, see David Brion Davis, *The Problem of Slavery in the Age of Revolution, 1770–1823* (Ithaca, N.Y.: Cornell University Press, 1975), 185–95; Elsa Goveia, *A Study on the Historiography of the British West Indies to the End of the Nineteenth Century* (Washington, D.C.: Howard University Press, 1980 [originally published in Mexico City in 1956]); Olwyn M. Blouet, "Bryan Edwards, F.R.S., 1743–1800," *Notes and Records of the Royal Society of London* 54 (2000): 215–22; Olwyn M. Blouet, "Bryan Edwards and the Haitian Revolution," in *The Impact of the Haitian Revolution in the Atlantic World*, ed. David Geggus (Columbia: University of South Carolina Press, 2001), 44–58; Elizabeth A. Bohls, *Slavery and the Politics of Place: Representing the Colonial Caribbean, 1770–1833* (Cambridge: Cambridge University Press, 2014), 103–19; and Edward B. Rugemer, *The Problem of Emancipation: The Caribbean Roots of the Civil War* (Baton Rouge: Louisiana State University Press, 2008). The wealth of the Edwards-Bayly clan was very large. Nathaniel Bayly died with an estate valued at £1,117,860. Personal communication, Karina Williamson, 17 July 2018.

54. Bryan Edwards, *The History, Civil and Commercial, of the West Indies, with a Continuation to the Present Time*, 3 vols. (London, 1793), 2:65–66.

55. Letter from Francis Treble, Kingston, to Caleb Dickinson, Bristol, 12 June 1760, Dickinson Mss., Somerset Heritage Centre, DDDN 4/1/25/59.

56. Bryan Edwards, *Poems Written Chiefly in the West Indies* (Kingston: Alexander Aikman, 1792), 67–68. Gordon K. Lewis's summary of Edwards's political philosophy is apt. He notes how Edwards reverted to a Hobbesian outlook on dealing with slave resistance when pushed. Edwards, Lewis argues, is "generally regarded as the moderate voice of West Indian planting." He was determined to show his humanitarian credentials, but these credentials faded during the crisis (for planters) of slave revolt in Saint-Domingue, where Edwards came to insist on "might is right" Hobbesian arguments. Gordon K. Lewis, *Main Currents in Caribbean Thought: The Historical Evolution of Caribbean Society in Its Ideological Aspects* (Baltimore: Johns Hopkins University Press, 1983), 113–16.

57. Bryan Edwards, *The History, Civil and Commercial, of the West Indies, with a Continuation to the Present Time* 5 vols. (London:, 1819), 3:9. For the color line and how race was tied to the exercise of individual power by whites, see Trevor Burnard and John Garrigus, *The Plantation Machine: Atlantic Capitalism in French Saint-Domingue and British Jamaica, 1748–1788* (Philadelphia: University of Pennsylvania Press, 2016); and John Garrigus, " 'Affranchis' and 'Coloreds': Why Were Racial Codes Stricter in Eighteenth-Century Saint-Domingue Than in Jamaica?" *Quaderni Storici* 148 (2015): 69–86.

58. Richard Follett, *The Sugar Masters: Planters and Slaves in Louisiana's Cane World, 1820–1860* (Baton Rouge: Louisiana State University Press, 2005); and Michael Wayne, *Death of an Overseer: Reopening a Murder Investigation from the Plantation South* (New York: Oxford University Press, 2001).

59. Edwards, *History, Civil and Commercial*, 3:7–11.

60. Ibid., 3:7.

61. Cited in Joan Dayan, *Haiti, History, and the Gods* (Berkeley: University of California Press, 1995), 213.

62. Christa Dierksheide, *Amelioration and Empire: Progress and Slavery in the Plantation Americas* (Charlottesville: University of Virginia Press, 2014).

63. Edwards, *History, Civil and Commercial*, 3:9–14.

64. For Machiavelli in British political thought, see Vicki B. Sullivan, *Machiavelli, Hobbes and the Formation of a Liberal Republicanism in England* (Cambridge: Cambridge University Press, 2004); and Sylvana Tomaselli, "The Spirit of Nations," in *The Cambridge History of Eighteenth-Century Political Thought*, ed. Mark Goldie and Robert Wokler (Cambridge: Cambridge University Press, 2006), 7–39.

65. Arno J. Mayer, *The Furies: Violence and Terror in the French and Russian Revolutions* (Princeton, N.J.: Princeton University Press, 2000), 99–101; and J. G. A. Pocock, *The Machiavellian Moment: Florentine Political Thought and the Atlantic Republican Tradition* (Princeton, N.J.: Princeton University Press, 1975).

66. Robert Travers, "Contested Despotism: Problems of Liberty in British India," in *Exclusionary Empire: English Liberty Overseas, 1600–1900*, ed. Jack P. Greene (New York: Cambridge University Press, 2010), 195–6

67. Quotations are taken from Thomas Nugent, trans., *The Spirit of the Laws*, 6th ed.(London: F. Wingrave, 1793), 112 A succinct summary of Montesquieu's political thought is Hilary Bok, "Baron de Montesquieu, Charles-Louis de Secondat," in *Stanford Encyclopedia of Philosophy* (Summer 2014 edition), http://plato.stanford.edu/archives/sum2014/entries/montesquieu. See also Paul Rahe, *Montesquieu and the Logic of Liberty* (New Haven, Conn.: Yale University Press, 2009); and Michael Curtis, *Orientalism and Islam: European Thinkers on Oriental Despotism and the Middle East and India* (Cambridge: Cambridge University Press, 2009), ch. 4.

68. Michel Foucault, *Discipline and Punish: The Birth of the Prison*, trans. Alan Sheridan, 2nd ed. (New York: Vintage, 1995), 3–31. The execution of Damiens evoked widespread controversy and may have accelerated the end of the ancien régime. Dale Van Kley, *The Damiens Affair and the Unraveling of the Old Regime* (Princeton, N.J.: Princeton University Press, 1984). For Montesquieu and moderate punishment, see David Carrithers, "Montesquieu's Philosophy of Punishment," *History of Political Thought* 19 (1978): 213–40.

69. Long, *History of Jamaica*, II: 442–44, 447, 460–75; Edwards, *Poems*, 67–68. For "spectacular terror," see Vincent Brown, "Spiritual Terror and Sacred Authority in Jamaican Slave Society," *Slavery and Abolition* 24 (2003): 27.

70. Trevor Burnard, "White West Indian Identity in the Eighteenth Century," in *Assumed Identities: Race and the National Imagination in the Atlantic World*, ed. John D. Garrigus and Christopher Morris (College Station: Texas A&M Press, 2010), 71–87.

71. Brown, "Spiritual Terror and Sacred Authority in Jamaican Slave Society," 27–29; Sarah Yeh, "Colonial Identity and Revolutionary Loyalty: The Case of the West Indies," in *The Oxford History of the British Empire: British North America in the Seventeenth and Eighteenth Centuries*, ed. Stephen Foster (Oxford: Oxford University Press, 2013), 203–5 (quote 204). See also Sarah Yeh, " 'A Sink of All Filthiness': Gender, Family, and Identity in the British Atlantic, 1688–1763," *Historian* 68 (2006): 66–88.

72. Edwards, *History, Civil and Commercial*, 2:169–70.

73. David Scott, "The Paradox of Freedom: An Interview with Orlando Patterson," *Small Axe* 17 (2013): 160.

74. Burgess, *British Political Thought*, 304, 314.

75. For hurricanes, see Michael Mulcahy, *Hurricanes and Society in the British Greater Caribbean, 1624–1783* (Baltimore: Johns Hopkins University Press, 2006); and Stuart B. Schwartz, *Sea of Storms:*

A History of Hurricanes in the Greater Caribbean from Columbus to Katrina (New Haven, Conn.: Yale University Press, 2015). For fears of foreign invasion, see Burnard and Garrigus, *Plantation Machine,* 206–7. For Maroons, see Michael Craton, *Testing the Chains: Resistance to Slavery in the British West Indies* (Ithaca, N.Y.: Cornell University Press, 1982), 81–98; and Kathleen Wilson, "The Performance of Freedom: Maroons and the Colonial Order in Eighteenth-Century Jamaica and the Atlantic Sound," *William and Mary Quarterly,* 3rd ser., 66 (2009): 45–86.

76. *Journals of the Assembly of Jamaica,* October 2, 7–8, 1762, 5:346–53.

77. Jack P. Greene, "The Jamaica Privilege Controversy, 1764–1766: An Episode in the Process of Constitutional Definition in the Early Modern British Empire," *Journal of Imperial and Commonwealth History* 22 (1994): 16–54.

78. Burnard and Garrigus, *Plantation Machine,* 257.

79. Madeleine Dobie, *Trading Places: Colonization and Slavery in Eighteenth-Century French Culture* (Ithaca, N.Y.: Cornell University Press, 2010), 256.

80. Jean-François de Saint-Lambert, "Ziméo," in, *Les Saisons, poème* (Amsterdam, 1769), 226–59.

81. Cited in Gabriel Debien, *Les esclaves aux Antilles françaises, XVIIe–XVIIIe siècles* (Basse-Terre: Société d'histoire de la Guadeloupe, 1974), 486.

82. Malick W. Ghachem, *The Old Regime and the Haitian Revolution* (New York: Cambridge University Press, 2012), 173, 196–202.

83. Cited in Burnard, *Mastery, Tyranny, and Desire,* 137.

84. Cited in Evelyne Camara, Isabelle Dion, and Jacques Dion, *Esclaves, Regards de blancs, 1672–1913, Collection Archives Nationales d'outre-mer* (Marseille: Images en Manoeuvres Editions, 2008), 134.

85. Lewis, *Main Currents in Caribbean Thought,* 166.

86. *The Humble Petition and Memorial of the Assembly of Jamaica to the King's Most Excellent Majesty in Council* (Philadelphia, 1774).

87. Berlin, *Generations of Captivity,* 97–158; and Egerton, *Death or Liberty.*

88. Johnson's quip is in [James Boswell], *Boswell's Life of Johnson,* ed. G. B. Hill, rev. L. F. Powell, 6 vols. (Oxford: Clarendon, 1934–64), 3:200. See also G. Basker, "'The Next Insurrection': Johnson, Race, and Rebellion," *Age of Johnson* 11 (2000): 39, 45, 47.

89. Trevor Burnard, "Powerless Masters: The Curious Decline of Jamaican Sugar Planters in the Foundational Period of British Abolition," *Slavery and Abolition* 32 (2011): 185–98.

90. This posture of abject helplessness and weakness was central to West Indian claims for relief following the devastating hurricanes of 1780. See Schwartz, *Sea of Storms*; and Mulcahy, *Hurricanes and Society in the British Greater Caribbean,* 164–88.

91. Stephen Conway, "From Fellow-Nationals to Foreigners: British Perceptions of the Americans Circa 1739–1783," *William and Mary Quarterly,* 3rd ser., 59 (2002): 65–100; Trevor Burnard, "Freedom, Migration and the Negative Example of the American Revolution: The Changing Status of Unfree Labor in the Second British Empire and the New American Republic," in *Empire and Nation: The American Revolution in the Atlantic World,* ed. Eliga H. Gould and Peter S. Onuf (Baltimore: Johns Hopkins University Press, 2004), 295–314.

92. Christopher Leslie Brown, *Moral Capital: The Foundations of British Abolitionism* (Chapel Hill: University of North Carolina Press, 2006); Seymour Drescher, "The Shocking Birth of British Abolitionism," *Slavery and Abolition* 33 (2012): 572–89.

93. For a sophisticated argument about West Indian and British North American identity, see Yeh, "Colonial Identity and Revolutionary Loyalty," 223–25. For Britain as "home," see Christer Petley, "'Home' and 'This Country': Britishness and Creole Identity in the Letters of a Transatlantic Slaveholder," *Atlantic Studies* 6 (2009): 43–61. For a similar argument about provincial identity in the eighteenth century, see Trevor Burnard, *Creole Gentlemen: The Maryland Elite, 1691–1776* (New York: Routledge, 2002).

94. Nathaniel Phillips to Hibbert, Fuhr, and Hibbert, 25 May 1788 and Robert Hibbert to George Hibbert, 14 May 1788, Stowe Papers, STB, 27 (33), Huntington Library.

95. Claudius Fergus, " 'Dread of Insurrection': Abolitionism, Security, and Labor in Britain's West Indian Colonies," *William and Mary Quarterly*, 3rd ser., 54 (2009): 757–80; and Matthews, *Caribbean Slave Revolts and the British Abolitionist Movement*.

96. For enslaved people as martyrs in the eighteenth century, see Boulukos, *Grateful Slave*, 233–45. For the nineteenth century, see Sarah N. Roth, *Gender and Race in Antebellum Popular Culture* (New York: Cambridge University Press, 2014), chs. 4 and 5.

97. Timothy Tackett, *The Coming of the Terror in the French Revolution* (Cambridge, Mass.: Harvard University Press, 2015); Richard Bourke, *Empire and Revolution: The Political Life of Edmund Burke* (Princeton, N.J.: Princeton University Press, 2015); and Tomaselli, "Spirit of Nations." In regard to the many books that insist that fear was a primary emotion shaping political change in the age of democratic revolutions, the cautionary words of David Bell in a review of Timothy Tackett and Adam Zamoyski, *Phantom Terror: Political Paranoia and the Creation of the Modern State, 1789–1848* (New York: Basic Books, 2015), are worth considering: we should be careful about taking the promoters of the politics of fear (then and now) too much at their word. The revolutionaries and counterrevolutionaries in the French Revolution and in the counterrevolutionary years of the early nineteenth century exaggerated the extent of conspiracies that they believed surrounded them. As Bell states, "Revolutionary leaders often deliberately exaggerated the dangers in the service of their own ambitions." David A. Bell, "Terror at the Dawn of Modern Europe," *Atlantic*, May 2015. https://www.theatlantic.com/magazine/archive/2015/05/terror-at-the-dawn-of-modern-europe/389552/

Chapter 2

1. Edward Long, *History of Jamaica . . .* , 3 vols. (London: T. Lowndes, 1774), 3:941.

2. Ibid., 3:941, 949.

3. James McClellan III and François Regoud, *The Colonial Machine: French Science and Overseas Expansion in the Old Regime* (Brepols, Belgium: Turnhout, 2011).

4. Long, *History of Jamaica*, 2:165–66.

5. Ibid.

6. Ibid., 1:402, 434.

7. For eighteenth-century Jamaica, see Trevor Burnard, *Planters, Merchants, and Slaves: Plantation Societies in British America 1650–1820* (Chicago: University of Chicago Press, 2015), ch. 4.

8. Long, *History of Jamaica*, 3:941–49.

9. Jack P. Greene, " 'Of Liberty and the Colonies': A Case Study of Constitutional Conflict in the Mid-Eighteenth-Century British American Empire," in *Creating the British Atlantic: Essays on Transplantation, Adaptation, and Continuity* (Charlottesville: University of Virginia Press, 2013), 140–207.

10. Long, *History of Jamaica*, 2:65–66.

11. Ibid., 3:372.

12. Burnard, *Planters, Merchants, and Slaves*, 110–20.

13. Gordon K. Lewis, *Main Currents in Caribbean Thought: The Historical Evolution of Caribbean Society in Its Ideological Aspects* (Baltimore: Johns Hopkins University Press, 1983), 109–14.

14. R. M. Howard, *Records and Letters of the Family of the Longs of Longville, Jamaica, and Hampton Lodge, Surrey*, 2 vols. (London,1925); Kenneth Morgan, "Long, Edward (1734–1813)," *Oxford Dictionary of National Biography*, Oxford University Press, 2004; online ed., May 2014, http://www.oxforddnb.com/view/article/16964, accessed 17 January 2016.

15. Stephen Saunders Webb, *The Governors-General: The English Army and the Definition of Empire, 1569–1681* (Chapel Hill: University of North Carolina Press, 1979), 230–31.

16. Elizabeth A. Bohls, "The Gentleman Planter and the Metropole: Long's History of Jamaica," in *The Country and the City Revisited: England and the Politics of Culture, 1550–1850*, ed. Gerald Maclean et al. (Cambridge: Cambridge University Press, 1999), 180–96; and Suman Seth, "Materialism, Slavery, and the *History of Jamaica*," *Isis* 105 (2014): 764–72.

17. Edward Long, *The Trial of Farmer Carter's Dog Porter for Murder . . . A Satire on the Game Laws* (London: T. Lowndes, 1757); Planter [Edward Long], *Candid Reflections upon the Judgement Lately Awarded by the Court of King's Bench in Westminster-Hall, on What Is Commonly Called the Negroe-Cause*

(London: T. Lowndes, 1772); Edward Long, *English Humanity No Paradox; or, An Attempt to Prove That the English Are Not a Nation of Savages* (London: T. Lowndes, 1778); Edward Long, *A Free and Candid Review, of a Tract Entitled "Observations on the Commerce of the American States'; Shewing the Pernicious Consequences Both to Great Britain and to the British Sugar Islands, of the Systems Recommended in That Tract* (London: T. Lowndes, 1784).

18. Elsa Goveia, *A Study on the Historiography of the British West Indies to the End of the Nineteenth Century* (Washington, D.C.: Howard University Press, 1980 [originally published in Mexico City in 1956]), 55–6.

19. Edward Braithwaite, *The Development of Creole Society in Jamaica, 1770–1820* (Oxford: Oxford University Press, 1971); and Michael Craton, "Reluctant Creoles: The Planter's World in the British West Indies," in *Strangers Within the Realm: Cultural Margins of the First British Empire*, ed. Bernard Bailyn and Philip D. Morgan (Chapel Hill: University of North Carolina Press, 1991), 314–62.

20. Long, *History of Jamaica*, 3:941.

21. Compare this work to earlier histories of Jamaica, notably James Knight's unpublished manuscript on the history of Jamaica. James Knight, "The Natural, Moral, and Political History of Jamaica and the Territories Thereon Depending," Long Family Papers, Add. Mss. 12, 418–19, British Library. Long had read this work and Charles Leslie, *A New and Exact Account of Jamaica* (Edinburgh: R. Fleming, ca. 1740). Long's work, however, is much more substantial than the works written by Knight or Leslie. The only work that compares with it in eighteenth-century British West Indian historiography is Bryan Edwards, *The History, Civil and Commercial, of the West Indies, with a Continuation to the Present Time* 5 vols. (London, 1819).

22. Long, *History of Jamaica*, I: 1–2, 6. Goveia, *Study on the Historiography of the British West Indies*, 55.

23. Christopher Leslie Brown, " 'Empire Without Slaves': British Concepts of Emancipation in the Age of the American Revolution," *William and Mary Quarterly*, 3rd ser., 56 (1999): 273–306; H. V. Bowen, "British Conceptions of Global Empire, 1756–1783," *Journal of Imperial and Commonwealth History* 26 (1998): 1–27; Richard Bourke, *Empire and Revolution: The Political Life of Edmund Burke* (Princeton, N.J.: Princeton University Press, 2015); Jennifer Pitts, *A Turn to Empire: The Rise of Liberal Imperialism in France and Britain* (Princeton, N.J.: Princeton University Press, 2005), pt. 1; and Patrick Griffin, *The Townshend Moment: The Making of Empire and Revolution in the Eighteenth Century* (New Haven, Conn.: Yale University Press, 2018).

24. For how empire became a growing topic of concern among a population ever more attuned to imperial issues, see Philip Lawson, " 'Arts and Empire Equally Extend': Tradition, Prejudice and Assumption in the Eighteenth Century Press Coverage of Empire," *Studies in History and Politics* 7 (1989): 119; Kathleen Wilson, *The Sense of the People: Politics, Culture, and Imperialism in England, 1715–1785* (Cambridge: Cambridge University Press, 1995); and Nicolas B. Dirks, *The Scandal of Empire: India and the Creation of Imperial Britain* (Cambridge, Mass.: Harvard University Press, 2009).

25. The best guide to this process is P. J. Marshall, *The Making and Unmaking of Empires: Britain, India and America c. 1760–1783* (Oxford: Oxford University Press, 2005). See also Eliga H. Gould, *The Persistence of Empire: British Political Culture in the Age of the American Revolution* (Chapel Hill: University of North Carolina Press, 2000).

26. Adam Anderson, *An Historical and Chronological Deduction of the Origin of Commerce . . .* , 4 vols. (London, 1787), 1:xliv.

27. Jack P. Greene, *Evaluating Empire and Confronting Colonialism in Eighteenth-Century Britain* (New York: Cambridge University Press, 2013), 84.

28. Thomas Pownall, *The Administration of the Colonies*, 4th ed. (London, 1768), 9–10.

29. Emma Rothschild, "Adam Smith in the British Empire," in *Empire and Modern Political Thought*, ed. Sankar Muthu (Cambridge: Cambridge University Press, 2012), 184–98.

30. A useful recent guide to this enormous literature is Jack P. Greene, *The Constitutional Origins of the American Revolution* (New York: Cambridge University Press, 2011), 67–186.

31. Arthur Young, *Political Essays Concerning the Present State of the British Empire* (London, 1772), 552; William Knox, *The Present State of the Nation, Particularly with Respect to Its Trade* (London, 1768), 39.

32. Long, *History of Jamaica*, 1:417, 491–509.

33. Ibid., 2:595–96.

34. Nick Bunker, *An Empire on the Edge: How Britain Came to Fight America* (New York: Vintage, 2014); and Stephen Conway, *The American Revolutionary War* (London: I. B. Tauris, 2013), ch. 1.

35. Jack P. Greene, *Evaluating Empire and Confronting Colonialism in Eighteenth-Century Britain* (New York: Cambridge University Press, 2013), chs. 4–5 (the quotes are the titles of each chapter).

36. Christopher Leslie Brown, *Moral Capital: Foundations of British Abolitionism* (Chapel Hill: University of North Carolina Press, 2006), 155–206; and Srividya Swaminathan, *Debating the Slave Trade: Rhetoric of British National Identity* (Farnham, Surrey: Ashgate, 2009), 46–170.

37. [William Burke], *An Account of the European Settlements in America*, 2 vols. (London, 1757), 2:124.

38. Trevor Burnard and John Garrigus, *The Plantation Machine: Atlantic Capitalism in French Saint-Domingue and British Jamaica, 1748–1788* (Philadelphia: University of Pennsylvania Press, 2016), 146, 194–95.

39. Adam Smith, *Theory of Moral Sentiments* (London, 1759), 402–3. See Trevor Burnard, "Slavery and the Enlightenment in Jamaica, 1760–1772: The Afterlife of Tackey's Rebellion," in *Enlightened Colonialism: Imperial Agents, Narratives of Progress and Civilizing Policies in the Eighteenth Century*, ed. Damien Tricoire (Basingstoke: Palgrave Macmillan, 2018), 227–46.

40. Granville Sharp, *A Representation of the Injustice and Dangerous Tendency of Tolerating Slavery; or of Admitting the Least Claim of Private Property in the Persons of Men, in England* (London, 1769), 47, 63, 149.

41. For the larger context, see Jack P. Greene, "Liberty, Slavery, and the Transformation of British Identity in the Eighteenth-Century West Indies," *Slavery and Abolition* 21 (2000): 1–31.

42. Long, *History of Jamaica*, 2:476.

43. Ibid., 2:483.

44. For a modern appreciation of Williams, see John Gilmore, "The British Empire and the Neo-Latin Tradition: The Case of Francis Williams," in *Classics and Colonialism*, ed. Barbara Goff (London: Duckworth, 2005), 92–106.

45. Anonymous, *Full and Free Inquiry into the Merits of the Peace . . .* (London: T. Payne, 1765), 1.

46. Long, *History of Jamaica*, 1:497, 515.

47. Ibid., 1:518.

48. Ibid., 1:96, 303.

49. Ibid., 1:336.

50. Roxann Wheeler, *The Complexion of Race: Categories of Difference in Eighteenth-Century British Culture* (Philadelphia: University of Pennsylvania Press, 2000), 260–87; Seymour Drescher, *From Slavery to Freedom: Comparative Studies in the Rise and Fall of Atlantic Slavery* (Basingstoke: Macmillan, 1999), 282–83, 285.

51. Important treatments include Winthrop D. Jordan, *White over Black: American Attitudes Towards the Negro, 1550–1812* (Chapel Hill: University of North Carolina Press, 1968); David Brion Davis, *The Problem of Slavery in the Age of Revolution, 1770–1823* (Ithaca, N.Y.: Cornell University Press, 1975); Wheeler, *Complexion of Race*. A useful brief guide is Andy Wells, "Race and Racism in the Global European World Before 1800," *History Compass* 13 (2015): 435–44.

52. Seymour Drescher, *From Slavery to Freedom: Comparative Studies in the Rise and Fall of Atlantic Slavery* (Basingstoke: Macmillan, 1999). See also Colin Kidd, *The Forging of Races: Race and Scripture in the Protestant Atlantic World, 1600–2000* (Cambridge: Cambridge University Press, 2006), 79–120.

53. Long, *History of Jamaica*, 1:433–34.

54. Ibid., 1:42.

55. Ibid, 1: 24. In fact, making sure that colonists stayed loyal to their imperial homelands was more of a problem for the French than for the British. Burnard and Garrigus, *Plantation Machine*, ch. 7.

56. Long, *History of Jamaica*, 1:23–24, 47. For government and law in Saint-Domingue, see Malick W. Ghachem, *The Old Regime and the Haitian Revolution* (New York: Cambridge University Press, 2012).

57. Long, *History of Jamaica*, 1:96.

58. Ibid., 1:40, 380; 2:595.

59. Ibid., 2:262, 280.

60. Ibid., 2:283. Trevor Burnard, *Mastery, Tyranny, and Desire: Thomas Thistlewood and His Slaves in the Anglo-Jamaica World* (Chapel Hill: University of North Carolina Press, 2004).

61. Long, *History of Jamaica*, 1:404.

62. Ibid., 1:420–26.

63. Burnard and Garrigus, *Plantation Machine*, 52.

64. Long, *History of Jamaica*, 1:404, 417; 2:283.

65. Ibid., 1:433.

66. Ibid., 1:413–14.

67. Ibid.

68. Ibid., 1:433. A crimp was a person or firm that engaged in illegal kidnapping through trickery or intimidation of people to serve in the merchant marine or the Royal Navy.

69. Kenneth Morgan, "Slave Women and Reproduction in Jamaica, c. 1776–1834," *History* 91 (2006): 231–53; B. W. Higman, *Plantation Jamaica 1750–1850: Capital and Control in a Colonial Economy* (Kingston: University of West Indies Press, 2005), 166–226.

70. Trevor Burnard, " 'Rioting in Goatish Embraces: Marriage and Improvement in Early British Jamaica, 1660–1780," *History of the Family* 11 (2006): 185–97.

71. Long, *History of Jamaica*, 2:246.

72. Ibid., 2:246–48.

73. Ibid., 2:246–47, 255.

74. Ibid., 2:411–13, 417–20.

75. Ibid., 2:253–59.

76. Ibid.

77. Ibid., 2:333.

78. Ibid., 1:133–36, 150, 2:317, 333–37.

79. David P. Geggus, "The French Slave Trade: An Overview," *William and Mary Quarterly*, 3rd ser., 58 (2001): 119–38.

80. Long, *History of Jamaica*, 1:403, 433, 518–20.

81. Ibid., 1:437, 555. For the ubiquity of the term "improvement" in British political discourse, see Paul Slack, *The Invention of Improvement: Information and Material Progress in Seventeenth-Century England* (Oxford: Oxford University Press, 2015).

82. Long, *History of Jamaica*, 2:440–41.

83. Ibid., 3:921.

84. Anonymous, *An Essay Concerning Slavery* . . . (London: Charles Corbett, 1746), 67, reprinted in *Exploring the Bounds of Liberty: Political Writings of Colonial British America from the Glorious Revolution to the American Revolution* ed. Jack P. Greene and Craig Yirush (Carmel, Ind.: Liberty Fund, 2018), 3 vols. This tract was probably written by Governor Edward Trelawney. It took a similar view to Long's about the need for an activist government. For the authorship of this tract by Trelawney, see Samuel Halkett and John Lang, *Dictionary of Anonymous and Pseudonymous English Literature*, new ed., 7 vols. (Edinburgh: Oliver and Boyd, 1926–34), 2: 188 and James Robertson, "*An Essay Concerning Slavery*: A Mid-Eighteenth Century Analysis from Jamaica," *Slavery and Abolition* 33 (2012): 68. . For activist government in early British America, see Steven Pincus, *The Heart of the Declaration: The Founders' Case for an Activist Government* (New Haven, Conn.: Yale University Press, 2016).

85. Long, *History of Jamaica*, 1:391, 404, 421, 440, 482; 2:28, 40, 68–69, 135, 165–66, 210, 258, 265; 3:912–13.

86. Ibid., 1:7, 96.

87. Ibid., 1:24, 114, 303, 404.

88. There is no indication that anyone subjected Long's ideas to serious scrutiny. We lack any study of the reception of Long in either Britain or Jamaica. Few newspapers from the period survive, and papers from people such as Thomas Fearon, a progressive chief justice much admired by Long, are similarly missing.

89. Long, *History of Jamaica*, 1:384–85.

90. Trevor Burnard, "'Passengers Only': The Extent and Significance of Absenteeism in Eighteenth-Century Jamaica," *Atlantic Studies* 1 (2004): 178–95; Trevor Burnard, "'The Countrie Continues Sicklie': White Mortality in Jamaica, 1655–1780," *Social History of Medicine* 12 (1999): 45–72.

91. Long, *History of Jamaica*, 2:269.

92. Burnard, *Planters, Merchants, and Slaves*, ch. 2.

93. David Eltis, Frank D. Lewis, and David Richardson, "Slave Prices, the African Slave Trade, and Productivity in the Caribbean, 1674–1807," *Economic History Review* 2nd ser. 4 (2005): 673–700.

94. Long, *History of Jamaica*, 2:410, 444.

95. The start of abolitionism led to more sustained attention to slave demography, aided by enslaved women's determination to create their own families. Sasha Turner, *Contested Bodies: Pregnancy, Childrearing and Slavery in Jamaica* (Philadelphia: University of Pennsylvania Press, 2017).

96. Burnard and Garrigus, *Plantation Machine*.

97. Long, *History of Jamaica*, 2:505. For transformations in medicine in the West Indies, see Mark Harrison, "'The Tender Frame of Man': Disease, Climate, and Racial Difference in India and the West Indies, 1760–1860," *Bulletin of the History of Medicine* 70 (1996): 68–93; Trevor Burnard and Richard Follett, "Caribbean Slavery, British Anti-Slavery and the Cultural Politics of Venereal Disease," *Historical Journal* 55 (2012): 427–51; and Katherine Paugh, "Yaws, Syphilis, Sexuality, and the Circulation of Medical Knowledge in the British Caribbean and the Atlantic World," *Bulletin of the History of Medicine* 88 (2014): 225–52.

98. Long, *History of Jamaica*, 2:542–43.

99. Ibid., 2:543. For "chaos of men, negroes and things," see [J. Hector St. John de Crèvecoeur], "Sketches of Jamaica and Bermudas and Other Subjects," ed. Dennis D. Moore (Athens: University of Georgia Press, 1995), 106–13. See also Christopher Iannini, "An Itinerant Man: Crèvecouer's Caribbean, Raynal's Revolution, and the Fate of Atlantic Cosmopolitanism," *William and Mary Quarterly*, 3rd ser., 61 (2004): 201–34.

100. Long, *History of Jamaica*, 2:543–44.

101. Michel-René Hilliard d'Auberteuil, *Considérations sur l'état présent de la colonie française de Saint-Domingue: Ouvrage politique et législatif présenté au ministre de la marine*, vol. 1 (Paris: Grangé, 1776), 149.

102. Cited in Burnard and Garrigus, *Plantation Machine*, 26. See also William Max Nelson, "Making Men: Enlightenment Ideas of Racial Engineering," *American Historical Review* 115 (2010), 1364–94.

103. Brown, "'Empire Without Slaves'"; Bowen, "British Conceptions of Global Empire"; Daniel K. Richter, "Native Americans, the Plan of 1764, and a British Empire That Never Was," in *Cultures and Identities in Colonial British America*, ed. Robert Olwell and Alan Tully (Baltimore: Johns Hopkins University Press, 2006), 269–92.

104. Hannah Weiss Muller, "Bonds of Belonging: Subjecthood and the British Empire," *Journal of British Studies* 53 (2014): 29–58; Sudipta Sen, "Imperial Subjects on Trial: On the Legal Identity of Britons in Late Eighteenth-Century India," *Journal of British Studies* 45 (2006): 532–55; Aaron Willis, "The Standing of New Subjects: Grenada and the Protestant Constitution After the Treaty of Paris (1763)," *Journal of Imperial and Commonwealth History* 42 (2014): 1–21; and Caitlin Anderson, "Old Subjects, New Subjects, and Non-Subjects: Silences and Subjecthood," in *War, Empire, and Slavery, 1770–1830*, ed. Jane Rendall et al. (Houndsmill: Palgrave, 2010), 201–17.

105. [Benjamin Franklin], *The Causes of the Present Distractions in America Explained in Two Letters to a Merchant in London* (New York, 1774), 2.

106. For the empire before 1763 as heavily contested, see Justin du Rivage, *Revolution Against Empire: Taxes, Politics, and the Origins of American Independence* (New Haven, Conn.: Yale University Press, 2017).

107. John Darwin, *Unfinished Empire: The Global Expansion of Empire* (London: Allen Lane, 2012), 161.

108. Andrew O'Shaughnessy, *An Empire Divided: The American Revolution and the British Caribbean* (Philadelphia: University of Pennsylvania Press, 2000), 44–45, 60–62, 85–86, 235–37; Louis P. Nelson, *Architecture and Empire in Jamaica* (New Haven, Conn.: Yale University Press, 2016).

109. Richard D. Brown, *Revolutionary Politics in Massachusetts: The Boston Committee of Correspondence and the Towns, 1772–1774* (Cambridge, Mass.: Harvard University Press, 1970).

110. Burnard and Garrigus, *Plantation Machine*, 89, 161, 165, 207.

111. Michael Duffy, *Soldiers, Sugar and Seapower: The British Expeditions to the West Indies and the War with Revolutionary France* (Oxford: Oxford University Press, 1987); P. J. Cain and A. G. Hopkins, "Gentlemanly Capitalism and British Expansion Overseas I: The Old Colonial System, 1688–1850," *Economic History Review* 39 (1986): 501–25.

112. For how the Second Hundred Years' War between France and Britain made empire a focus of nationalism, see David A. Bell, *The Cult of the Nation in France: Inventing Nationalism, 1680–1800* (Cambridge, Mass.: Harvard University Press, 2003); Linda Colley, *Britons: Forging the Nation 1707–1837* (New Haven, Conn.: Yale University Press, 1992); David Armitage, *The Ideological Origins of the British Empire* (Cambridge: Cambridge University Press, 2000); Gould, *Persistence of Empire*; P. J. Marshall, *"A Free Though Conquering People": Eighteenth-Century Britain and Its Empire* (Aldershot: Ashgate, 2003).

113. For the French Atlantic colonial system, see Jean Tarrade, *Le commerce colonial de la France à la fin de l'ancien régime: L'évolution du régime de l'exclusif de 1763 à 1789*, 2 vols. (Paris: Presses Universitaires de France, 1972); Gilles Havard and Cécile Vidal, *Histoire de l'Amérique Française* (Paris: Flammarion, 2003); Silvia Marzagalli, "The French Atlantic World in the Seventeenth and Eighteenth Centuries," in *The Oxford Handbook of the Atlantic World, 1450–1850*, ed. Philip D. Morgan and Nicholas Canny (Oxford: Oxford University Press, 2011), 235–51; Paul Cheney, *Revolutionary Commerce: Globalization and the French Monarchy* (Cambridge: Mass.: Harvard University Press, 2010); Cécile Vidal and François-Joseph Ruggiu, eds., *Sociétés, colonisation et esclavages dans le monde Atlantique: Historigraphie des sociétés américaines des XVIe–XIXe siècles* (Paris: Perseides, 2015); and Christopher Hodson and Brett Rushforth, *Discovering Empire: France and the Atlantic World from the Age of the Crusades to the Rise of Napoleon* (New York: Oxford University Press, forthcoming).

114. J. R. Ward, *British West Indian Slavery, 1750–1834: The Process of Amelioration* (New York: Oxford University Press, 1988).

Chapter 3

1. Elsa Goveia, *A Study on the Historiography of the British West Indies to the End of the Nineteenth Century* (Washington, D.C.: Howard University Press, 1980 [originally published in Mexico City in 1956]), 53–62 (quote 55). See also Howard Johnson, "Introduction: Edward Long, Historian of Jamaica," in Edward Long, *The History of Jamaica . . .* (Kingston: Ian Randle, 2002), 1:i–xxv.

2. Edward Long, *History of Jamaica . . .*, 3 vols. (London: T. Lowndes, 1774), 2:262.

3. Ibid., 2:267, 269, 407, 502. The "sinews" metaphor is revealing. It suggests seeing enslaved people in biological rather than in human or even in animal terms—organisms that shrink, grow, and reproduce.

4. Ibid., 2:267, 441.

5. Ibid., 2:270–71.

6. Ibid., 2:399–400.

7. Sheila Lambert, ed., *House of Commons Sessional Papers of the Eighteenth Century* (Wilmington, Del.: Scholarly Resources, 1975), 69:280–84 (quote 282).

8. Long, *History of Jamaica*, 2:400.

9. Ibid.

10. Lambert, *House of Commons Sessional Papers*, 69:280–84 (quote 282).

11. Long, *History of Jamaica*, 2:441.

12. Ibid., 2:503.

13. Charles Leslie, *A New and Exact Account of Jamaica* (Edinburgh: R. Fleming, 41; [Edward Trelawney], *An Essay Concerning Slavery and the Danger Jamaica Is Expos'd to from the Too Great Number of Slaves* (London: C. Corbett, 1746) reprinted in *Exploring the Bounds of Liberty: Political Writings of Colonial British America from the Glorious Revolution to the American Revolution* ed. Jack P. Greene and Craig Yirush (Carmel, Ind.: Liberty Fund, 2018), 3 vols., II:1134–64.

14. Michael Craton, "Hobbesian or Panglossian? The Two Extremes of Slave Conditions in the British Caribbean, 1783–1834," *William and Mary Quarterly*, 3rd ser., 35 (1978): 324–56; Richard S. Dunn, "A Tale of Two Plantations: Slave Life in Mesopotamia in Jamaica and Mount Airy in Virginia," *William and Mary Quarterly*, 3rd ser., 34 (1977): 64. See also Trevor Burnard and John Garrigus, *The Plantation Machine: Atlantic Capitalism in French Saint-Domingue and British Jamaica, 1748–1788* (Philadelphia: University of Pennsylvania Press, 2016). The scholarly consensus that Jamaican slavery was especially brutal in the period before the American Revolution is very strong. See Michael Craton, *Searching for the Invisible Man: Slaves and Plantation Life in Jamaica* (Cambridge, Mass.: Harvard University Press, 1978); B. W. Higman, *Plantation Jamaica 1750–1850: Capital and Control in a Colonial Economy* (Kingston: University of West Indies Press, 2005); Richard S. Dunn, *A Tale of Two Plantations: Slave Life and Labor in Jamaica and Virginia* (Cambridge, Mass.: Harvard University Press, 2014); Justin Roberts, *Slavery and the Enlightenment in the British Atlantic, 1750–1807* (New York: Cambridge University Press, 2013); Richard B. Sheridan, *Doctors and Slaves: A Medical and Demographic History of Slavery in the British West Indies, 1680–1834* (Cambridge: Cambridge University Press, 1985); J. R. Ward, *British West Indian Slavery, 1750–1834: The Process of Amelioration* (New York: Oxford University Press, 1988); and David Beck Ryden, *West Indian Slavery and British Abolition, 1783–1807* (New York: Cambridge University Press, 2009).

15. Srividhya Swaminathan, "Developing the West Indian Proslavery Position After the *Somerset* Decision," *Slavery and Abolition* 24 (2003): 40–60.

16. Michael Craton, *Empire, Enslavement and Freedom in the Caribbean* (Kingston: Ian Randle, 1997), 164.

17. Trevor Burnard, Laura Panza, and Jeffrey Williamson, "Living Costs, Real Incomes and Inequality in Colonial Jamaica," *Explorations in Economic History* 71 (2019): 55–71.

18. Long, *History of Jamaica*, 2:269.

19. T. C. E. Cheng and S. Podolsky, *Just-in-Time Manufacturing—an Introduction* (London: Chapman and Hall, 1993). The best description of slave management practices in the late eighteenth-century British West Indies is Roberts, *Slavery and the Enlightenment*.

20. For increasing slave prices, see Trevor Burnard, "A Serious Business: Slave Prices in Jamaica, 1674–1784," in *La société de plantation esclavagiste: Caribe Anglophone, francophone, hispanophone (18e–19e s.)*, ed. Jacques de Cauna and Cécile Révauger (Paris: Indes Savantes, 2013),59–80; and David Eltis and David Richardson, "Prices of African Slaves Newly Arrived in the Americas, 1673–1865: New Evidence on Long-Run Trends and Regional Differences," in *Slavery in the Development of the Americas*, ed. David Eltis et al. (Cambridge: Cambridge University Press, 2004), 181–218.

21. Philip D. Morgan, "Slavery in the British Caribbean," in *The Cambridge World History of Slavery*, vol. 3, *A.D. 1420–A.D. 1804*, ed. David Eltis and Stanley L. Engerman (Cambridge: Cambridge University Press, 2011), 387; David Eltis, Frank D. Lewis, and David Richardson, "Slave Prices, the African Slave Trade," *Economic History Review* 58 (2005): 690, 692.

22. David Beck Ryden, "Does Decline Make Sense? The West Indian Economy and the Abolition of the British Slave Trade," *Journal of Interdisciplinary History* 31 (2001): 347–74; and Christer Petley,

"Slaveholders and Revolution: The Jamaican Planter Class, British Imperial Politics, and the Ending of the Slave Trade, 1775–1807," *Slavery and Abolition* 39 (2018): 53–79.

23. The debate is usefully summarized in J. R. Ward, "The Amelioration of British West Indian Slavery: Anthropological Evidence," *Economic History Review* 71 (2018): 1199–1226.. For arguments that doubt improvements in health and food, see A. Jabour, "Slave Health and Health Care in the British Caribbean: Profits, Racism and the Failure of Amelioration," *Journal of Caribbean History* 28 (1994): 1–26; and J. E. Candow, "A Reassessment of the Provision of Food to Enslaved Persons, with Special Reference to Salted Cod in Barbados," *Journal of Caribbean History* 43 (2009): 265–81. For a recent survey of Jamaica in this period, see Audrey A. Diptee, *From Africa to Jamaica: The Makings of an Atlantic Slave Society, 1775–1807* (Gainesville: University Press of Florida, 2010).

24. Lambert, *House of Commons Sessional Papers*, vol. 82.

25. Ward, "Amelioration of British West Indian Slavery."

26. Trevor Burnard, *Planters, Merchants, and Slaves: Plantation Societies in British America 1650–1820* (Chicago: University of Chicago Press, 2015), ch. 2.

27. James Ramsay had worked in St. Kitts as a clergyman in the 1770s and had a good working knowledge of slave conditions, which he described to devastating effect in Ramsay, *An Essay on the Treatment and Conversion of African Slaves in the British Sugar Colonies* (London, 1784). See, for Ramsay, Christopher Leslie Brown, *Moral Capital: The Foundations of British Abolitionism* (Chapel Hill: University of North Carolina Press, 2006), 226–53. For the purposes of this chapter, I have examined the evidence presented by witnesses to the House of Commons committee on the slave trade in 1791, which came from people with experience of Jamaican conditions and which have been collected in volume 82 of Lambert, *House of Commons Sessional Papers*. The testimonies principally relied on for this analysis, and which will be cited without page attribution, are by William Fitzmaurice, James Towne, Captain Thomas Lloyd, Captain Hall, Dr. Jackson, Robert Ross, Henry Coor, Lieutenant Baker Davidson, Mark Cook, Hercules Ross, and Thomas Irving.

28. Thomas Clarkson, *The History of the Rise, Progress, and Accomplishment of the Abolition of the Slave Trade by the British Parliament*, 2 vols. (London: J. Phillips, 1808). For wider context, see Roger Anstey, *The Atlantic Slave Trade and British Abolition, 1760–1810* (London: Macmillan, 1975); and J. R. Oldfield, *Popular Politics and British Anti-Slavery: The Mobilization of Public Opinion Against the Slave Trade, 1787–1807* (London: Frank Cass, 1998).

29. In general, the parliamentarians stuck to a set of similar concerns: that the large number of nonresident slave owners was particularly devastating within a system already marked by cruelty and immorality. It steered witnesses to agreeing that a hard-driving slave system and a lack of concern for pregnant women caused a decline in slave numbers. They tested witnesses about the moral character of enslaved people and slave managers, seeking confirmation of their belief that Jamaica was immoral in all ways.

30. Roberts, *Slavery and the Enlightenment in the British Atlantic*; Randy Browne, *Surviving Slavery in the British Caribbean* (Philadelphia: University of Pennsylvania Press, 2017), 162–65.

31. For work outside sugar by enslaved people on sugar plantations, see Justin Roberts, "Working Between the Lines: Labour and Agriculture on Two Barbadian Sugar Plantations, 1796–97," *William and Mary Quarterly*, 3rd. ser., 63 (2006): 551–86.

32. Mary Turner, "The 11 O'Clock Flog: Women, Work and Labour Law in the British Caribbean," *Slavery and Abolition* 20 (1999): 38–58.

33. Selwyn Carrington, *The Sugar Industry and the Abolition of the Slave Trade, 1775–1810* (Gainesville: University Press of Florida, 2002), 257–64; and Ryden, *West Indian Slavery and British Abolition*, 229–30.

34. Justin Roberts, "The 'Better Sort' and the 'Poorer Sort': Wealth Inequalities, Family Formation and the Economy of Energy on British Caribbean Sugar Plantations," *Slavery and Abolition* 35 (2014): 458–73; and Browne, *Surviving Slavery*, 1.

35. Roberts, " 'Better Sort.' "

36. Ibid. See also John Campbell, "Reassessing the Consciousness of Labour and the Role of the 'Confidentials' in Slave Society: Jamaica, 1750–1834," *Jamaican Historical Review* 21 (2001): 23–32.

37. Matthew Mulcahy, *Hurricanes and Society in the British Greater Caribbean, 1624–1783* (Baltimore: Johns Hopkins University Press, 2006), 180–86.

38. Simon Taylor to Chaloner Arcedeckne, 5 July 1789, Vanneck Mss., Bundle 2/15. Cambridge University Library; Hector McNeill, *Observations on the Treatment of the Negroes, in the Island of Jamaica* (London: G. G. J. and J. Robinson, 1788), 39; "Second Report: Presented the 12th Day of November," in *Two Reports from the Committee of the Honourable House of Assembly of Jamaica, on the Subject of the Slave Trade* (London: Stephen Fuller, Agent for Jamaica, 1789), 13–15.

39. Cited in Roberts, " 'Better Sort,' " 467.

40. William Beckford, *A Descriptive Account of the Island of Jamaica* (London: T. and J. Egerton, 1788), 27–28.

41. Roberts, " 'Better Sort,' " 468. J. R. Ward estimates that the percentage of adult enslaved people on Jamaican sugar estates who were African was 78 percent. Ward, *British West Indian Slavery*, 129.

42. Robert E. Gallman and Ralph V. Anderson, "Slaves as Fixed Capital: Slave Labor and Southern Economic Development," *Journal of American History* 64 (1977): 24–46.

43. Patrick Kein, *An Essay upon Pen-Keeping and Plantership* (Kingston, 1786); Roberts, *Slavery and the Enlightenment*, 149–50; Nicholas Radburn and Justin Roberts, " 'Gold Versus Life': Jobbing Gangs and Sugar Planting in the British Caribbean," *William and Mary Quarterly* 3d ser. 76 (2019): 223–56; William Sutherland to William Philip Perrin, 14 January 1798, Blue Mountain, D239/M/E/17483, Fitzherbert Mss. Derby Record Office, Matlock; Beckford, *A Descriptive Account of Jamaica*, 2:345.

44. Richard B. Sheridan, "The Crisis of Slave Subsistence in the British West Indies During and After the American Revolution," *William and Mary Quarterly*, 3rd ser., 33 (1976): 615–41. See also Selwyn Carrington, *The British West Indies During the American Revolution* (Dordrecht: Foris, 1988).

45. For increasing debt, see Burnard, *Planters, Merchants, and Slaves*, 197–99.

46. David Eltis, Frank D. Lewis, and David Richardson, "Slave Prices, the African Slave Trade, and Productivity in the Caribbean, 1674–1807," *Economic History Review* 4 (2005), 673–700.

47. Although absenteeism was high by British North American standards, many Jamaican proprietors were residents. Moreover, attorneys could be highly effective managers. Trevor Burnard, " 'Passengers Only': The Extent and Significance of Absenteeism in Eighteenth-Century Jamaica," *Atlantic Studies* 1 (2004): 178–95; Higman, *Plantation Jamaica*, 21–29; and Douglas Hall, "Absentee-Proprietorship in the British West Indies to About 1850," *Jamaican Historical Review* 4 (1964): 15–35. Edward Braithwaite notes that it "was very much to be doubted" that if planters had been residents instead of "handling their properties over to the clumsy mercies of attorneys and overseers, things would have been more efficient, more humane, certainly different." Edward Braithwaite, *The Development of Creole Society in Jamaica, 1770–1820* (Oxford: Oxford University Press, 1971), 130–31. For the origins of a historiography that saw absenteeism as a problem, see Lowell. J. Ragatz, "Absentee Landlordism in the British Caribbean, 1750–1833," *Agricultural History* 5 (1931): 7–24.

48. For the plough as a symbol of modernity in the Caribbean, see Chris Evans, "The Plantation Hoe: The Rise and Fall of an Atlantic Commodity," *William and Mary Quarterly* 3d ser. 69 (2012): 96–98.

49. Richard B. Sheridan, *Sugar and Slavery: An Economic History of the British West Indies, 1623–1775* (Bridgetown, Barbados: University of the West Indies Press, 1974), 229–31.

50. Burnard, *Planters, Merchants, and Slaves*, 177. For the responsiveness of slave traders to changing economic circumstances in Jamaica, see Nicholas Radburn, "Guinea Factors, Slave Sales, and the Transatlantic Slave Trade in Late Eighteenth-Century Jamaica: The Case of John Tailyour," *William and Mary Quarterly* 3d ser. 72 (2015): 261. Slave imports jumped from 18,381 in Jamaica in 1792 to 27,135 in 1793 following the boom year of 1792. It slipped back to 15,015 in 1794 as 1793 proved less profitable than 1792 (the year after the implosion of Saint-Domingue).

51. Trevor Burnard, "From Periphery to Periphery: The Pennants' Jamaican Plantations, 1771–1812 and Industrialization in North Wales," in *Wales and Empire, 1607–1820*, ed. H. V. Bowen (Manchester: Manchester University Press, 2011), 114–42.

52. *Glasgow Courier*, 19 January 1792 (citation courtesy of David Ryden). For Taylor's income, see Richard B. Sheridan, "Simon Taylor, Sugar Tycoon of Jamaica, 1740–1813," *Agricultural History* 45 (1971): 295.

53. Burnard, *Planters, Merchants, and Slaves*, 181. See also Ahmed Reid and David B. Ryden, "Sugar, Land Markets, and the Williams Thesis: Evidence from Jamaica's Property Sales, 1750–1810," *Slavery and Abolition* 34 (2013): 401–24.

54. Simon Taylor to Chaloner Arcedeckne, 13 April 1771, in *Travel, Trade and Power in the Atlantic, 1765–1884*, ed. Betty Wood et al. (Cambridge: Royal Historical Society, 2002); Simon Taylor to John Taylor, 7 June 1774, 28 March 1775, Taylor Papers, II/A/9, 14, Institute of Commonwealth Studies, London. This episode is covered fully in Christer Petley, *White Fury: A Jamaican Slaveholder and the Age of Revolution* (Oxford: Oxford University Press, 2018). See also Trevor Burnard, "Et in Arcadia Ego: West Indian Planters in Glory, 1674–1784," *Atlantic Studies* 9 (2012): 65–83.

55. Trevor Burnard, *Mastery, Tyranny, and Desire: Thomas Thistlewood and His Slaves in the Anglo-Jamaica World* (Chapel Hill: University of North Carolina Press, 2004), 64.

56. Burnard, *Planters, Merchants, and Slaves*, 199.

57. S. D. Smith, *"Merchants and Planters* Revisited," *Economic History Review* 55 (2002): 434–65; S. D. Smith, *Slavery, Family and Gentry Capitalism in the British Atlantic: The World of the Lascelles, 1648–1834* (Cambridge: Cambridge University Press, 2006); and David Hancock, *Citizens of the World: London Merchants and the Integration of the British Atlantic Community, 1735–1785* (New York: Cambridge University Press, 1995).

58. Kenneth Morgan, ed., *The Bright-Meyler Papers: A Bristol–West India Connection 1732–1837* (Oxford: Oxford University Press, 2007), 99–102, 105–7.

59. Quotations are from Petley, *White Fury*, 48; and Higman, *Plantation Jamaica*, 197–200.

60. Burnard, *Planters, Merchants, and Slaves*, 191–92; and Ryden, *West Indian Slavery and British Abolitionism*, 235.

61. Trevor Burnard, " 'The Countrie Continues Sicklie': White Mortality in Jamaica, 1655–1780," *Social History of Medicine* 12 (1999): 45–72; and Stanley L. Engerman and B. W. Higman, "The Demographic Structure of the Caribbean Slave Societies in the Eighteenth and Nineteenth Centuries," in *General History of the Caribbean*, vol. 3, *The Slave Societies of the Caribbean*, ed. Franklin W. Knight (London: UNESCO, 2007), 45–104.

62. Burnard, "Serious Business"; Burnard, *Planters, Merchants, and Slaves*, 196.

63. Roberts, "Working Between the Lines."

64. Roberts, " 'Better Sort,' " 468.

65. Kenneth Morgan, "Slave Women and Reproduction in Jamaica, c. 1776–1834," *History* 91 (2006): 231–53.

66. Sasha Turner, *Contested Bodies: Pregnancy, Childrearing and Slavery in Jamaica* (Philadelphia: University of Pennsylvania Press, 2017).

67. Roberts, *Slavery and the Enlightenment*; Mark Overton, *The Agricultural Revolution in England: The Transformation of the Agrarian Economy, 1500–1850* (Cambridge: Cambridge University Press, 1996); and Jane Humphries, "Child Labour and the Industrial Revolution," *Economic History Review* 66 (2013): 395–418.

68. Robert W. Fogel, "Economic Growth, Population Theory, and Physiology: The Bearing of Long-Term Processes and the Making of Economic Policy," *American Economic Review* 84 (1994): 373–74.

69. David Meredith and Deborah Oxley, "Food and Fodder: Feeding England, 1700–1900," *Past and Present* 222 (2014): 213.

70. Planter [Edward Long], *Candid Reflections upon the Judgement Lately Awarded by the Court of King's Bench in Westminster-Hall, on What Is Commonly Called the Negroe-Cause* (London, 1772).

71. Jonathan Hersh and Hans-Joachim Voth, "Sweet Diversity: Colonial Goods and the Rise of European Living Standards After 1492." 17 July 2009. Available at SSRN: https://ssrn.com/abstract = 1402322 or http://dx.doi.org/10.2139/ssrn.1402322.

72. Kenneth Pomeranz, *The Great Divergence: China, Europe and the Making of a Modern World Economy* (Princeton, N.J.: Princeton University Press, 2000).

73. Robert William Fogel, *Without Consent or Coercion: The Rise and Fall of American Slavery* (New York: W. W. Norton, 1989), 11–15.

74. E. P Thompson, "Time, Work-Discipline and Industrial Capitalism," *Past and Present* 38 (1967): 56–97; Jane Humphries, *Childhood and Child Labour in the British Industrial Revolution* (Cambridge: Cambridge University Press, 2010).

75. J. B. Moreton, *West India Customs and Manners*... (London, 1793), 61.

76. Burnard, Panza, and Williamson, "Living Costs."

77. Robert Allen, "The Great Divergence in European Wages and Prices from the Middle Ages to the First World War," *Explorations in Economic History* 38 (2001): 411–47. See also A. Arroyo, L. Elwyn Davies, and J. L. van Zanden, "Between Conquest and Independence: Real Wages and Demographic Change in Spanish America, 1530–1820," *Explorations in Economic History* 49 (2012): 149–66.

78. Burnard, Panza, and Williamson, "Living Costs."

79. Robert Dirks, "Resource Fluctuations and Competitive Transformations in West Indian Slave Societies," in *Extinction and Survival in Human Populations*, ed. Charles E. Laughlin Jr. and Ivan A. Brady (New York: Columbia University Press, 1978), 141–42.

80. Ibid., 163–7; Ward, *British West Indian Slavery*, 149–51.

81. John Komlos, "The Secular Trend in the Biological Standard of Living in the United Kingdom, 1730–1860," *Economic History Review* 46 (1993): 115–44.

82. Ward, *British West Indian Slavery*, 129–36; and Michael Craton, "Jamaican Slave Mortality: Fresh Light from Worthy Park, Longville, and the Tharp Estates," *Journal of Caribbean History* 3 (1971): 1–27.

83. P. Morgan, "Slavery in the British Caribbean," 383–84; Sheridan, *Sugar and Slavery*, 246–47. Michael Craton estimates depletion at 3 percent in the first half of the eighteenth century and 2.5 percent in the period between 1750 and 1770. Ward suggests, however, that Craton's data reflects conditions on more mature, fully developed estates. Craton, "Jamaican Slave Mortality," 24; Ward, *British West Indian Slavery*, 121. For this reason, the more mathematically sound figures of Morgan are preferred.

84. Ward's estimates are preferred here. He explains that attrition during "seasoning" may have been greater in the early eighteenth century but that greater efficiencies in the slave trade, making enslaved people healthier on arrival, and better management of "new Negroes" reduced losses during this period. Higher estimates remained current, however, owing to the twin needs for humanitarian reformers to emphasize how the horrors of "seasoning" showed how bad the slave trade was while planters insisted that "seasoning" helped explain much of annual attrition rates, emphasizing therefore the poor quality of the African enslaved people they bought from slave ships more than how plantation labor caused ill health. Ward, *British West Indian Slavery*, 124–29.

85. P. Morgan, "Slavery in the British Caribbean," 383; Selwyn Carrington, "Management of Sugar Estates in the British West Indies at the End of the Eighteenth Century," *Journal of Caribbean History* 33 (1999): 30–43; Ward, *British West Indian Slavery*, ch. 5. Morgan's figures show more dramatic changes in Barbados, with an increase from 2.8 percent annual attrition between 1725 and 1750 to 4.8 percent attrition between 1751 and 1775 followed by a massive decline in attrition rates to 0.8 percent per annum between 1776 and 1800 and then a small amount of annual natural increase after 1800. The worst area of the British Caribbean for slave mortality between 1751 and 1775 was the newly acquired Windward Islands, including Grenada, where annual attrition was 11.3 percent, falling to 5.9 percent annual decline in the next quarter century.

86. Dunn, *Tale of Two Plantations*, 415–17, 431–32, 435–36.

87. Ryden, *West Indian Slavery and British Abolitionism*, 147–49; Burnard, *Mastery, Tyranny, and Desire*, 204; and Burnard, " 'Impatient of Sub-Ordination and Liable to Sudden Transports of Anger': White Masculinity and Homosocial Relations with Black Men in Eighteenth-Century Jamaica," in *New Men: Manliness in Early America*, ed. Thomas S. Foster (New York: New York University Press, 2010), 134–54.

88. Burnard, *Planters, Merchants, and Slaves,* 196. There was a counterdiscourse, less prevalent than the practice of investing in men rather than women, where pronatalists in both Britain and Jamaica argued for policies that favored reforming reproduction policies so that women would "breed" more easily and thus mitigate the need for the Atlantic slave trade. Katherine Paugh, "The Politics of Child-rearing in the British Caribbean and the Atlantic World During the Age of Abolition," *Past and Present* 221 (2013): 119–60.

89. Ward, "Amelioration of British West Indian Slavery," 5, 23; John Komlos, "The Height of Runaway Slaves in Colonial America, 1720–1770," in *Stature, Living Standards and Economic Development: Essays in Anthropometric History,* ed. John Komlos (Chicago: Chicago University Press, 1994), 93–116; F. Cinnirella, "Optimists or Pessimists? A Reconsideration of Nutritional Status in Britain, 1740–1865," *European Review of Economic History* 12 (2008): 325–52; A. E. Challú, "The Great Decline: Biological Well-Being and Living Standards in Mexico, 1730–1840," in *Living Standards in Latin American History: Height, Welfare and Development, 1750–2000,* ed. R. D. Salvatore et al. (Cambridge, Mass.: Harvard University Press, 2010), 23–68.

90. Kenneth F. Kiple, *The Caribbean Slave: A Biological History* (Cambridge: Cambridge University Press, 1984), 69–70.

91. Joyce E. Chaplin, *An Anxious Pursuit: Agricultural Innovation and Modernity in the Lower South, 1730–1815* (Chapel Hill: University of North Carolina Press, 1993).

92. Christer Petley, " 'Home' and 'This Country': Britishness and Creole Identity in the Letters of a Transatlantic Slaveholder," *Atlantic Studies* 6 (2009): 43–61; and Christer Petley, " 'Devoted Islands' and 'That Madman Wilberforce': British Proslavery Patriotism During the Age of Abolition," *Journal of Imperial and Commonwealth History* 39 (2011): 393–415.

93. Higman, *Plantation Jamaica,* 218.

94. Ibid, 197–220. For demographic data, see Betty Wood and T. R. Clayton, "Slave Birth, Death and Disease on Golden Grove Plantation, Jamaica, 1765–1810," *Slavery and Abolition* 6 (1985): 99–121. For technology and improvements in slave management, see Heather Cateau, "Conservatism and Change Implementation in the British West Indian Sugar Industry 1750–1810," *Journal of Caribbean History* 29 (1995): 1–36; and Veront Satchell, "Early Use of Steam Power in the Jamaican Sugar Industry, 1768–1810," *Transactions of the Newcomen Society* 67 (1995–96): 222–31.

95. Higman, *Plantation Jamaica,* 225.

96. Sheridan, "Crisis of Slave Subsistence," 632–33.

97. Simon Taylor to Chaloner Arcedeckne, 5 July 1789, Vanneck Mss., bundle 2/10, Cambridge University Library.

98. This phrase is the sole example of empathy Thomas Thistlewood expressed for an enslaved person—in this case his mistress, whose plight on being parted from him he lamented in his diary in 1757. Burnard, *Mastery, Tyranny, and Desire,* 239.

99. The phrase is from Nell Irvin Painter, *Southern History Across the Color Line* (Chapel Hill: University of North Carolina Press, 2002), ch. 1. See also Saidiya Hartman, *Scenes of Subjection: Terror, Slavery, and Self-Making in Nineteenth-Century America* (Oxford: Oxford University Press, 1997).

100. Radburn and Roberts, " 'Gold Versus Life' "; Browne, *Surviving Slavery,* 3; Kristin A. Shuler, "Life and Death on a Barbadian Sugar Plantation," *International Journal of Osteoarchaeology* 21 (2011): 66–81.

101. Browne, *Surviving Slavery,* 32; Trevor Burnard, "A Voice for Slaves: The Office of the Fiscal in Berbice and the Beginnings of Protection in the British Empire," *Pacific Historical Review* 87 (2018): 30–53; Trevor Burnard and Randy Browne, "Husbands and Fathers: The Family Experiences of Enslaved Men in Berbice," *New West India Guide/Nieuwe West-Indische Gids* 91 (2017): 193–222; and Paul Cheney, *Cul de Sac: Patrimony, Capitalism, and Slavery in French Saint-Domingue* (Chicago: Chicago University Press, 2017), 179–80, 189.

102. David Watts, *The West Indies: Patterns of Development, Culture and Environmental Change Since 1492* (Cambridge: Cambridge University Press, 1987), 127–228; J. R. McNeill, *Mosquito Empires: Ecology and War in the Greater Caribbean, 1620–1914* (New York Cambridge University Press, 2010), 47–52; Stephen Behrendt, "Ecology and Seasonality in the Slave Trade," in *Soundings in*

Atlantic History: Latent Structures and Intellectual Currents, ed. Bernard Bailyn and Patricia Denault (Cambridge, Mass.: Harvard University Press, 2009), 54; and Pablo Gómez, *The Experiential Caribbean: Creating Knowledge and Healing in the Early Modern Atlantic* (Chapel Hill: University of North Carolina Press, 2017), 45.

103. Golfo Alexopoulos, *Illness and Inhumanity in Stalin's Gulag* (New Haven, Conn.: Yale University Press, 2017).

104. J. R. McNeill, "The Changing Disease Environment of the Caribbean to 1850," in *Caribbean Environmental History to 1850*, ed. Philip D. Morgan (unpublished mss., forthcoming).

105. Emily Mendenhall, "Syndemics: A New Path for Global Health Research," *Lancet* 389 (4 March 2017): 889–91; and Merrill Singer et al., "Syndemics and the Biosocial Conception of Health," *Lancet* 389 (4 March 2017): 941–50.

106. See Burnard, "'Countrie Continues Sicklie.'"

107. Niklas Thode Jensen, *For the Health of the Enslaved: Slaves, Medicine, and Power in the Danish West Indies, 1803–1848* (Copenhagen: Museum Tusculanum Press, 2012). Alex van Stipriaan, *Surinaams contrest: Roojbouw en oveileven in een Caraïbiscbe plantage-kolonie, 1750–1863* (Leiden: KITLV Uitgeverij, 1993), 316–18, shows that mortality on sugar estates in Surinam in the second half of the eighteenth century was 50–100 percent higher than on coffee estates. For Martinique, see Geneviève Leti, *Santé et société esclavagiste à la Martinique (1802–1848)* (Paris: L'Harmattan, 1998).

108. Londa Schiebinger, *Secret Cures of Slaves: People, Plants and Medicine in the Eighteenth-Century Atlantic World* (Stanford, Calif.: Stanford University Press, 2017), 156; Karen Bourdier, *Vie quotidienne et conditions sanitaires sur les grandes habitations sucrières du nord de Saint-Domingue à la veille de l'insurrection d'août 1791* (Lille: Atelier National de Reproduction des Thèses, 2008), 75–83, 232–309; Karen Bourdier, "Les conditions sanitaires sur les habitations sucrières de Saint-Domingue à la fin du siècle," *Dix-huitième Siècle* 43 (2011): 360–61; and S. Turner, *Contested Bodies*, 160–61.

109. P. Morgan, "Slavery in the British Caribbean," 384.

110. Ward, "Amelioration of British West Indian Slavery," 1–2, 25–26.

111. Mary Turner, "Planter Profits and Slave Rewards: Amelioration Reconsidered," in *West Indies Accounts: Essays on the History of the British Caribbean and the Atlantic Economy in Honour of Richard Sheridan*, ed. Roderick A. McDonald (Kingston: University of the West Indies Press, 1996), 232–52; and S. Turner, *Contested Bodies*.

112. S. Turner, *Contested Bodies*; Schiebinger, *Secret Cures of Slaves*, 9, 24, 46, 131–32; and Pratik Chakrabarti, *Materials and Medicine: Trade, Conquest and Therapeutics in the Eighteenth Century* (Manchester: Manchester University Press, 2010), ch. 2.

113. Ward, *British West Indian Slavery*; Christa Dierksheide, *Amelioration and Empire: Progress and Slavery in the Plantation Americas* (Charlottesville: University of Virginia Press, 2014), 155–210; Trevor Burnard and Kit Candlin, "Sir John Gladstone and the Debate over Amelioration in the British West Indies in the 1820s," *Journal of British Studies* 57 (2018): 760–82.

Chapter 4

1. Robert Middlekauff, *The Glorious Cause: The American Revolution, 1763–1789* (New York: Oxford University Press, 1982). A pioneering reevaluation of periodization is Jack P. Greene, "A Posture of Hostility: A Reconsideration of Some Aspects of the Origins of the American Revolution," *American Antiquarian Society Proceedings* 87 (1977): 27–68. For an argument in favor of a longer American Revolution, see Alan Taylor, *American Revolutions: A Continental History, 1750–1804* (New York: W. W. Norton, 2016).

2. Stephen Conway, *The American Revolutionary War* (London: I. B. Tauris, 2013), 2.

3. P. J. Marshall, *The Making and Unmaking of Empires: Britain, India and America c. 1760–1783* (Oxford: Oxford University Press, 2005).

4. Stephen Conway, "'A Joy Unknown for Year's Past': The American War, Britishness and the Celebration of Rodney's Victory at the Saints," *History* 86 (2001): 180–99.

5. Andrew Jackson O'Shaughnessy, *The Men Who Lost America: British Leadership, the American Revolution, and the Fate of the Empire* (New Haven, Conn.: Yale University Press, 2013), 316.

6. The Sugar Act was intended to combat the more efficient French sugar industry. French producers, however, were able to easily absorb the loss of the new duty and reduced prices on their sugar-derived products. Andrew Jackson O'Shaughnessy, *An Empire Divided: The American Revolution and the British Caribbean* (Philadelphia: University of Pennsylvania Press, 2000), 65–67.

7. I have used Wager rather than Apongo in this chapter. It is by this name that he is recognized in most accounts—all written by Europeans.

8. Philip Wright, "War and Peace with the Maroons, 1730–1739," *Caribbean Quarterly* 16 (1970): 5–27; Helen McKee, "From Violence to Alliance: Maroons and White Settlers in Jamaica, 1739–1795," *Slavery and Abolition* 39 (2018): 27–52.

9. All references to Thistlewood's diaries are by dates only. The diaries are in Thomas Thistlewood's Papers, James Marshall and Marie-Louise Osborn Collection, Beinecke Rare Book and Manuscript Library, Yale University.

10. 24 October 1760.

11. O'Shaughnessy, *Empire Divided*, 130–32. Jack P. Greene sees more convergence between the concerns of mainland and island colonists. Jack P. Greene, "The Jamaica Privilege Controversy, 1764–1766: An Episode in the Process of Constitutional Definition in the Early Modern British Empire," *Journal of Imperial and Commonwealth History*, 22 (1994): 16–54; and Jack P. Greene, *The Constitutional Origins of the American Revolution* (New York: Cambridge University Press, 2011), 67–186.

12. *The Humble Petition and Memorial of the Assembly of Jamaica to the King's Most Excellent Majesty in Council* (Philadelphia, 1774); George Metcalf, *Royal Government and Political Conflict in Jamaica, 1729–1783* (London: Longman, 1965), 189; Jack P. Greene, "Liberty, Slavery, and the Transformation of British Identity in the Eighteenth-Century West Indies," *Slavery and Abolition* 21 (2000): 1–31; and Sarah Yeh, "Colonial Identity and Revolutionary Loyalty: The Case of the West Indies," in *The Oxford History of the British Empire: British North America in the Seventeenth and Eighteenth Centuries*, ed. Stephen Foster (Oxford: Oxford University Press, 2013), 218–19.

13. Claudius Fergus, "'Dread of Insurrection': Abolitionism, Security, and Labor in Britain's West Indian Colonies," *William and Mary Quarterly*, 3rd ser., 66 (2009): 757–80; Maria Alessandra Bollettino, "'Of Equal or of More Service': Black Soldiers and the British Empire in the Mid-Eighteenth-Century Caribbean," *Slavery and Abolition* 38 (2017): 510–33; and Siân Williams, "The Royal Navy and Caribbean Colonial Society During the Eighteenth Century," in *The Royal Navy and the British Atlantic World, c. 1750–1820*, ed. John McAleer and Christer Petley (London: Palgrave Macmillan, 2016), 27–50.

14. Edward Long, *History of Jamaica . . .* , 3 vols. (London: T. Lowndes, 1774), 2:442; Fergus, "'Dread of Insurrection,'" 769.

15. Long, *History of Jamaica*, 2:410, 444.

16. Fergus, "'Dread of Insurrection,'" 761–62, 769–70, 778–79; James Ramsay, *An Essay on the Treatment and Conversion of African Slaves in the British Sugar Colonies* (London: J. Phillips, 1784), 89. Katherine Paugh stresses the importance of demographic information in the politics of reproduction that were crucial for abolitionists in the debates over the abolition of the slave trade. Katherine Paugh, *The Politics of Reproduction: Race, Medicine, and Fertility in the Age of Abolition* (Oxford: Oxford University Press, 2017).

17. Fergus, "'Dread of Insurrection,'" 760.

18. David Barry Gaspar, *Bondmen and Rebels: A Study of Master-Slave Relations in Antigua with Implications for Colonial British America* (Baltimore: Johns Hopkins University Press, 1985).

19. Marjoleine Kars, "Dodging Rebellion: Politics and Gender in the Berbice Slave Uprising of 1763, *American Historical Review* 121 (2016): 47. For what constituted the Akan people, see Rebecca Shumway, *The Fante and the Transatlantic Slave Trade* (Rochester, N.Y.: University of Rochester Press, 2011), 17–21; and Randy Sparks, *Where the Negroes Are Masters: An African Port in the Era of the Slave Trade* (Cambridge, Mass.: Harvard University Press, 2014). For Coromantee ideology, see John K. Thornton, "The Coromantees: An African Cultural Group in Colonial North America and the Caribbean," *Journal of Caribbean History* 32 (1998): 161–78; John K. Thornton, "War, the State and Religious Norms in 'Coromantee' Thought: The Ideology of an African American Nation," in *Possible*

Pasts: Becoming Colonial in Early America, ed. Robert Blair St. George (Ithaca, N.Y.: Cornell University Press, 2000), 181–200; and Robin Law, "Ethnicities of Enslaved Africans in the Diaspora: On the Meanings of Mina (Again), *History in Africa* 32 (2005): 247–67. For Coromantees' propensity to rebellion, see C. Roy Reynolds, "Tacky and the Great Slave Rebellion of 1760," *Jamaica Journal* 6 (1970): 5–8; and Vincent Brown, *The Coromantee Wars: An Archipelago of Insurrection* (Cambridge Mass.: Harvard University Press, 2020).

20. James Knight, "The Natural, Moral, and Political History of Jamaica and the Territories Thereon Depending," Long Family Papers, Add. Mss., 12, 418, British Library.

21. Maria Alessandra Bollettino, "Slavery, War, and Britain's Atlantic Empire: Black Soldiers, Sailors, and Rebels in the Seven Years' War" (PhD diss., University of Texas, Austin, 2009), 192.

22. Trevor Burnard, *Mastery, Tyranny, and Desire: Thomas Thistlewood and His Slaves in the Anglo-Jamaican World* (Chapel Hill: University of North Carolina Press, 2004), 137.

23. Emilia Viotti da Costa, *Crowns of Glory, Tears of Blood: The Demerara Slave Rebellion of 1823* (New York: Oxford University Press, 1994), 205; Richard B. Sheridan, "The Jamaican Slave Insurrection Scare of 1776 and the American Revolution," *Journal of Negro History* 3 (1975): 300.

24. Bollettino, "Slavery, War and Britain's Atlantic Empire," ch. 5.

25. Burnard, *Mastery, Tyranny, and Desire*, 166.

26. Arthur Forrest was a long-serving naval officer who served in the siege of Cartagena in 1741 and took command of the *Wager* in 1745, operating first in Boston and then in the West Indies. His most prominent action was at Cap François at Saint-Domingue in 1757, where through what Richard Pares sniffs was "irresponsible zeal" he allowed France to escape a blockade. Forrest returned to England in 1759, where he remained during Tacky's Revolt, returning as senior officer briefly in 1761–62 and more permanently in 1769 as commander in chief of the naval squadron, He died in Jamaica on 26 May 1770. J. K. Laughton, rev. Ruddock McKay, "Forrest, Arthur, d. 1770," *Oxford Dictionary of National Biography* https://doi-org.ezp.lib.unimelb.edu.au/10.1093/ref:odnb/9885; and Richard Pares, *War and Trade in the West Indies, 1739–1763* (London: Cass, 1936), 281.

27. 7 June, 18 December 1760, 3 October 1761.

28. 25, 26, 28 May; 7, 19 June, 29 July, 21 August, 2 September, 6 October, 18 December 1760.

29. Francis Treble to Caleb Dickinson, 12 June 1760, Caleb Dickinson Letters, Somerset Record Office, Taunton, Somerset, DD/DN/218.

30. David Miller to Brigadier General William Lewis, 20 November 1765, Lyttleton Papers, Box 15, Clements Library, Ann Arbor, Michigan.

31. Long, *History of Jamaica*, 2:446–47.

32. Trevor Burnard, *Planters, Merchants, and Slaves: Plantation Societies in British America, 1650–1820* (Chicago: University of Chicago Press, 2015), 161.

33. See, inter alia, Michael Craton, *Searching for the Invisible Man: Slaves and Plantation Life in Jamaica* (Cambridge, Mass: Harvard University Press, 1978); Vincent Brown, "Spiritual Terror and Sacred Authority in Jamaican Slave Society," *Slavery and Abolition* 24 (2003): 24–53; Orlando Patterson, *The Sociology of Slavery: An Analysis of the Origins, Development, and Structure of Negro Slave Society in Jamaica* (Cranbury, N.J.: Farleigh Dickinson University Press, 1969); and Richard D. E. Burton, *Afro-Creole: Power, Opposition, and Play in the Caribbean* (Ithaca, N.Y.: Cornell University Press, 1997).

34. Charles Leslie, *A New and Exact Account of Jamaica* (Edinburgh: R. Fleming, ca. 1740), 41, 336–38.

35. Knight, "Natural, Moral, and Political History." f.81.

36. [Edward Trelawney], *An Essay Concerning Slavery, and the Danger Jamaica Is Expos'd to from the Too Great Number of Slaves* (London: C. Corbett, 1746), reprinted in *Exploring the Bounds of Liberty: Political Writings of Colonial British America from the Glorious Revolution to the American Revolution*, ed. Jack P. Greene and Craig Yirush (Carmel, Ind.: Liberty Fund, 2018), 1131–65. For discourses around slavery, see James Robertson, "An Essay Concerning Slavery: A Mid-Eighteenth-Century Analysis," *Slavery and Abolition* 33 (2012): 65–85; and James Robertson, "A 1748 'Petition of Negro Slaves' and the Local Politics of Slavery in Jamaica," *William and Mary Quarterly*, 3rd ser., 67 (2010): 319–46.

37. [Trelawney], *Essay Concerning Slavery*, 1135, 1144.

38. Ibid, 1134–36, 1144–45, 1158.

39. Ibid.,

40. Ibid., 1145–46.

41. Ibid., 1144–47.

42. Fergus, "'Dread of Insurrection,'" 764; Eric Williams, *The British West Indies at Westminster, 1789-1823: Extracts from the Debates in Parliament* (1954; rept. Westport, Conn.: Greenwood, 1970), 23.

43. [Trelawney], *Essay Concerning Slavery*, 1136.

44. Justin du Rivage, *Revolution Against Empire: Taxes, Politics, and the Origins of American Independence* (New Haven, Conn.: Yale University Press, 2017), 26–36.

45. Trevor Burnard, "Thomas Thistlewood and the Problem of *Petit Marronage* in Eighteenth-Century Jamaica," in *The Precariousness of Freedom*, ed. Charmaine Nelson (New York: Routledge University Press, 2020).

46. 27–28 December 1752, 6 January 1753.

47. John Campbell, "Reassessing the Consciousness of Labour and the Role of the 'Confidentials' in Slave Society: Jamaica, 1750–1834," *Jamaican Historical Review* 21 (2001): 23–32.

48. Trevor Burnard and John Garrigus, *The Plantation Machine: Atlantic Capitalism in French Saint-Domingue and British Jamaica, 1748-1788* (Philadelphia: University of Pennsylvania Press, 2016), 122–36, 138–47; Trevor Burnard, "Slavery and the Enlightenment in Jamaica, 1760–1772: The Afterlife of Tackey's Rebellion," in *Enlightened Colonialism: Imperial Agents, Narratives of Progress and Civilizing Policies in the Eighteenth Century*, ed. Damien Tricoire (Basingstoke: Palgrave Macmillan, 2017), 227–46; Burnard, *Mastery, Tyranny, and Desire*, 170–74.

49. For other accounts, see Long, *History of Jamaica*, 2:447–65; Vincent Brown, *The Reaper's Garden: Death and Power in the World of Atlantic Slavery* (Cambridge, Mass.: Harvard University Press, 2008); "Slave Revolt in Jamaica, 1760–1761: A Cartographic Narrative" http://revolt.axismaps.com/project.html; and Bollettino, "Slavery, War, and Britain's Atlantic Empire," 191–256.

50. For scholarship that raises problems of evidence about slave rebellions, see Kars, "Dodging Rebellion"; Jason Sharples, "Discovering Slave Conspiracies: New Fears of Rebellion and Old Paradigms of Plotting in Seventeenth-Century Barbados," *American Historical Review* 120 (2015): 811–43; Michel-Rolph Trouillot, *Silencing the Past: Power and the Production of History* (Boston: Beacon, 1995); *Hearing Slaves' Voices: African and Indian Slave Testimony in British and French America, 1700-1848*, ed. Trevor Burnard and Sophie White (New York: Routledge, 2020); and Keith Mason, "The Absentee Planter and the Key Slave: Privilege, Patriarchalism, and Exploitation in the Early Eighteenth-Century Caribbean," *William and Mary Quarterly*, 3rd ser., 70 (2013): 79–102.

51. For theoretical perspectives, see Carolyn Steedman, *Dust* (Manchester: Manchester University Press, 2002); Ann Laura Stoler, *Along the Archival Grain: Epistemic Anxieties and Colonial Common Sense* (Princeton, N.J.: Princeton University Press, 2009); Nicola J. Aljoe, "Caribbean Slave Narratives," in *The Oxford Handbook of Slave Narratives*, ed. John Ernest (Oxford: Oxford University Press, 2014), 362–70; Henrice Altink, *Representations of Slave Women in Discourses on Slavery and Abolition, 1780-1838* (New York: Routledge, 2007); Tony Ballantyne, "Colonial Knowledge," in *The British Empire: Themes and Perspectives*, ed. Sarah Stockwell (Oxford: Blackwell, 2008), 177–99; and Marisa J. Fuentes and Brian Connolly, "Slavery in the Archives," *History of the Present* 6 (2016): 105–16.

52. Kenneth Bilby, "Swearing by the Past, Swearing to the Future: Sacred Oaths, Alliances and Treaties Among the Guianese and Jamaican Maroons," *Ethnohistory* 44 (1997): 655–89.

53. Burnard and Garrigus, *Plantation Machine*, 127–31; Brown, *Reaper's Garden*; G. Basker, "'The Next Insurrection': Johnson, Race, and Rebellion," *Age of Johnson* 11 (2000): 37–51; Thomas Day and John Bicknell, *The Dying Negro, a Poetical Epistle . . .* , 3rd ed. (London: W. Flexney, 1775); and Diana Paton, "Punishment, Crime and the Bodies of Slaves in Eighteenth-Century Jamaica," *Journal of Social History* 34 (2001): 931, 938.

54. Jason T. Sharples, "Hearing Whispers, Casting Shadows: Jailhouse Conversation and the Production of Knowledge During the Antigua Slave Conspiracy of 1736," in *Buried Lives: Incarcerated in Early America*, ed. Michele Lise Tarter and Richard Bell (Athens: University of Georgia Press, 2012),

35–59; Jill Lepore, *New York Burning: Liberty, Slavery and Conspiracy in Eighteenth-Century Manhattan* (New York: Random House, 2005); John Garrigus, *"Macandal Is Saved!": Disease, Conspiracy, and the Coming of the Haitian Revolution* (forthcoming); Richard C. Wade, "The Vesey Plot: A Reconsideration," *Journal of Southern History* 30 (1964): 143–61; and Steven Hahn, " 'Extravagant Expectations' of Freedom: Rumour, Political Struggle and the Christmas Insurrection Scare of 1865," *Past and Present* 157 (1997): 122–58. For controversy over whether the plan by Denmark Vesey for rebellion was real, see Douglas Egerton, *He Shall Go Out Free: The Lives of Denmark Vesey* (Madison: University of Wisconsin Press, 1999); Michael P. Johnson, "Denmark Vesey and His Co-Conspirators," *William and Mary Quarterly*, 3rd ser., 58 (2001): 915–76; and James O'Neill Spady, "Power and Confession: On the Credibility of the Earliest Reports of the Denmark Vesey Slave Conspiracy," *William and Mary Quarterly*, 3rd ser., 68 (2011): 287–304.

55. 3, 29–30 July, 1, 3 August, 4, 12, 18, 20 December 1760; 12 July 1760, *Pennsylvania Gazette*; and Long, *History of Jamaica*, 2:452–53.

56. Tony Ballantyne, "The Changing Shape of the Modern British Empire and Its Historiography," *Historical Journal* 53 (2010): 429–52.

57. Moore's official correspondence can be traced for Jamaica in C.O. [Colonial Office] 137/31–32; and for New York in C.O. 5/1072–74, 1098, National Archives, Kew, London. See also Add. Mss. 12,440; 22,469; Egerton Mss. 3,490, British Library, London.

58. Jack P. Greene, " 'Of Liberty and the Colonies': A Case Study of Constitutional Conflict in the Mid-Eighteenth-Century British American Empire," in *Creating the British Atlantic: Essays on Transplantation, Adaptation, and Continuity,* (Charlottesville: University of Virginia Press, 2013), 140–207; Jack P. Greene, "Jamaica Privilege Controversy."

59. Joseph Tiedemann, "Moore, Sir Henry (1713–1769)," *Oxford Dictionary of National Biography*, https://doi-org.ezp.lib.unimelb.edu.au/10.1093/ref:odnb/19116; Joseph Tiedemann, *Reluctant Revolutions: New York City and the Road to Independence, 1763–1776* (Ithaca, N.Y.: Cornell University Press, 1997).

60. In the longer run, the victor in the contest was Wager. The death of Jamaican rebels was memorialized in Britain though the iconography of Christian martyrdom and formed an important precondition for the start of the abolitionist movement in Britain. Brown, *Reaper's Garden*, 152–56. Wager's memory was passed down. The Jamaican planter Simon Taylor told his cousin in 1807 when he feared another rebellion was breaking out that "all new negroes know of the insurrection of 40 years ago." That knowledge, he believed was why rebellion was always possible: "If something were not going on, for what reason would they tell these New Negroes who have not been four months in the island of what happened before any of the negroes sent there were born?" Simon Taylor to Robert Taylor, October 24, 1807, Taylor Papers, I/I/44, Institute of Commonwealth Studies, London.

61. Cotes to Clevland, 19 April 1760, ADM 1/235, National Archives.

62. Leonard Stedman to William Vassall, June 7, 1760, Vassal Papers, bMsAm 1250/84, Houghton Library, Harvard University; "List of rebels taken and killed," Moore to Board of Trade, 14 July 1760, C.O. 137/32.

63. The rate of exchange with sterling was 1.4:1. Long Papers, Add. Mss. 18,275, British Library; C.O. 137/22/60–62; C.O. 140/23/486–87; C.O. 137/27/40–52.

64. Dr. John Lindsay to Dr. William Robertson, 6 August 1776, Robertson-Macdonald Letters, MS 3942, f. 262, National Library of Scotland; James Robertson, "Tackey Plus 5? The Slave Uprising in St. Mary's in 1765: The Experience and Imagination of a Slave Revolt in Jamaica" (unpublished paper); Michael Craton, *Testing the Chains: Resistance to Slavery in the British West Indies* (Ithaca: Cornell University Press, 1982), 139, 172–79; B. W. Higman, *Proslavery Priest: The Atlantic World of John Lindsay, 1729–1788* (Kingston: University of the West Indies Press, 2011); O'Shaughnessy, *Empire Divided*, 151–54; and Sheridan, "Jamaican Slave Insurrection Scare of 1776," 300 (quote).

65. David Miller to Brigadier General William Lewis. 20 November 1765, Lyttleton Papers, Box 15, Clements Library, Ann Arbor, Michigan.

66. William Lewis to Lyttleton, Westmoreland, 5 December 1760, Lyttleton Papers, Box 15, Clements Library.

67. Zachary Bayly to Caleb Dickinson, 1 June 1760, Caleb Dickinson Letters, DD/DN/218, Somerset Record Office.

68. Bayly to Lyttleton, 25, 26 November 1765; William Patrick Brown to Lyttleton, 30 November, Lyttleton Papers, Box 15, Clements Library.

69. "Testimony of Creole Cuffee, ca. December 1765"; Bayly to Lyttleton, December 1765; "An Account of What Negroes Were Concerned in the Late Insurrection . . . ," Lyttleton Papers, Box 15, Clements Library.

70. Long, *History of Jamaica*, 2:465.

71. Sir Simon Clarke to Benjamin Lyon, 23 July 1776, C.O. 137/71/256; Dr. John Lindsay to Dr. William Robertson, 6 August 1776, Robertson-Macdonald Letters, MS 3942, f. 262, National Library of Scotland; Sheridan, "Jamaican Slave Insurrection Scare of 1776," 303–4.

72. "List of the Impeached Estates in the Parish of Hanover and the Number of Negroes in Them . . . as of the 28th July 1776," C.O. 137/71/27.

73. Keith to Lord George Germain, 6 August 1776, C.O. 137/71/230–31.

74. Ibid.

75. *Journals of the Assembly of Jamaica . . . 1663–[1826]*, 14 vols. (Kingston, 1811–29) 6:634–60, 692–93; C.O. 137/71/230–308; Sheridan, "Jamaican Slave Insurrection," 297, 304.

76. Orlando Patterson, "Slavery and Slave Revolts: A Socio-Historical Analysis of the First Maroon War, 1655–1740," *Social and Economic Studies* 19 (1970): 289–325; Barbara Klamon Kopytoff, "Jamaican Maroon Political Organisation: The Effect of the Treaties," *Social and Economic Studies* 25 (1976): 87–107; and Barbara Klamon Kopytoff, "Colonial Treaty as Sacred Charter of the Jamaican Maroons," *Ethnohistory* 26 (1979): 45–64.

77. Long, *History of Jamaica*, 2:348–49; Kathleen Wilson, "The Performance of Freedom: Maroons and the Colonial Order in Eighteenth-Century Jamaica and the Atlantic Sound," *William and Mary Quarterly*, 3rd ser., 66 (2009): 46–47.

78. O'Shaughnessy, *Empire Divided*, 41.

79. Robert C. Dallas, *The History of the Maroons* (London, 1803), 1:129; O'Shaughnessy, *Empire Divided*, 147; Craton, *Testing the Chains*, 211–15.

80. McKee, "From Violence to Alliance." For customary and contemporary examples of colonial-indigenous relationships scarred by violence, see Peter Silver, *Our Savage Neighbors: How Indian War Transformed Early America* (New York: W. W. Norton, 2008); Krista Camenzind, "Violence, Race, and the Paxton Boys," in *Friends and Enemies in Pennow's Woods: Indians, Colonists, and the Racial Construction of Pennsylvania*, ed. William Pencak and Daniel K. Richter (University Park: Penn State University Press, 2004), 201–37; Bain Atwood, *The Good Country: The Djadja Wurrung, the Settlers and the Protectors* (Melbourne: Monash University Press, 2017); and Hal Langfur, *The Forbidden Lands: Colonial Identity, Frontier Violence, and the Persistence of Brazil's Eastern Indians* (Stanford, Calif.: Stanford University Press, 2006).

81. McKee, "From Violence to Alliance," 28, 38–40.

82. Burnard, *Planters, Merchants, and Slaves*, ch. 5.

83. Greene, "Liberty, Slavery."

84. For the global turn in early American studies, see Rosemarie Zagarri, "The Significance of the 'Global Turn' for the Early American Republic," *Journal of the Early Republic* 31 (2011): 1–57; Linda Colley, "The Difficulties of Empire: Present, Past and Future," *Historical Research* 29 (2006): 367–82; and Trevor Burnard, "Empire Matters? The Historiography of Imperialism in Early America, 1492–1830," *History of European Ideas* 33 (2007): 87–107.

85. Aaron Graham, "The Colonial Sinews of Imperial Power: The Political Economy of Jamaican Taxation, 1768–1838," *Journal of Imperial and Commonwealth History* 45 (2017): 188–209.

86. Max M. Edling, *A Hercules in the Cradle: War, Money, and the American State, 1783–1867* (Chicago: University of Chicago Press, 2014), chs. 1–2.

87. "Report Relative to a Provision for the Support of Public Credit," in *The Papers of Alexander Hamilton*, ed. Harold C. Syrett, 27 vols. (New York: Columbia University Press, 1961–87), 6:68.

88. Edling, *Hercules in the Cradle*, 240; John J. McCusker, "Estimating Early American Gross Domestic Product, 1650–1800," *Historical Methods* 33 (2000): 135–62.

89. David Beck Ryden, *West Indian Slavery and British Abolition, 1783–1807* (New York: Cambridge University Press, 2009), 293–94.

90. Christer Petley, "Slaveholders and Revolution: The Jamaican Planter Class, British Imperial Politics, and the Ending of the Slave Trade, 1775–1807," *Slavery and Abolition* 39 (2018): 53–79.

91. Michael Duffy, "The French Revolution and British Attitudes to the West Indian Colonies," in *A Turbulent Time: The French Revolution and the Greater Caribbean*, ed. David Barry Gaspar and David Patrick Geggus (Bloomington: Indiana University Press, 1997), 87.

92. David Geggus, "The Cost of Pitt's Caribbean Campaigns, 1793–1798," *Historical Journal* 2 (1983): 699–706; and McKee, "From Violence to Alliance."

93. Graham, "Colonial Sinews," 202.

94. See Petley, "Slaveholders and Revolution"; Fergus, " 'Dread of Insurrection,' " 764, 768–69; and David Geggus, "Slavery, War, and Revolution in the Greater Caribbean" in *A Turbulent Time: The French Revolution and the Greater Caribbean*, ed. David Barry Gaspar and David Patrick Geggus (Bloomington: Indiana University Press, 1997), 1–50. Christer Petley argues that white Jamaicans believed that they could weather the crisis of abolitionism and that if Parliament had passed the abolition of the slave trade before the 1790s they "might have encountered severe, even violent, opposition from slaveholders." Petley, "Slaveholders and Revolution," 63. J. R. Ward argues that planter-led efforts at amelioration in the 1790s answered abolitionist criticisms about the material living conditions of slaves. J. R. Ward, "The Amelioration of British West Indian Slavery: Anthropological Evidence," *Economic History Review* 71 (2018): 1199–1216.

95. In November 1781, Campbell replaced John Dalling, who "had unfortunately lost the confidence of the people." "By arts of address and liberality," he became "one of the most successful governors of the period." Metcalf, *Royal Government and Political Conflict in Jamaica*, 220.

96. Justin Girod-Chantrans, *Voyage d'un suisse dans les colonies d'Amérique*, ed. Pierre Pluchon (Paris: J. Tallandier, 1980), 245–46; Burnard and Garrigus, *Plantation Machine*, 206.

97. Archibald Campbell, *Memoir Relative to the Island of Jamaica Shewing the Nature and Strength of the Colony, the Situation of the Retreats and Military Posts, Together with the Dispositions and General Plan of Defence Established in 1782* (London, 1783), 3.

98. Ibid., 3–6, 13, 18, 35, 79–82. See also Alexander Dirom, *Thoughts on the State of the Militia of Jamaica* (Kingston: A. Aikman, 1783).

99. Long, *History of Jamaica*, 1:42.

100. Williamson to Henry Dundas, January 15, 1792, C.O. 137/90; David Geggus, "The Enigma of Jamaica in the 1790s: New Light on the Causes of Slave Rebellions," *William and Mary Quarterly*, 3rd ser., 44 (1987): 274–99; Bryan Edwards, *The History, Civil and Commercial, of the British Colonies in the West Indies*, 5 vols., 5th ed. (London, 1819), 4:1–240.

101. Cited in McKee, "From Violence to Alliance," 41.

102. Balcarres to Portland, March 26, 1796, C.O. 137/96/117; Dallas, *History of the Maroons*; and Mavis Campbell, *The Maroons of Jamaica, 1655–1796: A History of Resistance, Collaboration and Betrayal* (Trenton, N.J.: Africa World Press, 1990).

103. Craton, *Testing the Chains*, 223.

104. Quote from Patrick Richardson, *Empire and Slavery* (London: Longmans, 1968), 45; Petley, "Slaveholders and Revolution"; Fergus, " 'Dread of Insurrection.' "

105. David Geggus, *Slavery, War, and Revolution: The British Occupation of Saint Domingue, 1793–1798* (Oxford: Clarendon Press, 1982); Kit Candlin, *The Last Caribbean Frontier, 1795–1815* (London: Palgrave Macmillan), 2012.

106. Duffy, "French Revolution and British Attitudes to the West Indian Colonies," 95–96. See also Ronald Hyam, "British Imperial Expansion in the Late Eighteenth Century," *Historical Journal* 10 (1967): 113–31; C. A Bayly, *The Birth of the Modern World, 1780–1914* (Oxford: Blackwell, 2004); Robin Blackburn, "Haiti, Slavery, and the Age of Democratic Revolution," *William and Mary Quarterly*, 3rd ser., 63 (2006): 643–74; Kit Candlin, "The Role of the Enslaved in the "Fedon Rebellion" of 1795,"

Slavery and Abolition 39 (2018): 685–707; and Nicholas Draper, "The Rise of a New Planter Class? Some Counter-Currents from British Guiana and Trinidad, 1807–1834," *Atlantic Studies* 9 (2012): 65–83.

Chapter 5

This chapter was originally published, in French, as " 'Une Véritable Nuisance pour la Communauté': La Place ambivalent des libres de couleur dans la société libre de la Jamäique au XVIIIe siècle," in *Sortir de l'esclavage: Stigmates, assimilations et recompositions identitaires du XVe au XXe siècle (Méditerranée, Europe, Amériques, Afriques)*, ed. Dominique Rogers and Boris Lesueur (Paris: Karthala, collection esclavages, 2018), 176–200.

1. Bryan Edwards, *The History, Civil and Commercial, of the West Indies, with a Continuation to the Present Time* 5 vols. (London:, 1819), 2:7.

2. For an example of such slavery, see Elaine G. Breslaw, *Tituba: Reluctant Witch of Salem; Devilish Indians and Puritan Fantasies* (New York: New York University Press, 1995), 3–64.

3. Daniel Livesay, *Children of Uncertain Fortune: Mixed-Race Jamaicans in Britain and the Atlantic Family, 1733–1833* (Chapel Hill: University of North Carolina Press, 2018), 25.

4. For mixed-race people in the Caribbean generally, see Jerome S. Handler, *The Unappropriated People: Freedmen in the Slave Society of Barbados* (Baltimore: Johns Hopkins University Press, 1974); Edward L. Cox, *Free Coloreds in the Slave Societies of St. Kitts and Grenada, 1766–1833* (Knoxville: University of Tennessee Press, 1984: Melanie J. Newton, *The Children of Africa in the Colonies: Free People of Color in Barbados in the Ae of Emancipation* (Baton Rouge: Louisiana University Press, 2008); and Emily Clark, *The Strange History of the American Quadroon: Free Women of Color in the Revolutionary Atlantic World* (Chapel Hill: University of North Carolina Press, 2013).

5. The first laws enacted dealing with free people, apart from private bills, came in 1711 and 1712. 10 Anne, c.4 of 19 May 1711, and 11 Anne, c.3 of 14 November 1712, *The Laws of Jamaica* (St. Jago de la Vega, Jamaica, 1792).

6. Gad Heuman, *Between Black and White: Race, Politics, and the Free Coloreds in Jamaica, 1792–1865* (Westport, Conn.: Greenwood Press, 1981), 7.

7. The phrase comes from a 1774 petition of the Assembly. Jamaica Assembly to King, 23 December 1774, Colonial Office [C.O.] Series, Class 140, vol. 46, 569–70, Public Record Office, Kew.

8. "Return of Militia, 1788," C.O. 137/87/32.

9. "General Return of the Militia in the Island of Jamaica," 26 November 1813, *Journals of the Assembly of Jamaica . . . 1663–[1826]*, 14 vols. (Kingston, 1811–29), vol. 12 (henceforth *JAJ*).

10. A. E. Furness, "The Maroon War of 1795," *Jamaican Historical Review* 5 (1965): 30–49; Roger Norman Buckley, *Slaves in Red Coats: The British West Indian Regiments, 1795–1815* (New Haven: Yale University Press, 1979).

11. Heuman, *Between Black and White*, 7.

12. Edwards, *History, Civil and Commercial*, 2:22–23.

13. John Stewart, *A View of the Past and Present State of the Island of Jamaica* (Edinburgh: Oliver and Boyd, 1823), 332, 334.

14. See Heuman, *Between Black and White*, 3–54.

15. Alfred D. Crosby notes that Queensland is the only place in the tropics where British have prospered, and then only after settlement in the late nineteenth century. Alfred D. Crosby, *Ecological Imperialism: The Biological Expansion of Europe, 900–1900*, 2nd ed. (Cambridge: Cambridge University Press, 2004).

16. One society where whites came to an accommodation with browns is nineteenth-century Brazil. See Carl Degler, *Neither Black nor White: Slavery and Race Relations in Brazil and the U.S.* (Madison: University of Wisconsin Press, 1986); and Thomas Skidmore, *Black into White: Race and Nationality in Brazilian Thought* (New York: Oxford University Press, 1974).

17. Winthrop D. Jordan, *White over Black: American Attitudes Towards the Negro, 1550–1812* (Chapel Hill: University of North Carolina Press, 1968), 126–27, 176–77; Arnold Sio, "Race, Colour and Miscegenation: The Free Coloured of Jamaica and Barbados," *Caribbean Studies* 16 (1976): 5–21;

Samuel J. Hurwitz and Edith F. Hurwitz, "A Token of Freedom: Private Bill Legislation for Free Negroes in Eighteenth Century Jamaica," *William and Mary Quarterly*, 3rd ser., 24 (1967): 423–31; 30 March, 15 April 1733, 18 July 1738, *JAJ*, 3:123, 141, 455.

18. [Edward Trelawney], *An Essay concerning Slavery, and the Danger Jamaica is expos'd from the Too Great Number of Slaves* . . . (London: C. Corbett, 1746), reprinted in *Exploring the Bounds of Liberty: Political Writings of Colonial British America from the Glorious Revolution to the American Revolution* ed. Jack P. Greene and Craig Yirush (Carmel, Ind.: Liberty Fund, 2018), 3 vols., II: 1134–1135.

19. Livesay, *Children of Uncertain Fortune*, 35–45. For Iberia, see Ann Twinam, *Purchasing Whiteness: Pardos, Mulattos, and the Quest for Social Mobility in the Spanish Indies* (Stanford: Stanford University Press, 2015).

20. Livesay, *Children of Uncertain Fortune*, 39, 65, 89.

21. Ibid., 48–50. These ambiguities about the precise status of wealthy but racially suspect free people of color in Jamaica were mirrored by similar tensions over the links between colonists and their mixed-race offspring in India. See Margot Finn, "Anglo-Indian Lives in the Later Eighteenth and Early Nineteenth Centuries," *Journal for Eighteenth-Century Studies* 33 (2010): 49–50.

22. 2 George III, c.8 of 19 December 1761, *Laws of Jamaica*.

23. See David Brion Davis, *Inhuman Bondage: The Rise and Fall of Slavery in the New World* (New York: Oxford University Press, 2006), 48–76.

24. Private acts are listed in *Laws of Jamaica*.

25. Census of 1730—C.O. 137/19 (pt. 2)/48; Edward Long, *History of Jamaica* . . . , 3 vols. (London: T. Lowndes, 1774), 2:337; Census of 1774—C.O. 137/70/88; Census of 1788—C.O. 137/87; "Mulattoes, Quads, Negroes Able to Bear Arms," Add. Mss. 12,435, British Museum, London; St. Andrew Parish Register, 1666–1780, St. Catherine Parish Register, 1667–1764, Manumissions, vols. 5 and 7, Jamaica Archives; Wills, vols. 1–60, Island Record Office; Kingston Parish Register, 1722–74, Island Record Office Armoury, Spanish Town, Jamaica.

26. The Bartholomew family can be traced in the St. Catherine Parish Register.

27. 16 November 1762, *JAJ*, 5:376–77.

28. Manumissions, vols. 5 and 7 (vols. 1–4, the records of manumissions before 1747 have been lost); Kingston Register of Free Persons, 1761–95, Kingston Vestry Records, Jamaica Archives.

29. Hurwitz and Hurwitz, "Token of Freedom," 425.

30. Ibid., 426.

31. 19 April 1747, *JAJ*, 4:67; 21 George II, c.7 of 13 August 1748, *Laws of Jamaica*.

32. These are listed in *Laws of Jamaica* (1792). Edward Braithwaite notes that between 1772 and 1796, sixty-seven petitions for 512 free people were passed. Edward Braithwaite, *The Development of Creole Society in Jamaica, 1770–1820* (Oxford: Oxford University Press, 1971), 172.

33. Livesay, *Children of Uncertain Fortune*, 88.

34. Heuman, *Between Black and White*, 5.

35. Long, *History of Jamaica*, 2:475–85.

36. Ibid., 2: 479.

37. 20–21 November 1724, *JAJ*, 2:512–14.

38. "Petition of Francis Williams to the Board of Trade," ca. 1730, C.O. 137/19/29–30.

39. Long, *History of Jamaica*, 478.

40. John Garrigus, "Blue and Brown: Contraband Indigo and the Rise of a Free Colored Planter Class in French Saint-Domingue," *Americas* 50 (1993): 257–61; and John Garrigus, "Colour, Class and Identity on the Eve of the Haitian Revolution: Saint-Domingue's Free Coloured Elite as *Colons Americains*," *Slavery and Abolition* 17 (1996): 26.

41. 2 George III c.7 of 19 December 1761, *Laws of Jamaica*.

42. Long, *History of Jamaica*, 2: 328; An Act to Intitle Anna Petronella Woodart Spinster a free Mulatto to the same Rights & privileges with English Subjects born of White Parents, 14 November 1760, Jamaica Acts, 1760, CO 139/21/14. For the complete record of Woodart's life, see Anonymous, *The Vicissitudes of a Mid-18th Century Female Jamaica Slave: From Slavery to Freedom, to Marriage with a White Englishman and Ownership of a Plantation* (London, 1756–69). See also Brooke N. Newman,

"Contesting 'Black' Liberty and Subjecthood in the Anglophone Caribbean, 1730s-1780s," *Slavery & Abolition* 32 (2011): 169–83.

43. Livesay, *Children of Uncertain Fortune*, 68.

44. Ibid.

45. Long, *History of Jamaica*, 2:323–33. In encouraging free people to be trades people, Long rejected a previously common opinion that freed tradesmen were a restraint on the opportunity of white artifices and prevented more sizable immigration into Jamaica of white tradespeople. See Florentius Vassall to Rose Fuller, 1 January 1750, RF17/19/J65, Fuller Papers, East Sussex Record Office; and, [Trelawney], *Essay Concerning Slavery*, 41–42.

46. "Account of Houses, Annual Rents etc in St. Jago de la Vega July August 1754," Fuller Papers, RF17/XVIII/J21; "List of Landholders in Jamaica, 1750 [misdated for 1754]," Add. Mss. 12,436, British Museum.

47. "Mulattoes, Quads, and Negroes Able to Bear Arms."

48. C.O. 137/19/30.

49. 2 September 1756, *JAJ*, 4:584.

50. Douglas Hall, "Jamaica," in *Neither Slave nor Free: The Freedmen of African Descent in the Slave Societies of the New World*, ed. David W. Cohen and Jack P. Greene (Baltimore: Johns Hopkins University Press, 1972), 198.

51. Stewart, *View of Past and Present State of the Island of Jamaica*, 327.

52. The Augiers can be traced in the Kingston Parish Register and in the will of John Augier, Wills (1722), Island Record Office, Jamaica. See also private acts of 1738 and 1747, *Laws of Jamaica*.

53. Heuman, *Between Black and White*, 10–11.

54. Richard Hill, *Lights and Shadows of Jamaican History* (Kingston: Ford and Gall, 1859), 104.

55. 2 November 1726, *Weekly Jamaican Courant*.

56. 1 November 1759, *JAJ*, 5:149; 1 George III, c.23, 1760, *Laws of Jamaica*.

57. "Petition of Thomas Edwards, Late Slave to Thomas Fuller Esquire of Jamaica," 1744, C.O. 137/48/92.

58. Lyttleton to Earl of Egremont, n.d. [1763], C.O. 137/61/116.

59. Edwards, *History, Civil and Commercial*, 2:310.

60. [Trelawney], *Essay Concerning Slavery*, 42.

61. Long, *History of Jamaica*, 2:331–35.

62. Stewart, *View of Past and Present State of the Island of Jamaica*, 331–35.

63. Jack P. Greene, "Independence, Improvement and Authority: Toward a Framework for Understanding the Histories of the Southern Backcountry During the Era of the American Revolution," in *An Uncivil War: The Southern Backcountry During the American Revolution*, ed. Ronald Hoffman, Thad W. Tate, and Peter J. Albert (Charlottesville: University of Virginia Press, 1985), 16–21.

64. "Journal of an Officer [Lord Adam Gordon] Who Travelled in America and the West Indies in 1764 and 1765," in *Travels in the American Colonies*, ed. Newton D. Mereness (New York: Macmillan, 1916), 377–79.

65. Long, *History of Jamaica*, 1:374.

66. Samuel J. Hurwitz and Edith Hurwitz, "The New World Sets an Example for the Old: The Jews of Jamaica and Political Rights, 1661–1831," *American Jewish Historical Quarterly* 55 (1965): 37–56.

67. John Fothergill, *Considerations Relative to the North American Colonies* (London, 1765), 41–42.

68. Long, *History of Jamaica*, 2:271, 280.

69. *The Works of James Houston, M.D.* (London: S. Bladon, 1753), 293.

70. John R. McNeill, *Mosquito Empires: Ecology and War in the Greater Caribbean, 1620–1914* (Cambridge: Cambridge University Press, 2012).

71. David Eltis, "Free and Coerced Transatlantic Migrations: Some Comparisons," *American Historical Review* 87 (1982): 255.

72. For the significance of sojourning in the late eighteenth century, see Alan L. Karras, *Sojourners in the Sun: Scottish Migrants in Jamaica and the Chesapeake, 1740–1800* (Ithaca: Cornell University Press, 1992).

73. Stewart, *View of the Past and Present State of the Island of Jamaica*, 332.

74. Anonymous, *Marly: or, A Planter's Life in Jamaica* (Glasgow: Richard Griffin, 1828), 219.

Chapter 6

1. Eliga H. Gould, *Among the Powers of the Earth: The American Revolution and the Making of a New World Empire* (Cambridge, Mass.: Harvard University Press, 2012), 55.

2. Full details of the background to the case and details on legal proceedings can be found in Steven Wise, *Though the Heavens May Fall: The Landmark Case that Led to the End of Human Slavery* (Cambridge, Mass.: Harvard University Press, 2005). Somerset's master, Charles Steuart, was an established Scottish and Atlantic figure. He was paymaster of the customs office in Boston and a friend of at least two governors, William Franklin (son of Benjamin) in New Jersey and James Johnstone, governor of Quebec. Emma Rothschild, *The Inner Life of Empires: An Eighteenth-Century History* (Princeton, N.J.: Princeton University Press, 2011), 92.

3. James Oldham, "New Light on Mansfield and Slavery," *Journal of British Studies* 27 (1988): 45–68.

4. For Mansfield's career and life, see Norman Posner, *Lord Mansfield: Justice in the Age of Reason* (Montreal: McGill-Queen's University Press, 2013). For his friendship with Beckford, see Perry Gauci, *William Beckford: First Prime Minister of the London Empire* (New Haven, Conn.: Yale University Press, 2013), 24, 114.

5. *London Evening Post*, 23 May 1772. Mansfield, however, had other connections with the West Indies, most famously in having a black servant, Elizabeth Dido Lindsay, the daughter of his nephew, Sir John Lindsay, whom he freed in his will and to whom he gave an annuity of £500. Paula Byrne, *Belle: The Slave Daughter and the Lord Chief Justice* (London: Harper Collins, 2014).

6. Cited in Follarin Shyllon, *Black Slaves in Britain* (Oxford: Oxford University Press, 1974), 52.

7. Benjamin Franklin to Anthony Benezet, 22 August 1772, in *The Papers of Benjamin Franklin*, ed. Leonard W. Labaree et al., 41 volumes to date (New Haven, Conn.: Yale University Press, 1959–), 19:269. Franklin was being disingenuous. He understood that while Mansfield had done little to advance abolitionism in England, he had upheld his belief in imperial constitutionalism and the importance of Parliament being sovereign by showing himself ready to adjudicate over a matter that most Americans and West Indians believed was outside the sphere of English knowledge and influence. Franklin realized that *Somerset* was as much about empire and nationality as about antislavery. He understood that metropolitan proposals to extend subjecthood to slaves went hand in hand with an unacceptable subjection of their masters to British parliamentary dictates. David Waldstreicher, *Runaway America: Benjamin Franklin, Slavery, and the American Revolution* (New York: Hill and Wang, 2005), 198–201.

8. Cited in *A Letter to Philo-Africanus…* (London, 1789).

9. Shyllon, *Black Slaves in Britain*, ch. 12; David Brion Davis, *The Problem of Slavery in the Age of Revolution, 1770–1823* (Ithaca, N.Y.: Cornell University Press, 1975), 499–500.

10. Ruth Paley, "After *Somerset*: Mansfield, Slavery and the Law in England, 1772–1830," in *Law, Crime and English Society, 1660–1830*, ed. Norma Landau (Cambridge: Cambridge University Press, 2002), 165–84.

11. Many scholars have been misled by relying on a fragmentary and partial law report on the case, drawn up four years after the case. As James Walvin notes, "The Somerset case in fact provides us with a good example of historical myth displacing truth and developing an importance of its own." James Walvin, *England, Slaves and Freedom, 1776–1838* (Jackson: University Press of Mississippi, 1986), 41. The best legal account of the case is Oldham, "New Light on Mansfield and Slavery."

12. William R. Cotter, "The *Somerset* Case and the Abolition of Slavery in England," *History* 79 (1994): 31–56.

13. Seymour Drescher intriguingly suggests that by deterring planters from importing "insecure" capital into England and thus reducing or at least not increasing the number of imported blacks in the country, *Somerset* may have helped defer resolution of whether slavery was legal in Britain until a later date, after 1783, when a fresh infusion of black refugees entered the country. Seymour Drescher, *From Slavery to Freedom: Comparative Studies in the Rise and Fall of Atlantic Slavery* (Basingstoke: Macmillan, 1999), 19.

14. Cited in Vincent Caretta, *Equiano the African: Biography of a Self-Made Man* (Athens: University of Georgia Press, 2005), 208, 212.

15. Ira Berlin, *The Long Emancipation: The Demise of Slavery in the United States* (Cambridge, Mass.: Harvard University Press, 2015), ch. 1.

16. Oldham, "New Light on Mansfield and Slavery"; James Oldham, *English Common Law in the Age of Mansfield* (Chapel Hill: University of North Carolina Press, 1992), 305–23; George Van Cleve, "*Somerset's Case* and Its Antecedents in Imperial Perspective," *Law and History Review* 24 (2006): 601–45; David Waldstreicher, *Slavery's Constitution: From Revolution to Ratification* (New York: Hill and Wang, 2009), ch. 1; and Paley, "After *Somerset*."

17. Thomas Jefferson, for example, who became convinced that all forms of despotism and corruption were British in origin, leading him in 1776 to try and include an unconvincing rant in the first draft of the Declaration of Independence against the British for responsibility for their "cruel war against human nature" in forcing the slave trade on a reluctant American population, was not especially concerned about *Somerset* but was more concerned about the Privy Council overturning laws by Maryland and Virginia to tax and regulate slave imports. Peter S. Onuf, "Federalism, Democracy, and Liberty in the New American Nation," in *Exclusionary Empire: English Liberty Overseas, 1600–1900*, ed. Jack P. Greene (New York: Cambridge University Press, 2010), 155.

18. Planter [Edward Long], *Candid Reflections upon the Judgement Lately Awarded by the Court of King's Bench in Westminster-Hall, on What Is Commonly Called the Negroe-Cause* (London: T. Lowndes, 1772); Samuel Estwick, *Considerations on the Negroe Cause Commonly So Called, Addressed to the Right Honourable Lord Mansfield*, 2nd ed. (London: J. Dodsley, 1773); and Samuel Martin Sr., *A Short Treatise on the Slavery of Negroes in the British Colonies* (Antigua: Robert Mearns, 1775).

19. Lorena S. Walsh, *From Calabar to Carter's Grove: The History of a Virginia Slave Community* (Charlottesville: University of Virginia Press, 1997); Allan Kulikoff, "A 'Prolifick' People: Black Population Growth in the Chesapeake Colonies, 1700–1790," *Southern Studies* 16 (1977): 391–428; and Kenneth Morgan, "Slave Women and Reproduction in Jamaica, c. 1776–1834," *History* 91 (2006): 231–53.

20. William W. Wiecek, "Somerset: Lord Mansfield, and the Legitimacy of Slavery, *University of Chicago Law Review* 42 (1974): 86–146. ; and Daniel J. Hulseboch, "Nothing but Liberty: "Somerset's Case" and the British Empire," *Law and History Review* 24 (2006): 647–58.

21. The *Somerset* decision is not noted, for example, in Andrew Jackson O'Shaughnessy, *An Empire Divided: The American Revolution and the British Caribbean* (Philadelphia: University of Pennsylvania Press, 2000).

22. For the suddenness in the realization of West Indians that the abolitionist campaign was a serious threat to their welfare, see Bryan Edwards, *A Speech Delivered at a Free Conference Between the Honourable Council and the Assembly of Jamaica . . . on the Subject of Mr. Wilberforce's Propositions in the House of Commons* (Kingston: A. Aikman, 1789).

23. George Metcalf, *Royal Government and Political Conflict in Jamaica, 1729–1783* (London: Longman, 1965), 167.

24. Jack P. Greene, "The Jamaica Privilege Controversy, 1764–1766: An Episode in the Process of Constitutional Definition in the Early Modern British Empire," *Journal of Imperial and Commonwealth History* 22 (1994): 16–54; and Trevor Burnard, "Et in Arcadia Ego: West Indian Planters in Glory, 1674–1784," *Atlantic Studies* 9 (2012): 65–83.

25. Edward Long, *History of Jamaica . . .*, 3 vols. (London: T. Lowndes, 1774), 2:63.

26. Metcalf, *Royal Government and Political Conflict in Jamaica*, 167; Long, *History of Jamaica*, 1:76–77, 2:263; Michael Craton and James Walvin, *A Jamaican Plantation: The History of Worthy Park*,

1670–1970 (London: W. H. Allen, 1970), 157. Not everything was rosy in the West Indies in 1772. That year saw the start of a major credit crisis (which affected the eastern Caribbean more than the Greater Antilles) in which some planters and West Indian merchants were severely affected. As it was mostly a crisis created in Scotland, those parts of the empire, like Virginia and Grenada, that were highly exposed to Scottish credit were especially badly hurt by the crash of 1772. See Richard B. Sheridan, "The British Credit Crisis of 1772 and the American Colonies," *Journal of Economic History* 20 (1960): 1161–86; S. D. Smith, *Slavery, Family and Gentry Capitalism in the British Atlantic: The World of the Lascelles, 1648–1834* (Cambridge: Cambridge University Press, 2006), 131–32, 220; and Nick Bunker, *An Empire on the Edge: Now Britain Came to Fight America* (New York: Vintage, 2014), ch. 3.

27. Gauci, *William Beckford.*

28. Sarah Scott, *The History of Sir George Ellison* (1766) as cited in Brycchan Carey, *British Abolitionism and the Rhetoric of Sensibility* (London: Palgrave Macmillan, 2005), 51; Evangeline Andrews and Charles M. Andrews, *Journal of a Lady of Quality* (New Haven, Conn.: Yale University Press, 1921), 103–6 (quote 103); Natalie Zacek, "Cultivating Virtue: Samuel Martin and the Paternal Ideal in the Eighteenth-Century English West Indies," *Wadabagei* 10 (2007): 8–31; and Richard B. Sheridan, "Samuel Martin, Innovating Sugar Planter of Antigua, 1730–1775," *Journal of Economic History* 34 (1960): 126–39.

29. O'Shaughnessy, *Empire Divided*; Smith, *Slavery, Family and Gentry Capitalism*; and Trevor Burnard, "From Periphery to Periphery: The Pennants' Jamaican Plantations, 1771–1812 and Industrialization in North Wales," in *Wales and Empire, 1607–1820*, ed. H. V. Bowen (Manchester: Manchester University Press, 2011), 114–42.

30. Cited in Lowell J. Ragatz, *The Fall of the Planter Class in the British Caribbean, 1763–1833* (New York: Century, 1928), 50.

31. J. R. Ward, "The Profitability of Sugar Planting in the British West Indies, 1650–1834," *Economic History Review* 31 (1978): 206; David Beck Ryden, *West Indian Slavery and British Abolition, 1783–1807* (New York: Cambridge University Press, 2009), 226; Ahmed Reid and David Ryden, "Sugar, Land Markets, and the Williams Thesis: Evidence from Jamaica's Property Sales, 1750–1810," *Slavery and Abolition* 34 (2013): 401–24; and Seymour Drescher, *Econocide: British Slavery in the Era of Abolition*, 2nd ed. (Chapel Hill: University of North Carolina Press, 2010).

32. The number of West Indians living in Britain was greater than the number of Americans living there, though the great majority of wealthy West Indians remained in the tropics, where they expended most of their money. "Absenteeism" (itself a pejorative word that means it should be used with care by historians to describe West Indians in Britain) was not inconsiderable but generally has been a phenomenon overemphasized by contemporaries and by subsequent generations of historians. Trevor Burnard, " 'Passengers Only': The Extent and Significance of Absenteeism in Eighteenth-Century Jamaica," *Atlantic Studies* 1 (2004): 178–95. See also Julie Flavell, *When London Was Capital of America* (New Haven, Conn.: Yale University Press, 2010).

33. O'Shaughnessy, *Empire Divided*, 15–17, 88–89, 106–7.

34. See Trevor Burnard, "Harvest Years: Reconfigurations of Empire in Jamaica, 1756–1807," *Journal of Commonwealth and Imperial History* 40 (2012): 533–55; and Trevor Burnard, "Powerless Masters: The Curious Decline of Jamaican Sugar Planters in the Foundational Period of British Abolition," *Slavery and Abolition* 32 (2011): 185–98.

35. Sarah Yeh, " 'A Sink of All Filthiness': Gender, Family, and Identity in the British Atlantic, 1688–1763," *Historian* 68 (2006): 66–83; and Wylie Sypher, "The West-Indian as a 'Character' in the Eighteenth Century," *Studies in Philology* 36 (1939): 503–20.

36. Brendan Simms, *Three Victories and a Defeat: The Rise and Fall of the First British Empire, 1714–1783* (London: Allen Lane, 2007); P. J. Marshall, *The Making and Unmaking of Empires: Britain, India and America c. 1750–1783* (Oxford: Oxford University Press, 2005).

37. Important works include Linda Colley, *Britons: Forging the Nation 1707–1837* (New Haven, Conn.: Yale University Press, 1992); David Armitage, *The Ideological Origins of the British Empire* (Cambridge: Cambridge University Press, 2000); Eliga H. Gould, *The Persistence of Empire: British*

Political Culture in the Age of the American Revolution (Chapel Hill: University of North Carolina Press, 2000); P. J. Marshall, *"A Free Though Conquering People": Eighteenth-Century Britain and Its Empire* (Aldershot: Ashgate, 2003); and Eliga H. Gould, "Zones of Law, Zones of Violence: The Legal Geography of the British Atlantic," *William and Mary Quarterly*, 3rd ser., 60 (2003): 471–510.

38. Jack P. Greene, *Evaluating Empire and Confronting Colonialism in Eighteenth-Century Britain* (New York: Cambridge University Press, 2013), xi.

39. Ibid., xii–xiii. Greene echoes and amplifies points made in P. J. Marshall, "The Moral Swing to the East: British Humanitarianism, India and the West Indies," in *"A Free Though Conquering People,"* ch. 9.

40. Richard Bourke, *Empire and Revolution: The Political Life of Edmund Burke* (Princeton: Princeton University Press, 2017); Hannah Weiss Muller, *Subjects and Sovereign: Bonds of Belonging in the Eighteenth-Century British Empire* (New York: Oxford University Press, 2017); Dana Rabin, " 'In a Country of Liberty?': Slavery, Villeinage and the Making of Whiteness in the Somerset Case (1772)," *History Workshop Journal* 72 (2011): 5–29; Brooke N.. Newman, *A Dark Inheritance: Blood, Race, and Sex in Colonial Jamaica* (New Haven, Conn.: Yale University Press, 2018); H. V. Bowen, "British Conceptions of Global Empire, 1756–1783," *Journal of Imperial and Commonwealth History* 26 (1998): 1–27; Greene, *Evaluating Empire and Confronting Colonialism*; and Burnard, "Harvest Years." Marshall's work is the indispensable background to any exploration of Britain's encounter with non-whites in imperial context. See the four presidential addresses to the Royal Historical Society, 1998–2001, collected as P. J. Marshall, "Britain and the World in the Eighteenth Century," *Transactions of the Royal Historical Society*, vol. 8 (1998): 1–18; vol. 9 (1999): 1–16; vol. 10 (2000): 1–16; and vol. 11 (2001): 1–15.

41. Heather Freund, "Who Should Be Treated 'with Every Degree of Humanity'? Debating Rights for Planters, Soldiers, and Caribs/Kalinago on St. Vincent, 1763–1773," *Atlantic Studies* 13 (2016): 125–43; J. Paul Thomas, "The Caribs of St. Vincent: A Study in Imperial Maladministration, 1763–1773," *Journal of Caribbean History* 18 (1983): 60–73; Christopher Taylor, *The Black Carib Wars: Freedom, Survival, and the Making of the Garifuna* (Jackson: University of Mississippi Press, 2012); and Brooke N. Newman "Identity Articulated: British Settlers, Black Caribs, and the Politics of Indigeneity on St. Vincent, 1763–1797," in *Native Diasporas: Indigenous Identities and Settler Colonialism in the Americas*, ed. Gregory D. Smithers and Brooke N. Newman (Lincoln: University of Nebraska Press, 2014), 109–49.

42. Davis, *Problem of Slavery in the Age of the American Revolution*, 394–95.

43. Granville Sharp, *A Representation of the Injustice and Dangerous Tendency of Tolerating Slavery; or of Admitting the Least Claim of Private Property in the Persons of Men, in England* (London, 1769), 13.

44. Sharp's attitudes toward America (more than his relationship to the West Indies, which was consistently hostile) were mixed. He was a "friend of America" insofar as he supported colonial claims that in the absence of direct representation in Britain they had the right to legislate for themselves. But he was also concerned about what he called "the natural rights of mankind." Americans should have the right to make laws for themselves, but those laws should include abolishing slavery. If they did not do this, then colonial legislators were no better than the British parliamentarians he despised. One could dismiss their complaints against tyranny because American liberty "has so little right to that sacred name, that it seems to differ from the arbitrary power of monarchies only in one circumstance; viz. that it is a *many-headed monster of tyranny*, which entirely subverts our excellent constitution." As Christopher Brown notes, Sharp "used the budding conflict to reshape how the British thought about the oppression of Africanism to reorient anti-slavery sentiment in a way increasingly common in revolutionary America," where there was a willingness "to inspect their own institutions, to conceive the guilt of slavery as a common possession of the American people." In short, Sharp supported American independence but only if Americans, as occurred in Pennsylvania and New England, rethought the colonial relationship with slavery and started to get rid of the institution on American soil. Sharp, *Representation*, 81–82; Christopher Brown, *Moral Capital: Foundations of British Abolitionism* (Chapel Hill: University of North Carolina Press, 2006), 161–82 (quote 170).

45. Sharp, *Representation*, 92, 110. See also Catherine Molineux, *Faces of Perfect Ebony: Encountering Atlantic Slavery in Imperial Britain* (Cambridge, Mass.: Harvard University Press, 2012).

46. For the racialization of whiteness, see Deirdre Coleman, "Janet Schaw and the Complexions of Empire," *Eighteenth-Century Studies* 36 (2003): 169–93.

47. Prince Hoare, *Memoirs of Granville Sharp* (London, 1820), iv; Katherine Paugh, "The Curious Case of Mary Hylas: Wives, Slaves, and the Limits of British Abolitionism," *Slavery and Abolition* 35 (2014): 629–51.

48. Sharp, *Representation*, 133–34; Rabin, " 'In a Country of Liberty?' " 5, 7, 18–19, 24.

49. Gretchen Holbrook Gerzina, *Black London: Life Before Emancipation* (New Brunswick, N.J.: Rutgers University Press, 1995), 104–5.

50. Cited in Brown, *Moral Capital*, 94.

51. Cited in Wise, *Though the Heavens May Fall*.

52. Shyllon, *Black Slaves in Britain*, 82–164.

53. Sharp, *Representation*, 65, 68.

54. Ibid., 82, 163.

55. For slave traders' complaints about *Somerset*, see An African Merchant [John Peter Demarin], *A Treatise upon the Trade from Great Britain to Africa* (London, 1772); and Thomas Thompson, *The African Trade for Negro Slaves Shewn to be Consistent with Principles of Humanity and the Laws of Revealed Religion* (Canterbury, 1772). These are usefully analyzed in Greene, *Evaluating Empire*, 180–83.

56. [Long], *Candid Reflections*, 4, 13.

57. Expositions of the arguments of Estwick, Long, and Martin (the latter being much less substantial than the first two pamphlets) can be found in Srividhya Swaminathan, "Developing the West Indian Proslavery Position After the *Somerset* Decision," *Slavery and Abolition* 24 (2003): 40–60; and Greene, "Liberty, Slavery and the Transformation of British West Indian Liberty."

58. Among a large literature, see Davis, *Problem of Slavery in the Age of Revolution*; David Brion Davis, *The Problem of Slavery in the Age of Emancipation* (New York: Alfred A. Knopf, 2014); Robin Blackburn, *The American Crucible: Slavery, Emancipation and Human Rights* (London: Verso, 2011); and Seymour Drescher, *Abolition—a History of Slavery and Antislavery* (New York: Cambridge University Press, 2009).

59. This is a point made in a different context by Kenneth Pomeranz, in his discussion of "ghost acres" as a way of determining the amount of land that would have been needed to turn over to the production of food with sufficient calories to match tropical agriculture. Kenneth Pomeranz, *The Great Divergence: China, Europe and the Making of the Modern World* (Princeton, N.J.: Princeton University Press, 2000).

60. Christa Dierksheide, *Amelioration and Empire: Progress and Slavery in the Plantation Americas* (Charlottesville: University of Virginia Press, 2014).

61. Estwick, *Considerations on the Negroe Cause*, xvi; Srividhya Swaminathan, *Debating the Slave Trade: Rhetoric and British National Identity* (Farnham: Ashgate, 2009), 158. For a sophisticated treatment of the various ways in which geography and place shaped understandings of the West Indies, see Elizabeth A. Bohls, *Slavery and the Politics of Place: Representing the Colonial Caribbean, 1770–1833* (Cambridge: Cambridge University Press, 2014).

62. Martin, *Short Treatise*, 5; Estwick, *Considerations on the Negroe Cause*, 80, 82.

63. Sharp, *Representation*, 46–49, 75, 81.

64. [Long], *Candid Reflections*, 41.

65. Estwick, *Considerations on the Negroe Cause*, 94; Martin, *Short Treatise*, 12.

66. Suman Seth, "Materialism, Slavery, and the *History of Jamaica*," *Isis* 105 (2014): 764–72.

67. Elizabeth Bohls, "The Gentleman Planter and the Metropole: Long's History of Jamaica," in *The Country and the City Revisited: England and the Politics of Culture, 1560–1840*, ed. Donna Landry, Gerald MacLean, and Joseph Ward (New York: Cambridge University Press, 1999), 180–96.

68. Long, *History of Jamaica*.

69. Ibid., 2:476–85; Estwick, *Considerations on the Negroe Cause*, 36, 51.

70. Long thought the French had a better imperial policy than the British, stating that "the *French* have shewn much sagacity, and great attention to the true interest of their American colonies, in this and in many other regulations affecting them, [which are] not unworthy of being copied by other great trading nations." Ibid., 46, 48, 49.

71. Ibid., 48.

72. Ibid., 49.

73. Ibid., 42, 49, 59, 62.

74. Ibid., 55.

75. Michael Craton, "Reluctant Creoles: The Planters' World in the British West Indies," in *Strangers within the Realm: Cultural Margins of the First British Empire*, ed. Bernard Bailyn and Philip D. Morgan (Chapel Hill: University of North Carolina Press, 1991), 333–35; Long, *History of Jamaica*, 2:76–77.

76. Trevor Burnard, " 'Rioting in Goatish Embraces: Marriage and Improvement in Early British Jamaica, 1660–1780," *History of the Family* 114 (2006): 185–97.

77. Wise, *Though the Heavens May Fall*.

78. Richard Bourke, " 'Pocock and the Presuppositions of the New British History," *Historical Journal* 53 (2010): 747–70; J. G. A. Pocock, "The New British History in Atlantic Perspective: An Antipodean Commentary," *American Historical Review* 102 (1999): 490–500.

79. For a similar conclusions, focusing on British distaste for aspects of West Indian culture and character, including their love of hot food and tendency to overeat as well as their penchant for sexual dalliances with black women, see Christer Petley, "Gluttony, Excess and the Fall of the Planter Class," *Atlantic Studies* 9 (2012): 85–106; and Brooke N. Newman, "Gender, Sexuality, and the Formation of Racial Identities in the Eighteenth-Century Anglo-Caribbean World," *Gender History* 22 (2010): 585–602.

80. Kathleen Wilson, *This Island Race: Englishness, Empire and Gender in the Eighteenth Century* (London: Routledge, 2003), 130.

81. Swaminathan, *Debating the Slave Trade*, 46–170; David Lambert, "The Counter Revolutionary Atlantic: White West Indian Petitions and Proslavery Networks," *Social and Cultural Geography* 6 (2005): 405–20. See also Colin Kidd, *British Identities Before Nationalism: Ethnicity and Nationhood in the Atlantic World 1600–1800* (Cambridge: Cambridge University Press, 1999).

82. J. G. A. Pocock, "The Ideal of Citizenship since Classical Times," in *The Citizenship Debates: A Reader*, ed. Gershon Shafir (Minneapolis: University of Minnesota Press, 1998), 31–41. For the Jew Bill, see Thomas W. Perry, *Public Opinion, Propaganda and Politics in Eighteenth-Century England: A Study of the Jew Bill of 1753* (Cambridge, Mass.: Harvard University Press, 1962). For England as being tolerant to Jews, see Todd Endelman, "The Englishness of Jewish Modernity in England," in *Toward Modernity: The European Jewish Model*, ed. Jacob Katz (New Brunswick, N.J.: Transaction Books, 1987), 225–46. For the opposite view, see Frank Felsenstein, *Anti-Semitic Stereotypes: A Paradigm of Otherness in English Populist Culture, 1660–1830* (Baltimore: Johns Hopkins University Press, 1995). For issues of subjecthood in the New World, see Anderson, "Old Subjects, New Subjects and Non-Subjects; Brooke N. Newman, "Contesting 'Black Liberty' and Subjecthood in the Anglophone Caribbean, 1730s–1780s," *Slavery and Abolition* 32 (2011): 169–83; and Jacob Selwood, "Left Behind: Subjecthood, Nationality, and the Status of Jews After the Loss of English Surinam," *Journal of British Studies* 54 (2015): 578–601. For French America, see John D. Garrigus, "New Christians/'New Whites': Sephardic Jews, Free People of Color and Citizenship in French Saint-Domingue, 1760–1789," in *The Jews and the Expansion of Europe to the West, 1450–1800*, ed. Paolo Bernadini and Norman Fiering (New York, 2001), 314–27.

83. Rabin, " 'In a Country of Liberty?' " 15–21.

84. Hoare, *Memoirs of Granville Sharp*, 91.

85. Mark Harrison, *Climates and Constitutions: Health, Race, Environment and British Imperialism in India, 1600–1850* (Oxford: Oxford University Press, 2002); and Colin Kidd, *The Forging of Races* (Cambridge: Cambridge University Press, 2006).

86. Greene, *Exclusionary Empire*.

87. Daniel Livesay, *Children of Uncertain Fortune: Mixed-Race Jamaicans in Britain and the Atlantic Family, 1733–1833* (Chapel Hill: University of North Carolina Press, 2018), 123–29.

88. Roxann Wheeler, *The Complexion of Race: Categories of Difference in Eighteenth-Century British Culture* (Philadelphia: University of Pennsylvania Press, 2000), 260–87; Petley, "Gluttony and Excess'; Newman, *Dark Inheritance.*

89. Nicolas B. Dirks, *The Scandal of Empire: India and the Creation of Imperial Britain* (Cambridge, Mass.: Harvard University Press, 2009), 32–34.

90. James Ramsay, *An Essay on the Treatment and Conversion of African Slaves in the British Sugar Colonies* (Dublin, 1784), 55–6, 72; James Ramsay, "Memorial Suggesting Motives for the Improvement of the Sugar Colonies," Add. Mss. 27261, British Library.

91. See, for example, [James Tobin], *Cursory Remarks upon the Reverend Mr. Ramsay's Essay on the Treatment and Conversion of African Slaves in the British Sugar Colonies* (London: Wilkie, 1785).

92. Burnard, "Powerless Masters."

93. Kay Dian Kriz, *Slavery, Sugar, and the Culture of Refinement: Picturing the British West Indies, 1700–1840* (New Haven, Conn.: Yale University Press, 2008), 115.

Chapter 7

1. The main records of the case are "Documents Relating to the Ship *Zong*," REC/19, National Maritime Museum, Greenwich, London (I refer to this document as *Sharp Transcript*), and "Answers of William Gregson (January 1784) and James Kelsall (November 1783)," E112/1258/173, NA. All references from the court cases not otherwise noted in this chapter come from these sources. A useful compendium is Andrew Lewis, "Martin Dockray and the *Zong*: A Tribute in the Form of a Chronology," *Journal of Legal History* 28 (2007): 357–70.

2. The case led to laws being passed between 1788 and 1795 to prevent such an incident being tried in such a way ever again. The most important of these laws was the Dolben Act, named after its sponsor, Sir William Dolben, which in 1794 had an amendment passed specifically disallowing insurance claims when enslaved Africans were thrown overboard. Elizabeth Donnan, ed., *Documents Illustrative of the History of the Slave Trade to America*, 4 vols. (Washington, D.C.: Carnegie Institution of Washington, 1930–55), 2:586. The amendment to the act is 29 Geo. III c.66, XV.

3. It is likely that the *William* landed its slaves in Kingston. The *Zong* engaged Kingston merchants Coppell and Aguilar as factors in the sale of slaves. The Gregsons, the owners of the *Zong* and the *William*, overwhelmingly concentrated on Kingston as a place to discharge slaves in Jamaica. Of eighty-three voyages to Jamaica, landing 28,697 sales, only one voyage—that of the *Zong* in 1781—was not to Kingston. All data relating to slaving voyages, unless otherwise indicated, come from the website Transatlantic Slave Trade Data Base http://www.slavevoyages.org/tast/database/search.faces. Moreover, Luke Collingwood, despite being desperately ill, made his way immediately after the sale of slaves from the *Zong*, according to James Kelsall, to Kingston, in which town he quickly died. I have not found any record of Collingwood's death in the Kingston burial registers for this year. Kingston Burial Register, 1722–1834, Jamaica Archives, Spanishtown, Jamaica.

4. The best recent treatment is James Walvin, *The Zong: A Massacre, the Law, and the End of Slavery* (New Haven, Conn.: Yale University Press, 2011).

5. James Oldham, "Insurance Litigation Involving the *Zong* and Other British Slave Ships, 1780–1807," *Journal of Legal History* 28 (2007): 299–318; Tim Armstrong, "Slavery, Insurance and Sacrifice in the Black Atlantic," in *Sea Changes: Historicizing the Ocean*, ed. Bernhard Klein and Gesa Mackenthun (London: Routledge, 2004), 167–85; and Michael Lobban, "Slavery, Insurance and the Law," *Journal of Legal History* 28 (2007): 319–28.

6. Vincent Brown, *The Reaper's Garden: Death and Power in the World of Atlantic Slavery* (Cambridge, Mass.: Harvard University Press, 2008), ch. 5; Robin Blackburn, *The Overthrow of Colonial Slavery, 1776–1848* (London: Verso, 1988), ch. 4; J. R. Oldfield, *Popular Politics and British Anti-Slavery: The Mobilization of Popular Opinion Against the Slave Trade* (London: Routledge, 1998), ch. 1.

7. *Sharp Transcript*, 1–3, 20–21.

8. Trevor Burnard, "'A Prodigious Mine': The Wealth of Jamaica Before the American Revolution Once Again," *Economic History Review* 54 (2001): 505–23.

9. J. R. Ward, "The Profitability of Sugar Planting in the British West Indies, 1650–1834," *Economic History Historical Review* 31 (1978): 207–9; Richard B. Sheridan, "The Crisis of Slave Subsistence in the British West Indies During and After the American Revolution," *William and Mary Quarterly*, 3rd ser., 33 (1976): 615–41. For the destructive impact of the American Revolution, see Selwyn Carrington, *The Sugar Industry and the Abolition of the Slave Trade, 1775–1810* (Gainesville: University Press of Florida, 2002); and Andrew Jackson O'Shaughnessy, *An Empire Divided: The American Revolution and the British Caribbean* (Philadelphia: University of Philadelphia Press, 2000). For the 1780s as difficult but not terminal, see John J. McCusker, "The Economy of the British West Indies, 1763–1790: Growth, Stagnation, or Decline?" in *Essays on the Economic History of the Atlantic World* (New York: Routledge, 1997), 330.

10. Michael Mulcahy, *Hurricanes and Society in the British Greater Caribbean, 1624–1783* (Baltimore: Johns Hopkins University Press, 2006), 107–15.

11. Letter from Jeremiah Meyler to Meyler and Maxse, Westmoreland, 30 October 1780; Letter from Meyler to Richard Bright, Bristol, 5 November 1783, in *The Bright-Meyler Papers: A Bristol–West India Connection, 1732–1837*, ed. Kenneth Morgan (Oxford: Oxford University Press, 2007), 516, 521.

12. Ibid., 110–11, 165. See also John Fowler, *A General Account of the Calamities Occasioned by the Late Tremendous Hurricanes and Earthquakes in the West-India Islands* (London, 1781).

13. Trevor Burnard, *Mastery, Tyranny, and Desire: Thomas Thistlewood and His Slaves in the Anglo-Jamaican World* (Chapel Hill: University of North Carolina Press, 2004), 65–66.

14. O'Shaughnessy, *Empire Divided*, 166. Slave prices dropped severely in Jamaica. In 1774, 2,026 slaves in seven shipments were sold in Jamaica for an average price of £51.09. Between 1778 and 1783, 1,598 slaves in five shipments were sold in Jamaica for an average price of £40.81. The reduction in value was 20.1 percent. For Gold Coast slaves only, 1,401 slaves were sold in five shipments in 1774–75 for an average price of £55.86. Between 1776 and 1783, 1,012 slaves from three ships were sold for £39.71 each. The percentage reduction was 29 percent.

15. Lowell J. Ragatz, *The Fall of the Planter Class in the British Caribbean, 1763–1833* (New York: Century, 1928), 165–66.

16. Walvin thinks that the acquisition of the *Zong* was an opportunity for William Gregson to double his trade for a relatively small outlay. Walvin, *Zong*, 69. But there are other ways of reading the evidence. Gregson had not sent a ship to Africa for three years before 1781. Moreover, for the first time, this venture was a family enterprise that he shared with his two sons and son-in-law. His two sons had reentered the Atlantic slave trade a year before Gregson but with mixed success. They had sent the *Swallow* to Tortola in early 1781, where it had landed 186 Africans but had been shipwrecked on its return to Britain. Certainly, the rapidity by which the Gregsons insured the *Zong* after being told that Hanley had bought it, on 3 July 1781, and the speed by which they sued their insurers for nonpayment after their insurers had rejected their claim for £8,000 recompense in late 1782, suggest that they faced financial hardship. In addition, the two ship captains chosen to replace the deceased Hanley and Collingwood in bringing the *William* and the *Richard* back to Liverpool were never in charge of any other slave ship. The *William* arrived in Kingston, out of season and during a prolonged economic downturn, on 26 May 1781. Stephen Behrendt, "Ecology, Seasonality, and the Transatlantic Slave Trade," in *Soundings in Atlantic History: Latent Structures and Intellectual Currents*, ed. Bernard Bailyn and Patricia Denault (Cambridge, Mass.: Harvard University Press, 2009), 44–85. For the Gregsons, see Walvin, *Zong*, 56–66; Ian Baucom, *Specters of the Atlantic: Finance Capital, Slavery, and the Philosophy of History* (Durham, N.C.: Duke University Press, 2005), 48–49.

17. Mulcahy, *Hurricanes and Society*, 111–12. Both Collingwoods came from Newcastle. Luke was the son of a barber-surgeon, born sometime in the 1740s and living in Liverpool by 1764. Lewis, "Martin Dockray and the *Zong*," 358. Cuthbert Collingwood (1748–1810) was a younger son from a middle-class family. Max Adams, *Admiral Collingwood: Nelson's Own Hero* (London: Weidenfeld and Nicolson, 2005).

18. O'Shaughnessy, *Empire Divided*, 232–34.

19. Linda E. Sturtz, "The 1780 Hurricane Donation: 'Insult Offered Instead of Relief,'" *Jamaican Historical Review* 21 (2001): 38.

20. Petition of Free People of Color, Hanover, March 2, 1782, and from Westmoreland, April 16, 1782, *JHA* 2:460, 470.

21. Mulcahy, *Hurricanes and Society*, 165–66, 168–70,180–82; Burnard, *Mastery, Tyranny, and Desire*, 65.

22. Walvin, *Zong*, 56–66.

23. David Richardson, "Prices of Slaves in West and West-Central Africa: Toward an Annual Series, 1698–1807," *Bulletin of Economic Research* 43 (1991): 55.

24. Baucom, *Specters of the Atlantic*, 95. See also Anita Rupprecht, "Excessive Memories: Slavery, Insurance and Resistance," *History Workshop Journal* 64 (2008): 6–28.

25. Stephanie Smallwood, *Saltwater Slavery: A Middle Passage from Africa to American Diaspora* (Cambridge, Mass.: Harvard University Press, 2007).

26. Joseph Arnould, *A Treatise on the Law of Marine Insurance*, 2 vols. (London, 1849), 1:320. Lord Mansfield had ruled in 1761 that valued policies were not void as gaming contracts. In colonial commerce, the effect was to allow insurance to become a means of making speculative gains. Lobban, "Slavery, Insurance and the Law," 324.

27. Harold Raynes, *A History of British Insurance* (London, 1948), 188–89; Jeremy Krikler, "The *Zong* and the Lord Chief Justice," *History Workshop Journal* 64 (2007): 32.

28. Lobban, "Slavery, Insurance and the Law," 324–25.

29. Christopher Kingston, "Governance and Institutional Change in Marine Insurance, 1350–1850," *European Review of Economic History* 18 (2014): 1–18; and Oldham, "Insurance Litigation."

30. Oldham, "Insurance Litigation," 305–10.

31. *Sharp Transcript*, 52–53; John Weskett, *A Digest of the Theory, Laws and Practice of Insurance* (London: Frys, Couchman, and Collier, 1781), 11, 525.

32. Why would members of the crew be interested in maximizing the profit for their employers? Slave voyages went wrong frequently. Certainly William Gregson, as an experienced slave trader, had faced his share of disappointments. Over his long career he suffered eleven shipwrecks, four wartime seizures of his vessel, and three slave insurrections (in 1765, 1776, and 1789). Walvin, *Zong*, 56, 60. The counsel for the insurers suggested that one reason why Collingwood might have chosen to enact an insurance fraud was that he desired "to saddle a bad Market upon the Underwriters instead of the Owners." They added that Collingwood was afraid that his first command "would make a bad voyage for the Owners." *Sharp Transcript*, 13–14; "Petition of Gilbert Syndicate to William Pitt," E112/1258/173, NA. But the person on the *Zong* who most wanted Gregson's favor might have been Robert Stubbs. He presented himself as an honest broker, a man "unconnected with the all the contending parties," including the owners, and "the only person who could be said to be perfectly disinterested in this question." *Sharp Transcript*, 65–66. But this was disingenuous. For a start, Stubbs had spent several months living with the Gregsons' favored ship captain, Richard Hanley—Hanley had arrived at Anomabu just when Stubbs's career was exploding. Hanley arrived on 14 January 1781, just after Stubbs had been recalled and a week before he was suspended officially from his post. On 26 January 1781, Stubbs suffered the indignity of having his clothes pulled off him by an angry mob, exposing him in "a cruell and shameful manner there in that posture among a vast number of Blacks, both men and women." Letter, 1781, T70/146, cited in Walvin, *Zong*, 84. Hanley was Stubbs's only friend. A creature of ridicule in Africa, Stubbs had prospects in England that were so poor he decided to take the long way back on a slaver rather than a direct route home. When in Britain he faced financial hardship. At the same time as he was substantiating the case of the Gregsons in March 1783, he was also involved in a dispute over money with his previous employers, in which he claimed he was owed £444 rather than the actual £73. "Minutes of the African Company," T70/145/134, NA. The Gregsons were on the verge of a period of sustained prosperity—between 1781 and 1790 they shipped 8,018 Africans from the Gold Coast to Jamaica, including 2,427 from Anomabu. Walvin, *Zong*, 161. William Gregson was also in the process of transferring his active participation in the trade to his sons and son-in-law. One imagines that this new and somewhat inexperienced generation of slave traders might find useful a man

with considerable experience in the slave trade, including contacts at Anomabu. That Stubbs might want to ingratiate himself with the Gregsons might also explain another mystery. Stubbs insisted he was not involved with any of the murders, although he admitted that he "amused himself with seeing [the victims] out of the Cabin Windows plunging into the Sea." Yet among those murdered were four slaves that he intended to sell in Jamaica. But he showed a remarkable indifference to his slaves' fate: "I never knew whether they were dead or living till we got to Jamaica." As Walvin states, the credibility of Stubbs is extremely questionable when it comes to his enslaved property and not much more credible in other areas, either. Walvin, *Zong*, 149–50.

33. Lewis, "Martin Dockray and the *Zong*," 363. James Kelsall claimed that he had objected at the meeting to throwing Africans overboard on 29 November because of the "horrid brutality" of the proposal and because he thought there was sufficient water for a few days, during which time Kelsall thought they could wait to see if another ship passed by or if rain fell. This piece of evidence was probably why the insurers insisted on calling Kelsall to give evidence. Lobban, "Slavery, Insurance and the Law," 321–22.

34. Stubbs commanded the *Black Joke* in 1757. The trip was not a success, as his vessel was captured by the French and forced to disembark in Martinique. The vessel was notorious for another slave trade atrocity. Isaac Parker, a seaman, testified to the House of Commons how Thomas Marshall, captain of the *Black Joke* in a 1765 trip to Barbados, flogged and tortured a nine-month-old child to death. "Testimony of Isaac Parker, 1790," in *House of Commons Sessional Papers of the Eighteenth Century*, ed. Sheila Lambert (Wilmington, Del.: Scholarly Resources, 1975), 73:124–25, 130.

35. "Letter from William Llewellin, Gold Coast, 13 January 1781," T[reasury] 70/1695, National Archives, Kew, London.

36. Walvin, *Zong*, 76–87. George Gregory, *Essays, Historical and Moral* (London: J. Johnson, 1785), 304, 355–78.

37. Oldham, "Insurance Litigation," 305, 309–10.

38. *Sharp Transcript*, 50–52.

39. David Richardson, "Shipboard Revolts, African Authority, and the Atlantic Slave Trade," *William and Mary Quarterly*, 3rd ser., 57 (2001): 69–92; Eric Robert Taylor, *If We Must Die: Shipboard Insurrections in the Era of the Atlantic Slave Trade* (Baton Rouge: Louisiana State University Press, 2006); and Stephen D. Behrendt, David Eltis, and David Richardson, "The Costs of Coercion: African Agency in the Pre-Modern World," *Economic History Review* 54 (2001): 454–76.

40. Samuel Gamble faced the same problem on the *Sandown* in 1794. He put everyone on two-thirds rations for sixteen days before refilling water casks in Barbados. These restrictions led to sixteen of his crew absconding at Barbados. Bruce L. Mouser, ed., *A Slaving Voyage to Africa and Jamaica: The Log of the Sandown, 1793–1794* (Bloomington: Indiana University Press, 2002), 104, 107–9.

41. The standard slave-to-ton ration in the British transatlantic slave trade prior to the passing of the Dolben Act in 1788 was 1.60. See Herbert S. Klein, *The Atlantic Slave Trade* (New York, 1999), 133. David Eltis, *The Rise of Atlantic Slavery in the Americas* (New York: Cambridge University Press, 2000), 125–28.

42. Alexander Falconbridge, *An Account of the Slave Trade on the Coast of Africa* (London: James Phillips, 1788), 32.

43. Weskett, *Digest*, 525; Oldham, "Insurance Litigation," 303–5.

44. Emma Christopher, *Slave Ship Sailors and Their Captive Cargies, 1730–1807* (Cambridge: Cambridge University Press, 2006), 171.

45. Sharp was determined that the crew of the *Zong* be arraigned for murder and made his views publicly known. Walvin, *Zong*, 104. An interesting question is why Mansfield was determined not to widen the case into a murder investigation, as Sharp wanted and as counsel for the defendants asserted was true to the facts of the case. Krikler asserts that Mansfield's judicial strategy must be seen in the context of his determination to protect the emerging system of commercial law and to maintain a commitment to certainty in the law over principle. He was concerned that a major statement influencing the development of marine insurance law not be derailed by transforming an insurance case into a case about murder. His eye was fixed on the questions of absolute necessity and the general average,

and he did not want to weaken the import of these questions by being side-tracked into considering slaves not as property but as humans. Krikler, "*Zong* and the Lord Chief Justice."

46. Granville Sharp to Lords Commissioners of the Admiralty, n.p., 2, July 1783, *Tracts* 35 (Old Jewry London, MS), British Library cited in Michelle Faubert, *Granville Sharp's Uncovered Letter and the Zong Massacre* (Basingstoke: Palgrave Macmillan, 2018), 139.

47. By way of comparison, the log of the *Sandown* shows that on its return voyage to Britain from Kingston, it took twelve days to sail from the Grand Cayman Islands past Mattanzas on the north coast of Cuba to the Florida coastline. Mouser, *Slaving Voyage*, 118–19.

48. Sharp to Lords Commissioner, cited in Faubert, *Granville Sharp's Uncovered Letter*, 138–39.

49. There is some evidence that ship captains and factors were keenly aware of trading conditions in late 1781. See the following letter: "Capt. Thorburn, of London, touched at St. Kitts, with 450 Negroes; if he gets down to Jamaica, by the advises we have received, he will come to a very great market. Capt. Hanley's prize De Jong, Capt. Collingwood, was to sail with Thorburn, and there is not any hearing of her." Letter, John and Thomas Hodgson to Richard Miles, Liverpool, 19 January 1782, T/70/1545, National Archives, Kew, London.

50. Morgan, *Bright-Meyler Papers*, 32.

51. Burnard, *Mastery, Tyranny, and Desire*, 3–4.

52. Evidence of William James, Master of the *Hound*, in [John Ranby], *Observations on the Evidence Given Before the Committee of the Privy Council in Support of the Bill for Abolishing the Slave Trade* (London: J. Stockdale, 1791), 103–4.

53. Mouser, *Slaving Voyage*, 111–16.

54. Lewis, "Martin Dockray and the *Zong*," 365.

55. Mulcahy, *Hurricanes and Society*, 147–49, 167–68.

56. In 1788, Clarkson wrote of the *Zong* that it was an event "unparalleled in the memory of man . . . and of so black and complicated a nature, that were it to be perpetuated to future generations . . . it could not possibly be believed." Thomas Clarkson, *Essay on the Slavery and Commerce of the Human Species* (London: J. Phillips, 1788), 99. See also John Newton, *Thoughts upon the African Slave Trade* (London: J. Buckland and J. Johnson, 1788), 11. For his recapitulation of the event late in his life, see Thomas Clarkson, *The History of the Rise, Progress and Accomplishment of the African Slave Trade by the British Parliament*, 2 vols. (London: J. W. Parker, 1839), 2:377.

Chapter 8

1. The incongruity has also been noted by Vincent Brown, *The Reaper's Garden: Death and Power in the World of Atlantic Slavery* (Cambridge, Mass.: Harvard University Press, 2008), 253.

2. Simon Taylor to Chaloner Arcedeckne, Kingston, 5 June 1775, in *Travel, Trade and Power in the Atlantic, 1765–1884*, ed. Betty Wood et al. (Cambridge: Royal Historical Society, 2002), 148.

3. Simon Taylor to John Taylor. Kingston, 7 June 1774, Institute of Commonwealth Studies, Taylor Papers, II/A/9.

4. Simon Taylor's life and correspondence are covered in the following works, from which the quotations from Taylor are drawn: Wood and Lynn, *Travel, Trade and Power*; Christer Petley, "'Home' and 'This Country': Britishness and Creole Identity in the Letters of a Transatlantic Slaveholder," *Atlantic Studies* 6 (2009): 43–62; Christer Petley, "'Devoted Islands' and "That Madman Wilberforce': Proslavery Patriotism During the Age of Abolition," *Journal of Imperial and Commonwealth History* 39 (2011): 393–415; Richard B. Sheridan, "Simon Taylor, Sugar Tycoon of Jamaica, 1740–1813," *Agricultural History* 45 (1971): 285–96; Trevor Burnard, "Harvest Years: Reconfigurations of Empire in Jamaica, 1756–1807," *Journal of Commonwealth and Imperial History* 40 (2012): 546–47; Trevor Burnard and John Garrigus, *The Plantation Machine: Atlantic Capitalism in French Saint-Domingue and British Jamaica, 1748–1788* (Philadelphia: University of Pennsylvania Press, 2016), 173, 179. Taylor is the main protagonist in Christer Petley, *White Fury: A Jamaican Slaveholder and the Age of Revolution* (Oxford: Oxford University Press, 2018).

5. Trevor Burnard and Cécile Vidal, "Location and the Conceptualization of Historical Frameworks: Early North American History and Its Multiple Reconfigurations in the US and in Europe," in *You, the People: Historical Writing About the United States in Europe*, ed. Nicolas Barreyre, Michael Heale, Stephen Tuck, and Cécile Vidal (Berkeley: University of California Press, 2014), ch. 7; and Trevor Burnard, "The British Atlantic World," in *Atlantic History: A Critical Appraisal*, ed. Jack P. Greene and Philip D. Morgan (New York: Oxford University Press, 2009), 111–36.

6. David Armitage, "The First Atlantic Crisis: The American Revolution," in *Early North America in Global Perspective*, ed. Philip D. Morgan and Molly A. Warsh (New York: Routledge, 2014), 309–36.

7. Gordon S. Wood, *The Radicalism of the American Revolution* (New York: Alfred A. Knopf, 1992); Patrick Griffin, *America's Revolution* (New York: Oxford University Press, 2013); T. H. Breen, *American Insurgents, American Patriots* (New York: Hill and Wang, 2010); and Thomas Slaughter, *Independence: The Tangled Roots of the American Revolution* (New York: Hill and Wang, 2014).

8. Kathleen Duval, *Independence Lost: Lives on the Edge of the American Revolution* (New York: Random House,2015), 226–27. For the international context, see Eliga Gould, *Among the Powers of the Earth: The American Revolution and the Making of a New World Empire* (Cambridge, Mass.: Harvard University Press, 2012); David Armitage, *The Declaration of Independence: A Global History* (Cambridge, Mass.: Harvard University Press, 2007); Andrew Stockley, *Britain and France at the Birth of America: The European Powers and the Peace Negotiations of 1782–1783* (Exeter: University of Exeter Press, 2001); and Paul Mapp, "The Revolutionary War and Europe's Great Powers," in *Oxford Handbook of the American Revolution*, ed. Edward G. Gray and Jane Kamensky (New York: Oxford University Press, 2013), 311–26. An important collection is Simon Newman, ed., *Europe's American Revolution* (Basingstoke: Palgrave Macmillan, 2006).

9. Rosemarie Zagarri, "The Significance of the 'Global Turn' for the Early American Republic: Globalization in the Age of Nation-Building," *Journal of the Early Republic* 31 (2011): 1–37.

10. Nick Bunker, *An Empire on the Edge: How Britain Came to Fight America* (New York: Vintage, 2014).

11. Ibid., 368–69.

12. Andrew Jackson O'Shaughnessy, *The Men Who Lost America: British Leadership, the American Revolution, and the Fate of Empire* (New Haven, Conn.: Yale University Press, 2013), 17–46; Jeremy Black, *George III: America's Last King* (New Haven, Conn.: Yale University Press, 2006); John L. Bullion, "The *Ancien Regime* and the Modernizing State: George III and the American Revolution," *Anglican and Episcopal History* 68 (1999): 67–84; and Linda Colley, "The Apotheosis of George III: Loyalty, Royalty and the British Nation, 1760–1820," *Past and Present* 102 (1984): 94–129.

13. Thomas Jefferson described his one and only meeting with George III in the summer of 1786 as deeply unpleasant. Thirty-five years after this meeting, he described how "nothing could have been more ungracious" than the reception the king had given him and John Adams. *Autobiography of Thomas Jefferson 1743–1790*, ed. Paul Leicester Ford (New York: Putnam's, 1914), 94. He said nothing at the time, however, and some commentators think that Jefferson took offence at his treatment only late in life. Charles R. Ritcheson, "The Fragile Memory: Thomas Jefferson at the Court of George III," *Eighteenth-Century Life* 6 (1981): 1–16. Andrew O'Shaughnessy, however, is inclined to believe that George III was indeed very rude to Jefferson. O'Shaughnessy, *Men Who Lost America*, 366n4.

14. Andrew Jackson O'Shaughnessy, *An Empire Divided: The American Revolution and the British Caribbean* (Philadelphia: University of Philadelphia, 2000), ch. 1.

15. Perry Gauci, *William Beckford: First Prime Minister of the London Empire* (New Haven, Conn.: Yale University Press, 2013), 203, 205.

16. William Doyle comments that it is "not a complete exaggeration to suggest that the costs of upholding a colonial system that could only work through slavery were what ultimately brought down the Ancien Regime in France." William Doyle, "Slavery and Serfdom," in *The Oxford Handbook of the Ancien Régime*, ed. William Doyle (Oxford: Oxford University Press, 2012), 271.

17. O'Shaughnessy, *Empire Divided*. For brief treatments of the British Caribbean during the American Revolution, see Edward L. Cox, "The British Caribbean in the Age of Revolution," in *Empire and Nation: The American Revolution in the Atlantic World*, ed. Eliga H. Gould and Peter S. Onuf

(Baltimore: Johns Hopkins University Press, 2005), 274–94; and Matthew Mulcahy, *Hubs of Empire: The Southeastern Lowcountry and British Caribbean* (Baltimore: Johns Hopkins University Press, 2014), 205–14. See also Edward Rugemer, *The Politics of Atlantic Slavery: A Comparative History of Jamaica and South Carolina* (Cambridge, Mass.: Harvard University Press, 2018); Trevor Burnard, *Planters, Merchants, and Slaves: Plantation Societies in British America, 1650–1850* (Chicago: University of Chicago Press, 2015), ch. 5; and Burnard and Garrigus, *Plantation Machine*, chs. 7–9; Burnard, "Empires, the Age of Revolution and Plantation America," in *The Routledge History of Western Empires*, ed. Robert Aldrich and Kirsten McKenzie (London: Routledge, 2014), 46–58; and Trevor Burnard, "Slavery and the Causes of the American Revolution in Plantation British America," in *The World of the Revolutionary American Republic: Expansion, Conflict, and the Struggle for a Continent*, ed. Andrew Shankman (New York: Routledge, 2014), 54–76.

18. Christopher Leslie Brown, "The Problems of Slavery," in *Oxford Handbook of the American Revolution*, ed. Gray and Kamensky, 431. For the importance of the Battle of the Saintes see Stephen Conway, "'A Joy Unknown for Year's Past': The American War, Britishness and the Celebration of Rodney's Victory at the Saints," *History* 86 (2001): 180–99. Rodney's victory stopped a potential invasion of Jamaica by French and Spanish troops. This episode in the American Revolution, perhaps because it happened after the Battle of Yorktown, has been comprehensively ignored in revolutionary historiography, even though the number of men lost to disease—around seven thousand—in Saint-Domingue in the first five months of 1782 was easily the largest single loss of lives in any single campaign during the War for American Independence. Burnard and Garrigus, *Plantation Machine*, 212–15.

19. Nicholas Radburn, "Guinea Factors, Slave Sales, and the Profits of the Transatlantic Slave Trade in Eighteenth-Century Jamaica: The Case of John Tailyour," *William and Mary Quarterly*, 3rd ser., 72 (2015): 243–86.

20. Gray and Kamensky, *Oxford Handbook of the American Revolution*, 4, 188.

21. Colin G. Calloway, *The Scratch of a Pen: 1763 and the Transformation of North America* (Oxford: Oxford University Press, 2006).

22. O'Shaughnessy, *Empire Divided*, ch. 4; Mulcahy, *Hubs of Empire*, 206–7.

23. There is considerable debate over the significance of the 1774 petition by Jamaica. Andrew O'Shaughnessy argues that the petition was not a political manifesto of support for American independence but a peculiar and limited document, passed by a minority mercantile faction from Kingston and after the announcement of the nonimportation and nonexportation resolutions of the Continental Congress. Jack P. Greene, on the other hand, considers the petition both a strong endorsement of the American political position and an acknowledgement on how their reliance on slavery kept them unable to stand up for their own rights in the tradition of independent English people. See O'Shaughnessy, *Empire Divided*, 137–44; and Jack P. Greene, *Peripheries and Center: Constitutional Development in the Extended Polities of the British Empire and the United States, 1607–1788* (Athens: University of Georgia Press, 1986), 139; and Jack P. Greene, "Liberty and Slavery: The Transfer of British Liberty to the West Indies, 1627–1865," in *Exclusionary Empire: English Liberty Overseas, 1600–1900*, ed. Jack P. Greene (New York: Cambridge University Press, 2010), 69–70.

24. Cited in Breen, *American Insurgents, American Patriots*, 67.

25. O'Shaughnessy, *Empire Divided*, chs. 4 and 6.

26. Jack P. Greene, "Changing Identity in the British Caribbean: Barbados as a Case Study," in *Colonial Identity in the Atlantic World*, ed. Anthony Pagden and Nicholas Canny (Princeton, N.J.: Princeton University Press, 1987), 213–66; Karl Watson, *The Civilised Island, Barbados: A Social History, 1750–1816* (Bridgetown, Barbados: K. Watson, 1979); O'Shaughnessy, *Empire Divided*, 208.

27. Trevor Burnard, "'A Matron in Rank, a Prostitute in Manners . . .': The Manning Divorce of 1741 and Class, Race, Gender, and the Law in Eighteenth-Century Jamaica," in *Working Out Slavery, Pricing Freedom: Perspectives from the Caribbean, Africa and the African Diaspora*, ed. Verene Shepherd (London: St. Martin's, 2002), 133–52.

28. Jack P. Greene, *The Constitutional Origins of the American Revolution* (New York: Cambridge University Press, 2011), 56.

29. Sarah Yeh, "Colonial Identity and Revolutionary Loyalty: The Case of the West Indies," in *The Oxford History of the British Empire: British North America in the Seventeenth and Eighteenth Centuries*, ed. Stephen Foster (Oxford: Oxford University Press, 2013), 208–9. For a similar argument, see Greene, "Liberty and Slavery," 67–70.

30. Jack P. Greene, "Liberty, Slavery and the Transformation of British Identity in the Eighteenth-Century West Indies," *Slavery and Abolition* 21 (2000): 24–25.

31. Cited in George Metcalf, *Royal Government and Political Conflict in Jamaica, 1729–1783* (London: Longmans, 1965), 189.

32. Breen, *American Insurgents, American Patriots*; Slaughter, *Independence*. It is important to note, however, that Americans devoted virtually no attention to persuading West Indians to join in their rebellion. That so little time was given to trying to make the thirteen colonies into the fourteen colonies with the addition of Jamaica is surprising, both given the importance of Jamaica to British geopolitical thinking and also because the easiest way to destroy a European army, as New Englanders knew from the attack on Cartagena in 1741 and the siege of Havana in 1762, and as was to become evident in the fruitless and extraordinarily ruinous efforts of Britain and France to intervene in Haiti in the 1790s and 1800s, was to send it to the West Indies. The reason was that disease killed European soldiers in droves. It would have required Britain to have employed the entirety of its armed forces to defend Jamaica if Jamaica had decided to rebel. They could have counted on losing many tens of thousands of troops in the process. Burnard and Garrigus, *Plantation Machine*, ch. 8. For how European soldiers could not cope with the Caribbean disease environment, see J. R. McNeill, *Mosquito Empires: Ecology and War in the Greater Caribbean, 1620–1914* (New York: Cambridge University Press, 2010).

33. O'Shaughnessy, *Empire Divided*, ch. 6.

34. Thomas Iredell to James Iredell, 8 January 1775, St. Dorothy, Jamaica, in *The Papers of James Iredell*, vol. 1, *1767–77*, ed. Don Higginbotham (Raleigh: North Carolina Division of Archives and History, 1976), 280.

35. Andrew O'Shaughnessy, "The West India Interest and the Crisis of American Independence," in *West Indies Accounts: Essays on the History of the British Caribbean and the Atlantic Economy in Honour of Richard Sheridan*, ed. Roderick A. McDonald (Kingston: University of the West Indies Press, 1996), 126.

36. Dr. John Lindsay to Dr. William Robertson, 6 August 1776, Robertson-Macdonald Letters, MS 3942, f. 262, National Library of Scotland. See B. W. Higman, *Proslavery Priest: The Atlantic World of John Lindsay, 1729–1788* (Kingston: University of the West Indies Press, 2011).

37. Burnard and Garrigus, *Plantation Machine*, chs. 5 and 6.

38. Greene, "Liberty, Slavery and the Transformation of British Identity."

39. Gould, *Among the Powers of the Earth*, 55–58.

40. Alan Taylor, *The Internal Enemy: Slavery and War in Virginia, 1772–1832* (New York: Norton, 2013), 3–4; Woody Holton, *Forced Founders: Indians, Debtors, Slaves, and the Making of the American Revolution in Virginia* (Chapel Hill: University of North Carolina Press, 1999), 70–71.

41. Among a large literature, see Brown, "The Politics of Slavery," 433–41. For a thirteen-colonies–centered perspective that sees the American Revolution as a wasted opportunity for black freedom, see Gary B. Nash, "Sparks from the Altar of '76: International Repercussions and Reconsiderations of the American Revolution," in *The Age of Revolutions in Global Context c. 1760–1840*, ed. David Armitage and Sanjay Subrahmanyam (London: Palgrave Macmillan, 2010), 1–19. See also Burnard, *Planters, Merchants, and Slaves*, 255–58; and the provocative arguments of Gerald Horne, *The Counter-Revolution of 1776: Slave Resistance and the Origins of the United States of America* (New York: New York University Press, 2014).

42. Julie Flavell, *When London Was Capital of America* (New Haven, Conn.: Yale University Press, 2010).

43. Structural similarities between the eighteenth-century mainland and the West Indies in social and economic organization were more important than ongoing personal contacts between individual planters and merchants. See Mulcahy, *Hubs of Empire*; and Paul Pressley, *On the Rim of the Caribbean: Colonial Georgia and the British Atlantic World* (Athens: University of Georgia Press, 2013). The area

of what later became the United States where there were the most frequent and long-lasting interactions with the Caribbean was New Orleans. Cécile Vidal, "Caribbean Louisiana: Church, *Métissage*, and the Language of Race in the Mississippi Colony During the French Period," in *Louisiana: Crossroads of the Atlantic World* (Philadelphia: University of Pennsylvania Press, 2013), ch. 7.

44. Richard Pares, *Yankees and Creoles: The Trade Between North America and the West Indies Before the American Revolution* (London: Longmans, 1956); Richard Pares, *War and Trade in the West Indies 1739–1763* (Oxford: Oxford University Press, 1936); and Thomas Truxes, *Defying Empire: Trading with the Enemy in Colonial New York* (New Haven, Conn.: Yale University Press, 2008).

45. O'Shaughnessy, *Empire Divided*, 131.

46. See Maria Alessandra Bollettino, "'Of Equal or of More Service': Black Soldiers and the British Empire in the Mid-Eighteenth-Century Caribbean," *Slavery and Abolition* 38 (2017): 510–33. William Henry Lyttleton was one of several governors (and probably the most influential in imperial circles) who were either West Indians or had West Indian experience. Antiguans Thomas Oliver of Massachusetts and Josiah Martin of North Carolina were the last royal governors of their respective colonies.

47. Michael A. McDonnell, *The Politics of War: Race, Class and Conflict in Revolutionary Virginia* (Chapel Hill: University of North Carolina Press, 2007); Holton, *Forced Founders*.

48. Burnard, *Planters, Merchants, and Slaves*, 237–38; Philip D. Morgan, *Slave Counterpoint: Black Culture in the Eighteenth-Century Chesapeake and Lowcountry* (Chapel Hill: University of North Carolina Press, 1998), 58–101.

49. Burnard, *Planters, Merchants, and Slaves*, 145–47; Burnard and Garrigus, *Plantation Machine*, 231–33; Richard B. Sheridan, "The Crisis of Slave Subsistence in the British West Indies During and After the American Revolution," *William and Mary Quarterly*, 3rd ser., 33 (1976): 615–41; Richard S. Dunn, *A Tale of Two Plantations: Slave Life and Labor in Jamaica and Virginia* (Cambridge, Mass.: Harvard University Press, 2014).

50. Richard S. Dunn, "The Glorious Revolution," in *The Oxford History of the British Empire: Origins of Empire*, ed. Nicholas Canny (Oxford: Oxford University Press, 1998), 445–66.

51. James Otis, *The Rights of the British Colonies Asserted and Proved* (Boston, 1764), in *Pamphlets of the American Revolution, 1750–1776*, ed. Bernard Bailyn (Cambridge, Mass.: Harvard University Press, 1965), 1:435–36, 439–40.

52. Benjamin Franklin, *Observations on the Increase of Mankind*, in *The Papers of Benjamin Franklin*, ed. Leonard W. Labaree et al. (New Haven, Conn.: Yale University Press 1959–), 4:225–34. For a sensitive discussion of Franklin's views on slavery in this pamphlet, see David Waldstreicher, *Runaway America: Benjamin Franklin, Slavery, and the American Revolution* (New York: Hill and Wang, 2004), 136–39.

53. Josiah Tucker, *A Treatise Concerning Civil Government, in Three Parts* (London: T. Cadell, 1781), 169.

54. Burnard, "Slavery and the Causes of the American Revolution in Plantation British America."

55. Isaac de Pinto, *Letters on the American Troubles, translated from the French* (London: John Boosey, 1776), 35–46, 72, 83.

56. Bunker, *Empire on the Edge*, 234.

57. Ibid., 243–45.

58. Franklin, *Observations*, 374.

59. Burnard, *Planters, Merchants, and Slaves*, chs. 1 and 3.

60. P. J. Marshall, "Britain's American Problem: The International Perspective," in Gray and Kamensky, *Oxford Handbook of the American Revolution*, 15–29. For mercantilism, see Steven Pincus, "Rethinking Mercantilism: Political Economy, the British Empire, and the Atlantic World in the Seventeenth and Eighteenth Centuries," *William and Mary Quarterly*, 3rd ser., 69 (2012): 3–34; and Philip J. Stern and Carl Wennerlind, eds., *Mercantilism Reimagined: Political Economy in Early Modern Britain and Its Empire* (Oxford: Oxford University Press, 2014). For doubts about whether mercantilism captures the reality of eighteenth-century political economy, see Julian Hoppit, *Britain's Political Economies: Parliament and Economic Life, 1660–1800* (Cambridge: Cambridge University Press, 2017), 5, 6, 17, 165, 215, 247, 324.

61. Marshall, "Britain's American Problem: The International Perspective," in Gray and Kamensky, *Oxford Handbook of the American Revolution* 19. Of course, France recognized what Britain had done and devoted the years after 1763 to building up its navy and waiting until it had a chance to inflict global revenge on Britain and so reduce what it considered its dangerous hegemony in Europe and the world. Jonathan R. Dull, *The French Navy and American Independence: A Study of Arms and Diplomacy, 1774–1787* (Princeton, N.J.: Princeton University Press, 1975).

62. Subjecthood was, however, as Hannah Weiss Muller argues, a capacious category and between the 1760s and 1780s was redefined through numerous debates about colonial rights throughout the empire. Hannah Weiss Muller, *Subjects and Sovereign: Bonds of Belonging in the Eighteenth-Century British Empire* (New York: Oxford University Press, 2017), 212. Subjects, moreover, disagreed that loyalty to the Crown meant being silent. British subjects—new and old—were relentless in proclaiming their "rights" and "privileges," often through petitions to the monarch. Hannah Weiss Muller, "From Requête to Petition: Petitioning the Monarch Between Empires," *Historical Journal* 60 (2017): 659–86.

63. Marshall, "Britain's American Problem," 20–21. This interpretation of British foreign policy in the age of the American Revolution follows that of Daniel Baugh, *The Global Seven Years' War, 1754 to 1763* (London: Longman, 2011); Daniel Baugh, "Withdrawing from Europe: Anglo-French Maritime Geopolitics, 1750–1800," *International History Review* 20 (1998): 1–32; and H. M. Scott, *British Foreign Policy in the Age of the American Revolution* (Oxford: Oxford University Press, 1990). For a different view, that argues that British ministers, led by the Earl of Rochford, believed that "a forward policy in Europe best secured Britain's maritime predominance," see Brendan Simms, *Three Victories and a Defeat: The Rise and Fall of the First British Empire* (London: Allen Lane, 2007), 514–15.

64. Philip Lawson, *The East India Company: A History* (London: Longmans, 1993); P. J. Marshall, *East Indian Fortunes: The British in Bengal in the Eighteenth Century* (Oxford: Oxford University Press, 1976); and Robert Travers, *Ideology and Empire in Eighteenth-Century India: The British in Bengal* (Cambridge: Cambridge University Press, 2007). For the pre-Plassey period, see Philip J. Stern, *The Company-State: Corporate Sovereignty and the Early Modern Foundations of the British Empire in India* (Oxford: Oxford University Press, 2011). An important overview is Jon Wilson, *India Conquered: Britain's Raj and the Chaos of Empire* (New York: Simon and Schuster, 2017).

65. James M. Vaughn, *The Politics of Empire and the Accession of George III: The East India Company and the Crisis and Transformation of Britain's Imperial Rule* (New Haven, Conn.: Yale University Press, 2019).

66. Bunker, *Empire on the Edge*, ch. 6; Lucy S. Sutherland, *The East India Company in Eighteenth-Century Politics* (Oxford: Oxford University Press, 1962), chs. 8 and 9. For the business structure of the East India Company, see H. V. Bowen, *The Business of Empire: The East India Company and Imperial Britain* (Cambridge: Cambridge University Press, 2005).

67. Of course, the other model of colonization that shaped imperial thinking was that of Spain. Spain provided an example of how a very small population of Spaniards exerted rule over a vast indigenous majority in the Americas, although Britons were contemptuous of Spaniard's cruelty to Indians and their heterogeneous mixing of populations. Hugh Thomas, *World Without End: Spain, Philip II, and the First Global Empire* (New York: Random House, 2015); and J. H. Elliott, *Empires of the Atlantic World: Britain and Spain in America, 1492–1830* (New Haven, Conn.: Yale University Press, 2007).

68. Burnard and Garrigus, *Plantation Machine*, 147–48.

69. Burnard, *Planters, Merchants, and Slaves*, 118–25; C. A. Bayly, *Imperial Meridian: The British Empire and the World, 1780–1830* (London: Longman, 1989); P. J. Marshall, *The Making and Unmaking of Empires: Britain, India and America c. 1760–1783* (Oxford: Oxford University Press, 2005); and Alan Atkinson, *The Europeans in Australia: The Beginning* (Melbourne: Oxford University Press, 1997). An interesting counterpoint to how Australia developed is what happened in French Guiana in the early nineteenth century, where France established its own colonial prison; see Miranda Spieler, *Empire and Underworld: Captivity in French Guiana* (Cambridge, Mass.: Harvard University Press, 2012).

70. Gordon S. Wood, *The Americanization of Benjamin Franklin* (New York: Penguin, 2004), 105–52.

71. Jack P. Greene, *Negotiated Authorities: Essays in Colonial Political and Constitutional History* (Charlottesville: University of Virginia Press, 1994); and James Belich, *Replenishing the Earth: The British Empire and the World, 1780–1830* (Oxford: Oxford University Press, 2009).

72. Christopher Leslie Brown, " 'Empire Without Slaves: British Concepts of Emancipation in the Age of the American Revolution," *William and Mary Quarterly*, 3rd ser., 56 (1999): 273–306.

73. The most negative assessment of the European founding of Australia is Robert Hughes, *The Fatal Shore: The Epic of Australia's Founding* (New York: Alfred A. Knopf, 1987). For a more balanced view, see Grace Karskens, *The Colony: A History of Early Sydney* (Sydney: Allen and Unwin, 2009).

74. Christopher Leslie Brown, *Moral Capital: Foundations of British Abolitionism* (Chapel Hill: University of North Carolina Press, 2006); Nicholas B. Dirks, *The Scandal of Empire: India and the Creation of Imperial Britain* (Cambridge, Mass.: Harvard University Press, 2009); and Maya Jasanoff, *Liberty's Exiles: American Loyalists in the Revolutionary World* (New York: Alfred A. Knopf, 2011).

75. We have been misled by later accounts of the abolitionist struggle, notably that by Thomas Clarkson, where he likened the course of abolitionism to a river, picking up strength over time until it became an irresistible flood, to underplay the role of contingency in the crucial events leading to the abolition of the slave trade and of slavery in the British Empire. Abolitionism was a movement that made its significant achievements in fits and starts (with the years 1772, 1787–88, 1806–7, and 1831–33 seeing the most significant triumphs), with long periods in which the slavery interest maintained its power within British and colonial politics.

76. Roger Anstey, *The Atlantic Slave Trade and British Abolition, 1760–1810* (London: Macmillan, 1975); and David Beck's Ryden, *West Indian Slavery and British Abolition, 1783–1807* (New York: Cambridge University Press, 2009).

77. Douglas R. Egerton, *Death of Liberty: African Americans and Revolutionary America* (New York: Oxford University Press, 2009); Sylvia Frey, *Water from the Rock: Black Resistance in a Revolutionary Age* (Princeton, N.J.: Princeton University Press, 1991); Cassandra Pybus, *Epic Journeys of Freedom: Runaway Slaves of the American Revolution and Their Global Quest for Liberty* (Boston: Beacon, 2006).

78. Brown, "The Problems of Slavery," 431 (quote); Taylor, *Internal Enemy*; Eric Foner, *The Fiery Trial: Abraham Lincoln and American Slavery* (New York: W. W. Norton, 2010), chs. 7 and 8.

79. David Geggus, "The Caribbean in an Age of Revolution," in *The Age of Revolutions in Global Context, 1760–1840*, ed. David Armitage and Sanjay Subrahmanyam (Basingstoke: Palgrave Macmillan, 2010), 83–99; David Geggus, *Slavery, War and Revolution: The British Occupation of Saint-Domingue, 1793–1798* (Oxford: Oxford University Press, 1982).

80. Nicholas Draper, *The Price of Emancipation: Slave-Ownership, Compensation and British Society at the End of Slavery* (Cambridge: Cambridge University Press, 2010); and Catherine Hall et al., *Legacies of British Slaveholding: Colonial Slavery and the Formation of Victorian Britain* (Cambridge: Cambridge University Press, 2011).

81. For the strength of Britain in the 1780s, see Seymour Drescher, "The Shocking Birth of British Abolitionism," *Slavery and Abolition* 33 (2012): 572–89.

82. J. G. A. Pocock, "British History: A Plea for a New Subject: A Reply," *Journal of Modern History* 47 (1975): 627.

83. For the consequences of defeat, see Stephen Conway, *The British Isles and the War for American Independence* (Oxford: Oxford University Press, 2000); and Harry T. Dickinson, "The Impact of the War on British Politics," in Gray and Kamensky, *Oxford Handbook of the American Revolution*, 355–69.

84. H. V. Bowen, "British Conceptions of Global Empire, 1756–1783," *Journal of Imperial and Commonwealth History* 26 (1998): 1–27; and Eliga H. Gould, "A Virtual Nation: Greater Britain and the Imperial Legacy of the American Revolution," *American Historical Review* 104 (1999): 476–89.

85. Richard B. Morris, *The Peacemakers: The Great Powers and American Independence* (Boston: Northeastern University Press, 1983); Stockley, *Britain and France at the Birth of America*; Duval,

Independence Lost, 231–37; O'Shaughnessy, *Men Who Lost America*, 361; and P. J. Marshall, *Remaking the British Atlantic: The United States and the British Empire After American Independence* (Oxford: Oxford University Press, 2012), ch. 1.

86. Adam Rothman, *Slave Country: American Expansion and the Origins of the Deep South* (Cambridge, Mass.: Harvard University Press, 2005); and George Van Cleve, *A Slaveholder's Republic: Slavery, Politics, and the Constitution in Early America* (Chicago: University of Chicago Press, 2010).

87. Steven Sarson, *The Tobacco-Plantation South in the Early American Plantation World* (London: Palgrave Macmillan, 2011); Allan Kulikoff, "'Such Things Ought Not to Be': The American Revolution and the First National Great Depression," in Shankman, *World of the Revolutionary American Republic*, 134–64.

88. Simon Taylor to Sir John Taylor, 2 November 1783, Simon Taylor Letter Book, 1A, Institute of Commonwealth Studies, University of London.

89. Petley, "'Home' and 'This Country'"; Trevor Burnard and Richard Follett, "Caribbean Slavery, British Anti-Slavery and the Cultural Politics of Venereal Disease," *Historical Journal* 55 (2012): 427–51; Trevor Burnard, "Powerless Masters: The Curious Decline of Jamaican Sugar Planters in the Foundational Period of British Abolition," *Slavery and Abolition* 32 (2011): 185–98.

90. Gary B. Nash, *The Forgotten Fifth: African Americans in the Age of the American Revolution* (Cambridge, Mass.: Harvard University Press, 2005).

91. Maya Jasanoff, *Liberty's Exiles: American Loyalists in the Revolutionary World* (New York: Alfred A. Knopf, 2011).

92. Ira Berlin, *The Long Emancipation: The Demise of Slavery in the United States* (Cambridge, Mass.: Harvard University Press, 2015).

93. Shane White, "'It Was a Proud Day': African Americans, Festivals, and Parades in the North, 1741–1834," *Journal of American History* 81 (1994): 13–50; Edward Rugemer, "Emancipation Day Traditions in the Anglo-Atlantic World," in *The Routledge History of Slavery*, ed. Gad Heuman and Trevor Burnard (London: Routledge, 2010), 316–20; Van Gosse, "'As a Nation the English Are Our Friends': The Emergence of African American Politics in the British Atlantic World, 1772–1861," *American Historical Review* 113 (2008): 1003–28.

94. Peter S. Onuf, *Jefferson's Empire: The Language of American Nationhood* (Charlottesville: University of Virginia Press, 2000), 117, 128.

95. Bryan Edwards, *The History, Civil and Commercial, of the British Colonies in the West Indies*, 5 vols., 5th ed. (London, 1819), 2:340–41; Francis Wharton, ed. *The Revolutionary and Diplomatic Correspondence of the United States* 6 vols. (Washington, D.C.: Government Printing Office, 1889), 6: 533.

Chapter 9

1. Eric Williams, *Capitalism and Slavery* (Chapel Hill: University of North Carolina Press, 1944).

2. Barbara L. Solow, "Capitalism and Slavery in the Very Long Run," *Journal of Interdisciplinary History* 17 (1987): 732.

3. See Roger Anstey, "Capitalism and Slavery: A Critique," *Economic History Review, , 21 (1968): 307–20; and Stanley L. Engerman, "The Slave Trade and British Capital Formation in the Eighteenth Century: A Comment on the Williams Thesis," *Business History Review* 46 (1972): 430–43, for early criticisms. For a defense, see J. E. Inikori, "Market Structure and the Profits of the British Atlantic Trade in the Late Eighteenth Century," *Journal of Economic History* 41 (1981): 745–76. A mostly critical set of essays was Barbara L. Solow and Stanley L. Engerman, eds., *British Capitalism and Caribbean Slavery: The Legacy of Eric Williams* (New York: Cambridge University Press, 1987). For a more positive view, see Heather Cateau and Selwyn Carrington, eds., *"Capitalism and Slavery" Fifty Years Later: Eric Eustace Williams—a Reassessment of His Work* (New York: Peter Lang, 2000). An excellent survey of the historiography on the topic up until 2000 is Kenneth Morgan, *Slavery, Atlantic Trade and the British Economy, 1660–1800* (Cambridge: Cambridge University Press, 2000), 29–35, 47–50.

4. David Eltis and Stanley L. Engerman, "The Importance of Slavery and the Slave Trade to Industrializing Britain," *Journal of Economic History* 60 (2000): 125–27, 138.

5. See "Interchange: The History of Capitalism," *Journal of American History* 101 (2014): 503–36.

6. Sven Beckert, "Slavery and Capitalism," *Chronicle of Higher Education* 12 December 2014, https://www.chronicle.com/article/SlaveryCapitalism/150787; Greg Grandin, "Capitalism and Slavery," *Nation*, 1 May 2015, https://www.thenation.com/article/capitalism-and-slavery/; and Seth Rockman and Sven Beckert, "How Slavery Led to Modern Capitalism," *Bloomberg*, 25 January 2012, "Echoes" blog.

7. Beckert, "Slavery and Capitalism." See also Seth Rockman, "The Unfree Origins of American Capitalism," in *Capitalism and Econometrics in Early American Economic History*, ed. Cathy Matson (University Park: Pennsylvania State University Press, 2006), 335–62; Seth Rockman, "Slavery and Capitalism," *Journal of the Civil War Era* 2 (2012): https://journalofthecivilwarera.org/forum-the-future-of-civil-war-era-studies/; Seth Rockman, "What Makes the History of Capitalism Newsworthy?" *Journal of the Early Republic* 34 (2014): 439–66.

8. For institutions, see Douglass C. North and Barry Weingast, "Constitutions and Commitment: The Evolution of Institutions Governing Choice in Seventeenth-Century England," *Journal of Economic History* 49 (1989): 803–32; and Barry Weingast, "Constitutions as Governance Structures: The Political Foundations of Secure Markets," *Journal of Institutional and Theoretical Economics* 149 (1993): 286–312. For bourgeois dignity, see Deirdre McCloskey, *Bourgeois Dignity: Why Economics Can't Explain the Modern World* (Chicago: University of Chicago Press, 2010). For the state as an actor in promoting mercantilism, see Ronald Findlay and Kevin O'Rourke, *Power and Plenty: Trade, War and the World Economy* (Princeton, N.J.: Princeton University Press, 2007).

9. Rockman, "Unfree Origins of American Capitalism," 347.

10. Beckert, "Slavery and Capitalism."

11. Sven Beckert, *Empire of Cotton: A New History of Global Capitalism* (New York: Alfred A. Knopf, 2014), xv

12. Ibid., 171–72, 239.

13. Trevor Burnard, Laura Panza, and Jeffrey Williamson, "Living Costs, Real Incomes and Inequality in Colonial Jamaica," *Explorations in Economic History* 71 (2019): 55–71.

14. Philip D. Morgan, *Slave Counterpoint: Black Culture in the Eighteenth-Century Chesapeake and Lowcountry* (Chapel Hill: University of North Carolina Press, 1998), 468; J. R. Ward, "The Profitability of Sugar Planting in the British West Indies, 1650–1834," *Economic History Review* 31 (1978): 197–213; and David Richardson, "Profits in the Liverpool Slave Trade: The Accounts of William Davenport, 1757–1784," in *Liverpool, the Atlantic Slave Trade, and Abolition*, ed. Roger Anstey and P. E. H. Hair, Historic Society of Lancashire and Cheshire, Occasional Series (1976), vol. 2, 60–90.

15. Williams, *Capitalism and Slavery*, 52.

16. See also Catherine Hall et al., *Legacies of British Slave-Ownership: Colonial Slavery and Formation of Victorian Britain* (Cambridge: Cambridge University Press, 2014).

17. Lowell J. Ragatz, *The Fall of the Planter Class in the British Caribbean, 1763–1833: A Study of Social and Economic History* (New York: Century, 1928).

18. Williams, *Capitalism and Slavery*, 169. For the fate of West Indian planters after the American Revolution, see Christer Petley, ed., "Rethinking the Fall of the Planter Class," special issue, *Atlantic Studies* 9 (2012).

19. Hilary McD. Beckles, " 'The Williams Effect': Eric Williams' *Capitalism and Slavery* and the Growth of West Indian Political Economy," in Solow and Engerman, *British Capitalism and Caribbean Slavery*, 303–16; W. A. Darity, "Eric Williams and Slavery: A West Indian Viewpoint," *Callaloo* 20 (1998): 801–16. For an appreciation of Williams as a politician and an intellectual, see Tanya L. Shields, *The Legacy of Eric Williams: The Postcolonial Moment* (Jackson: University of Mississippi Press, 2015).

20. Cited in Richard B. Sheridan, *Sugar and Slavery: An Economic History of the British West Indies, 1623–1775* (Bridgetown, Barbados: University of the West Indies Press, 1974), 5–6.

21. S. D. Smith, "*Merchants and Planters* Revisited," *Economic History Review* 55 (2002): 434–65; and Emma Rothschild, "Adam Smith in the British Empire," in *Empire and Modern Political Thought*, ed. Sankar Muthu (Cambridge: Cambridge University Press, 2012), 184–98.

22. Williams, *Capitalism and Slavery*, 5–6.

23. For a recent survey of mercantilism, which leaves the term largely undefined, see Philip J. Stern and Carl Wennerlind, eds., *Mercantilism Reimagined: Political Economy in Early Modern Britain and Its Empire* (Oxford: Oxford University Press, 2014).

24. Williams, *Capitalism and Slavery*, 164–65.

25. P. J. Marshall, *Remaking the British Atlantic: The United States and the British Empire After American Independence* (Oxford: Oxford University Press, 2012).

26. David Geggus, *Slavery, War and Revolution: The British Occupation of Saint Domingue, 1793–1798* (Oxford: Oxford University Press, 1982).

27. Klas Rönnbäck, "On the Economic Importance of the Slave Plantation Complex to the British Economy During the Eighteenth Century: A Value-Added Approach," *Journal of Global History* 13 (2018): 308–27.

28. Selwyn Carrington, *The British West Indies During the American Revolution* (Dordrecht: Foris, 1988).

29. Nicholas Draper, "The Rise of a New Planter Class? Some Counter-Currents from British Guiana and Trinidad, 1807–1834," *Atlantic Studies* 9 (2012): 65–83.

30. Seymour Drescher, *Econocide: British Slavery in the Era of Abolition*, 2nd ed. (Chapel Hill: University of North Carolina Press, 2010).

31. David Beck Ryden, "Does Decline Make Sense? The West Indian Economy and the Abolition of the British Slave Trade," *Journal of Interdisciplinary History* 31 (2001): 347–74. See also G. Checkland, "Finance for the West Indies, 1780–1815," *Economic History Review*, 10 (1958): 461–49.

32. Reviews of the book can be found at http://svenbeckert.com/.

33. Beckert, *Empire of Cotton*, xv, 92.

34. Karl Marx, *Capital: A Critique of Political Economy* (New York: International, 1967), 1:361.

35. B. W. Higman, *Slave Populations of the British Caribbean 1807–1834* (Baltimore: Johns Hopkins University Press, 1984), 50. Beckert does not cite recent work on the production of West Indian cotton, such as Justin Roberts, *Slavery and the Enlightenment in the British Atlantic, 1750–1807* (New York: Cambridge University Press, 2013); or David Beck Ryden, " 'One of the Finest and Most Fruitful Spots in America': An Analysis of Eighteenth-Century Carriacou," *Journal of Interdisciplinary History* 43 (2013): 539–70. Instead, he uses an 1848 history of Barbados for information on ant invasions and an obscure article from 1944 on cotton production. Beckert also cites Michael M. Edwards, *The Growth of the British Cotton Trade, 1780–1815* (Manchester: Manchester University Press, 1967), 79, although Edwards does not support his argument; and an article from 1987 by Selwyn Carrington, but the pages cited (841–42) refer not to cotton production but to Britain banning American shipping from the West Indies. Carrington does mention rising cotton production in the 1780s on page 847 but provides no figures like those that Beckert quotes. Selwyn Carrington, "The American Revolution and the British West Indies Economy," *Journal of Interdisciplinary History* 17 (1987): 823–50.

36. Beckert, *Empire of Cotton*, 56–60 (quote 60).

37. Legacies of British Slave Holding project, https://www.ucl.ac.uk/lbs/person/view/10314.

38. For Beckert and the world systems theory undergirding his approach, see Beckert, *Empire of Cotton*, 450–51nn7, 12, 15, and 16. For a Marxist criticism of Wallerstein, see Robert Brenner, "The Origins of Capitalist Development: A Critique of Neo-Smithian Marxism," *New Left Review* 104 (1977): 25–92.

39. Beckert, *Empire of Cotton*, xv–xvi, 37–38, 52, 54, 92, 155, 165–66, 169, 171–73, 239.

40. Ibid., 92.

41. P. K. O'Brien, "European Economic Development: The Contribution of the Periphery," *Economic History Review*, 35 (1982): 1–18.

42. Sanjay Subrahmanyam, "Introduction," in *The Cambridge World History*, vol. 6, *The Construction of a Global World, 1400–1800 C.E.*, part 1, *Foundations*, ed. Jerry H. Bentley, Sanjay Subrahmanyam, and Merry E. Wiesner-Hanks (Cambridge: Cambridge University Press, 2015), 11.

43. Jan De Vries, "The Limits of Globalisation in the Early Modern World," *Economic History Review* 63 (2010): 710–33.

44. C. Knick Harley insists that while the Atlantic economy made a central contribution to the causes of the Industrial Revolution, the route through which this contribution came was through trade to the nonplantation colonies of British North America. These colonies wanted industrial goods from Britain and financed them by developing the burgeoning provision trade to the West Indies. Harley argues that "in the absence of slavery, the northern settlements would have found alternative goods to sell into the Atlantic economy and their growth, and their demand for British manufactures, seems unlikely to have been stifled." C. Knick Harley, "Slavery, the British Atlantic Economy, and the Industrial Revolution," in *The Caribbean and the Atlantic World Economy*, ed. A. B. Leonard et al. (London: Palgrave Macmillan, 2015), 182. For mercantilism and empire, see Jonathan Barth, "Reconstructing Mercantilism: Consensus and Conflict in the British Imperial Economy in the Seventeenth and Eighteenth Centuries," *William and Mary Quarterly*, 3rd ser., 73 (2016): 257–90.

45. Patrick O'Brien, "The Nature and Historical Evolution of an Exceptional Fiscal State and Its Possible Significance For The Precocious Commercialization and Industrialization of the British Economy from Cromwell to Nelson," *Economic History Review* 64 (2011): 415; De Vries, "The Limits of Globalisation in the Early Modern World."

46. O'Brien, "Nature and Historical Evolution," 415–16.

47. Ibid., 416; Peter Liberman, *Does Conquest Pay? The Exploitation of Occupied Industrial Societies* (Princeton, N.J.: Princeton University Press, 1996).

48. O'Brien, "Nature and Historical Evolution," 435–56, 439.

49. As Barbara Hahn states, there is a romantic and nostalgic view to precapitalist behavior. It is noticeable, for example, that Beckert's opening essay starting with cotton growing in Aztec Mexico is singular in not mentioning violence. Violence, it seems, arrived only with Columbus and Cortes. Barbara Hahn, "Review," *Agricultural History* 89 (2015): 482–86 (quote 484).

50. Giorgio Riello, *Cotton: The Fabric that Made the Modern World* (Cambridge: Cambridge University Press, 2013), 240–46.

51. Ibid., 4, 10; Eltis and Engerman, "Importance of Slavery."

52. There is a rich literature on the connections between war and capitalism. See Bartolmé Yun-Casalilla and Patrick K. O'Brien, eds., *The Rise of Fiscal States: A Global History, 1500–1914* (Cambridge: Cambridge University Press, 2012); John Brewer, *The Sinews of Power: War, Money and the English State* (New York: Alfred A. Knopf, 1989); Philippe Contamine, ed., *War and Competition Behind States* (Oxford: Oxford University Press, 2000); Rafael Torres-Sanchez, ed., *War, State and Development: Fiscal-Military States in the Eighteenth-Century* (Pamplona: Eunsa, 2007); and Jan Glete, *War and the State in Early Modern Europe: Spain, the Dutch Republic, and Sweden as Fiscal-Military States, 1500–1600* (New York: Routledge, 2002).

53. Trevor Burnard, *Planters, Merchants, and Slaves: Plantation Societies in British America, 1650–1820* (Chicago: University of Chicago Press, 2015).

54. Nuala Zahedieh, "Colonies, Copper, and the Market for Inventive Activity in England and Wales, 1680–1730," *Economic History Review* 66 (2013): 805–25.

55. Lorena S. Walsh, *Motives of Honor, Pleasure and Profit: Plantation Management in the Colonial Chesapeake, 1607–1763* (Chapel Hill: University of North Carolina Press, 2010); Burnard, *Planters, Merchants, and Slaves.*

56. Andrew Jackson O'Shaughnessy, "The West India Interest and the Crisis of American Independence," in *West Indies Accounts: Essays on the History of the British Caribbean and the Atlantic Economy in Honour of Richard Sheridan*, ed. Roderick A. McDonald (Kingston: University of the West Indies Press, 1996), 126–49.

57. Burnard, *Planters, Merchants, and Slaves*, ch. 5.

58. Simon Newman, *Free and Bound Labor in the British Atlantic World: Black and White Workers and the Development of Plantation Slavery* (Philadelphia: University of Pennsylvania Press, 2013); J. H. Elliott, *Empires of the Atlantic World: Britain and Spain in America, 1492–1830* (New Haven, Conn.: Yale University Press, 2007), chs. 3 and 5.

59. Carla Gardina Pestana, "English Character and the Fiasco of the Western Design," *Early American Studies* 3 (2005): 1–31.

60. Nuala Zahedieh, "Making Mercantilism Work: London Merchants and Atlantic Trade in the Late Seventeenth Century," *Transactions of the Royal Historical Society* (1999): 43–84.

61. Nuala Zahedieh, *The Capital and Colonies: London and the Atlantic Economy 1660–1700* (Cambridge: Cambridge University Press, 2010), 285, 292.

62. Richard S. Dunn, "The Glorious Revolution and America," in *The Oxford History of the British Empire*, vol. 1, *The Origins of Empire*, ed. Nicholas Canny (Oxford: Oxford University Press, 1988), 463–65 (quote 465). Britain did not dislodge France and Spain in the Caribbean and had only limited power over agents at the periphery. Nuala Zahedieh, "Commerce and Conflict: Jamaica and the War of the Spanish Succession," in Leonard et al., *Caribbean and the Atlantic World Economy*, 78–80. Moreover, the process of British state formation was a contested process of highly partisan politics in which compromises were repeatedly negotiated between various irreconcilable public priorities and private interests. Aaron Graham, *Corruption, Party, and Government in Britain, 1702–1713* (Oxford: Oxford University Press, 2015).

63. Burnard, *Planters, Merchants, and Slaves*; and Trevor Burnard and John Garrigus, *The Plantation Machine: Atlantic Capitalism in French Saint-Domingue and British Jamaica, 1748–1788* (Philadelphia: University of Pennsylvania Press, 2016). A useful summary of the politics of slavery over the *longue durée* is John Craig Hammond, "Slavery, Sovereignty, and Empires: North American Borderlands and the American Civil War, 1660–1860," *Journal of the Civil War Era* 4 (2014): 264–98. Still insightful is Richard Pares, *War and Trade in the West Indies 1739–1763* (Oxford: Oxford University Press, 1936).

64. Williams, *Capitalism and Slavery*, 115. For the Seven Years' War in the Caribbean, see Burnard and Garrigus, *Plantation Machine*; and Brendan Simms, *Three Victories and a Defeat: The Rise and Fall of the First British Empire, 1714–1783* (London: Allen Lane, 2007).

65. Walsh, *Motives of Honor*; Emory Evans, *A Topping People: The Rise and Decline of Virginia's Old Political Elite, 1680–1790* (Charlottesville: University of Virginia Press, 2009); and Nicholas Draper, *The Price of Emancipation: Slave-Ownership, Compensation and British Slavery at the End of Slavery* (Cambridge: Cambridge University Press, 2010).

66. Burnard, *Planters, Merchants, and Slaves*, ch. 5.

67. Manisha Sinha, *The Slave's Cause: A History of Abolition* (New Haven, Conn.: Yale University Press, 2016); and David Brion Davis, *The Problem of Slavery in the Age of Emancipation* (New York: Alfred A. Knopf, 2014).

68. Allan Kulikoff, "'Such Things Ought Not to Be': The American Revolution and the First National Great Depression," in *The World of the Revolutionary American Republic: Expansion, Conflict, and the Struggle for a Continent*, ed. Andrew Shankman (New York: Routledge, 2014), 134–64.

69. Peter H. Lindert and Jeffrey G. Williamson, "American Incomes Before and After the Revolution," *Journal of Economic History* 73 (2016): 725–65 (quotes 741, 752).

70. Burnard, *Planters, Merchants, and Slaves*, ch. 3.

71. Lindert and Williamson, "American Incomes Before and After The Revolution," 747; Peter C. Mancall, Joshua L. Rosenbloom, and Thomas Weiss, "Conjectural Estimates of Economic Growth in the Lower South, 1720 to 1800," in *History Matters: Economic Growth, Technology, and Population*, ed. William Sundstrom and Timothy Guinnane (Stanford, Calif.: Stanford University Press, 2003), 389–424.

72. Daron Acemoglu, Simon Johnson, and James Robinson, "Reversal of Fortune: Geography and Institutions in the Making of the Modern World Income Distribution," *Quarterly Journal of Economics* 117 (2002): 1231–94.

73. Lindert and Williamson, "American Incomes Before and After the Revolution," 741–42, 750–53; Acemoglu, Johnson, and Robinson, "Reversal of Fortune,"; and Kulikoff, "'Such Things Ought Not to Be.'"

74. David Brion Davis, "Slavery, Emancipation, and Progress," in *British Abolitionism and the Question of Moral Progress in History*, ed. Donald A. Yerxa (Columbia: University of South Carolina Press, 2012), 18–19.

75. Richard Follett et al., *Plantation Kingdom: The American South and Its Global Commodities* (Baltimore: Johns Hopkins University Press, 2016).

76. Steven Sarson, *The Tobacco-Plantation South in the Early American Plantation World* (London: Palgrave Macmillan, 2013).

77. Lindert and Williamson, "American Incomes Before and After the Revolution," 742.

78. Marshall, *Remaking the British Atlantic*, 77–78.

79. Christopher Leslie Brown, *Moral Capital: Foundations of British Abolitionism* (Chapel Hill: University of North Carolina Press, 2006); and John Craig Hammond and Matthew Mason, eds., *Contesting Slavery: The Politics of Bondage and Freedom in the New American Nation* (Charlottesville: University of Virginia Press, 2011).

80. Trevor Burnard, "Powerless Masters: The Curious Decline of Jamaican Sugar Planters in the Foundational Period of British Abolition," *Slavery and Abolition* 32 (2011): 185–98.

81. James Oakes, *Freedom National: The Destruction of Slavery in the United States, 1861–1865* (New York: Norton, 2013).

82. Beckert, *Empire of Cotton*, 112, 122.

83. Cited in R. J. M. Blackett, *Building an Antislavery Wall: Black Americans in the Atlantic Abolitionist Movement, 1830–1860* (Baton Rouge: Louisiana State University Press, 1983), 6.

84. Richard Huzzey, *Freedom Burning: Anti-Slavery and Empire in Victorian Britain* (Ithaca, N.Y.: Cornell University Press, 2012).

85. Ibid.; Marcus Cunliffe, *Chattel Slavery and Wage Slavery: The Anglo-American Context, 1830–1860* (Athens: University of Georgia Press, 1979); and W. Caleb McDaniel, *The Problem of Democracy in the Age of Slavery: Garrisonian Abolitionists and Transatlantic Reform* (Baton Rouge: Louisiana State University Press, 2013).

86. Alan Olmstead and Paul Rhode, *Creating Abundance: Biological Innovation and American Agricultural Development* (New York: Cambridge University Press, 2002), 64–97.

87. Amanda Foreman, *A World on Fire: Britain's Crucial Role in the American Civil War* (New York: Random House, 2010); R. J. M. Blackett, *Divided Hearts: Britain and the American Civil War* (Baton Rouge: Louisiana University Press, 2001).

88. Kwame Anthony Appiah, *The Honor Code: How Moral Revolutions Happen* (New York: Norton, 2010), 103–36; and Seymour Drescher, *The Mighty Experiment: Free Labor Versus Slavery in British Emancipation* (New York: Oxford University Press, 2002), 202–3, 231–37.

89. Jay Sexton, "Epilogue: The United States in the British Empire," in *British North America in the Seventeenth and Eighteenth Centuries*, ed. Stephen Foster (Oxford: Oxford University Press, 2013), 320–21, 334.

90. Frank Thistlethwaite, *The Anglo-American Connection in the Early Nineteenth Century* (Philadelphia: University of Pennsylvania Press, 1959), 3–5; and Williams, *Capitalism and Slavery*, 131–33.

91. Cited in Kinsley Brauer, "The United States and British Imperial Expansion, 1815–1860," *Diplomatic History* 12 (1988): 24.

92. Alan Taylor, *The Internal Enemy: Slavery and War in Virginia, 1772–1832* (New York: Norton, 2013).

93. Catherine Hall, "Gendering Property, Racing Capital," *History Workshop Journal* 78 (2014): 23.

94. Edward Long, *The History of Jamaica . . .*, 3 vols. (London: T. Lowndes, 1774), 1:493–94.

Epilogue

1. Eric Williams, *Capitalism and Slavery* (Chapel Hill: University of North Carolina Press, 1944). For Williams as man, prime minister, and scholar, see Tanya L. Shields, ed., *The Legacy of Eric Williams: The Postcolonial Moment* (Jackson: University Press of Mississippi, 2015). For the "newer" colonies, see Nicholas Draper, "The Rise of a New Planter Class? Some Counter-Currents from British Guiana and Trinidad, 1807–1833," *Atlantic Studies* 9 (2012): 65–83.

2. Williams, *Capitalism and Slavery*, 121–23.

3. Catherine Hall et al., *Legacies of British Slave-Ownership: Colonial Slavery and the Formation of Victorian Britain* (Cambridge: Cambridge University Press, 2014), 1–33.

4. Trevor Burnard, "Powerless Masters: The Curious Decline of Jamaican Sugar Planters in the Foundational Period of British Abolition," *Slavery and Abolition* 32 (2011): 185–98.

5. Trevor Burnard and John Garrigus, *The Plantation Machine: Atlantic Capitalism in French Saint-Domingue and British Jamaica, 1748–1788* (Philadelphia: University of Pennsylvania Press, 2016).

6. David Geggus, *Slavery, War, and Revolution: The British Occupation of Saint-Domingue 1793–1798* (Oxford: Oxford University Press, 1982).

7. Christer Petley, "Slaveholders and Revolution: The Jamaican Planter Class, British Imperial Politics, and the Ending of the Slave Trade, 1775–1807," *Slavery and Abolition* 39 (2018): 53–79.

8. Christer Petley, *White Fury: A Jamaican Slaveholder and the Age of Revolution* (Oxford: Oxford University Press, 2018), 114–16.

9. Ira Berlin pointed out long ago that slavery has a temporal as well as a spatial history. Ira Berlin, "Time, Space, and the Evolution of Afro-American Society on British Mainland North America," *American Historical Review* 85 (1980): 44–78. Nevertheless, most surveys of slavery stick closely to regional comparisons. See Ira Berlin, *Many Thousands Gone: The First Two Centuries of Slavery in North America* (Cambridge, Mass.: Harvard University Press, 1998); Philip D. Morgan, *Slave Counterpoint: Black Culture in the Eighteenth-Century Chesapeake and Lowcountry* (Chapel Hill: University of North Carolina Press, 1998); and Justin Roberts, *Slavery and the Enlightenment in the British Atlantic, 1750–1807* (New York: Cambridge University Press, 2013). For Jamaican slavery over time, see Trevor Burnard, "E Pluribus Plures: Ethnicities in Early Jamaica," *Jamaican Historical Review* 21 (2001): 8–22, 56–59; and Trevor Burnard, "Slaves and Slavery in Kingston, 1770–1815," in *Urban Slavery in the Age of Abolition, 1770–1930*, ed. Karwan Fatah-Black (Leiden: Brill, forthcoming).

10. Richard S. Dunn, *A Tale of Two Plantations: Slave Life and Labor in Jamaica and Virginia* (Cambridge, Mass.: Harvard University Press, 2014); J. R. Ward, "The Amelioration of British West Indian Slavery: Anthropological Evidence," *Economic History Review* 71 (2018): 1199–1226.

11. Trevor Burnard, "The British Atlantic World," in *Atlantic History: A Critical Appraisal*, ed. Jack P. Greene and Philip D. Morgan (New York: Oxford University Press, 2009), 111–36; Tehila Sasson et al., "Britain and the World: A New Field?" *Journal of British Studies* 57 (2018): 677–708.

12. Edward Braithwaite, *The Development of Creole Society in Jamaica, 1770–1820* (Oxford: Oxford University Press, 1971).

13. Patrick Spero and Michael Zuckerman, eds., *The American Revolution Reborn* (Philadelphia: University of Pennsylvania Press, 2016); Edward G. Gray and Jane Kamensky, eds., *The Oxford Handbook of the American Revolution* (New York: Oxford University Press, 2013); Stephen Conway, *The American Revolutionary War* (London: I. B. Tauris, 2013); and Alan Taylor, *American Revolutions: A Continental History, 1750–1804* (New York: Norton, 2016).

14. Kathleen Wilson, "Rethinking the Colonial State: Family, Gender, and Governmentality in Eighteenth-Century British Frontiers," *American Historical Review* 116 (2011): 1294–1322; Carla Pestana, "State Formation from the Vantage of Early English Jamaica: The Neglect of Edward Doyley," *Journal of British Studies* 56 (2017): 383–405; and Aaron Graham, "The Colonial Sinews of Imperial Power: The Political Economy of Jamaican Taxation, 1768–1838," *Journal of Imperial and Commonwealth History* 45 (2017): 188–209.

15. Trevor Burnard and Aaron Graham, "Security, Settlement and the British Imperial System: The Imperial and Colonial State in Jamaica, 1721–1782," *Early American Studies* (forthcoming).

16. Philip Wright, "War and Peace with the Maroons, 1730–1739," *Caribbean Quarterly* 16 (1970): 5–27; Kathleen Wilson, "The Performance of Freedom: Maroons and the Colonial Order in Eighteenth-Century Jamaica and the Atlantic Sound,' *William and Mary Quarterly*, 3rd ser., 66 (2009): 45–86.

17. Michael Craton, *Testing the Chains: Resistance to Slavery in the British West Indies* (Ithaca, N.Y.: Cornell University Press, 1982), 92.

18. Veridicus [Thomas Fearon], *The Merchants, Factors, and Agents Residing at Kingston at the Said Island, COMPLAINANTS, Against the Inhabitants of Spanish-Town . . . THE RESPONDENTS CASE* (London, 1754), 60–61, 65–66.

19. Kenneth Morgan, ed., "Robert Dinwiddie's Reports on the British American Colonies," *William and Mary Quarterly*, 3rd ser., 65 (2008): 305–46; Anonymous, *An Inquiry Concerning the Trade, Commerce, and Policy of Jamaica . . .* (London: T. Kinnersly and T. Woodfal, 1759). The first version of this pamphlet was published in Jamaica in 1757 by Curtis Brett.

20. Jack P. Greene, *Settler Jamaica in the 1750s: A Social Portrait* (Charlottesville: University of Virginia Press, 2016), 204–5; P. J. Marshall, *The Making and Unmaking of Empires: Britain, India and America c. 1760–1783* (Oxford: Oxford University Press, 2005), 76.

21. Graham, "Colonial Sinews of Imperial Power," 192, fig. 1.

22. Jamaica was not the only West Indian colony that nearly succumbed to slave revolt. In Berbice, slave rebels took over the government in 1763 and nearly drove whites from the colony. Marjoleine Kars, "Dodging Rebellion: Politics and Gender in the Berbice Slave Uprising of 1763," *American Historical Review* 121 (2016): 39–69.

23. Trevor Burnard, "A Failed Settler Society: Marriage and Demographic Failure in Early Jamaica," *Journal of Social History* 28 (1994): 63–82.

24. [Edward Trelawney], *An Essay Concerning Slavery, and the Danger Jamaica Is Expos'd to from the Too Great Number of Slaves . . .* (London: C. Corbett, 1746),reprinted in *Exploring the Bounds of Liberty: Political Writings of Colonial British America from the Glorious Revolution to the American Revolution* ed. Jack P. Greene and Craig Yirush (Carmel, Ind.: Liberty Fund, 2018), 3 vols., 2: 1134–66; Jack P. Greene, "Edward Trelawney's 'Grand Elixir': Metropolitan Weakness and Constitutional Reform in the Mid-Eighteenth-Century British Empire," in *West Indies Accounts: Essays on the History of the British Caribbean and the Atlantic Economy in Honour of Richard Sheridan*, ed. Roderick A. McDonald (Kingston: University of the West Indies Press, 1996), 87–100; Knowles to Board of Trade, November 21, 1741, C.O. 137/27/24, National Archives, Kew, London.

25. The Seven Years' War was also important, but to a large extent Jamaica avoided most of the actions in this war. The most important geopolitical event in this war was the capture of Havana in 1762, but this military action was undertaken by British and New England troops more than Jamaicans. For the Seven Years' War in Jamaica, see Burnard and Garrigus, *Plantation Machine*, chs. 4–5; and Elena Schneider, *The Occupation of Havana: Slavery, War, and Empire in the Eighteenth-Century World* (Chapel Hill: University of North Carolina Press, 2018).

26. Burnard and Garrigus, *Plantation Machine*, 226–28.

27. Cited in Petley, "Slaveholders and Revolution," 59.

28. Seymour Drescher, *Econocide: British Slavery in the Era of Abolition*, 2nd ed. (Chapel Hill: University of North Carolina Press, 2010); Ahmed Reid, "Sugar, Slavery and Productivity in Jamaica, 1750–1807," *Slavery and Abolition* 37 (2016): 159–82; and J. R. Ward, "The Profitability of Sugar Planting in the British West Indies, 1650–1834," *Economic History Review* 31 (1978): 206.

29. Seymour Drescher, "The Shocking Birth of British Abolitionism," *Slavery and Abolition* 33 (2012): 575–84.

30. Sarah Yeh, "Colonial Identity and Revolutionary Loyalty: The Case of the West Indies," in *The Oxford History of the British Empire: British North America in the Seventeenth and Eighteenth Centuries*, ed. Stephen Foster (Oxford: Oxford University Press, 2013), 199–216.

31. "Resolution," *Journals of the Assembly of Jamaica*, December 10, 1789. Petley argues that such was the radical nature and strident tone of the resolution that the agent for Jamaica in London, Stephen Fuller, did not formally present the petition to Parliament, fearing that MPs would associate Jamaican planters with the turbulent forces of revolution that were troubling them from events in France. Petley, "Slaveholders and Revolution," 61.

32. Petley, "Slaveholders and Revolution," 63.

33. Roger Anstey, *The Atlantic Slave Trade and British Abolition, 1760–1810* (Atlantic Highlands, N.J.: Humanities, 1975), 275–78.

34. David Beck Ryden, *West Indian Slavery and British Abolition, 1783–1807* (Cambridge: Cambridge University Press, 2009), 293–94

35. David Geggus, "The Cost of Pitt's Caribbean Campaigns, 1793–1798," *Historical Journal* 2 (1983): 699–706; and Michael Duffy, "The French Revolution and British Attitudes to the West

Indian Colonies," in *A Turbulent Time: The French Revolution and the Greater Caribbean*, ed. David Barry Gaspar and David Patrick Geggus (Bloomington: Indiana University Press, 1997), 87.

36. Duffy, "French Revolution," 89; Drescher, *Econocide*, 104–6.

37. For the impotence of Jamaican elites against abolitionist pressure in the push for the abolition of the slave trade in 1805–7 and their economic helplessness as sugar prices fell after 1807, see David Beck Ryden, "Does Decline Make Sense? The West Indian Economy and the Abolition of the British Slave Trade," *Journal of Interdisciplinary History* 31 (2001): 347–74; and David Beck Ryden, "Sugar, Spirits, and Fodder: The London West India Interest and the Glut of 180715," *Atlantic Studies* 9 (2012): 41–64.

38. Christer Petley, "Rethinking the Fall of the Planter Class: Introduction," *Atlantic Studies* 9 (2012): 1–17.

39. Trevor Burnard, *Planters, Merchants, and Slaves: Plantation Societies in British America, 1650–1820* (Chicago: University of Chicago Press, 2015), and Burnard and Garrigus, *Plantation Machine*.

40. James Otis, *The Rights of the British Colonies Asserted and Proved* (Boston, 1764), in *Pamphlets of the American Revolution, 1750–1776*, ed. Bernard Bailyn (Cambridge, Mass.: Harvard University Press, 1965), 1:435–36.

41. James Ramsay, "Motives for the Improvement of the Sugar Colonies," Add. Mss. 27261, British Library, fols. 44, 69.

42. B. W. Higman, *Plantation Jamaica 1750–1850: Capital and Control in a Colonial Economy* (Kingston: University of the West Indies Press, 2005), 293.

Index

abolitionism: about history of, 212, 215, 301n75; African Americans and, 235; American North and, 207–8, 215, 229, 231, 234; American Revolution and, 153, 204, 231; British society, 8, 9, 17–18, 41, 156, 231, 235, 285n7; British West Indian merchants, 41, 156; cotton plantations/products, 234; creolization and, 107; fear of slave revolts and, 106, 107, 130; free people and, 133, 134; humanitarianism and, 39, 106, 170, 221; loyalism and, 195, 196, 212; martyrdom of rebellious enslaved people, 117, 279n60; mercantilism, 41; plantation system and, 15, 41, 69, 196, 211; planters and, 9, 15, 69, 72, 117, 127, 156, 170, 172, 173, 195, 214, 281n94; poor whites/lower classes as enslaved by blacks, 161, 166; pronatalism and population data, 267n95, 276n16; Royal Navy and, 235; security issue and, 106–7, 113; slave registration system and, 101; slave system abolished in 1834 and, 2–3, 34, 134, 215, 221, 234, 235; slave system and, 16, 72, 100–101, 154, 165; slave trade abolished in 1807, 8, 86, 152, 222, 241, 245–47; *Somerset* case and, 8, 23, 154–56, 204, 285n7; United States and, 153, 288n44; xenophobia and, 167, 170; *Zong* case and, 8, 9, 192, 204, 295n56. *See also* slave trade

Abruco (alias Blackwell), 122

absentee planters, 49, 57, 63, 84, 86, 152, 157, 159, 163, 198, 271n47, 287n32

Acemoglu, Daron, 232

Adams, John, 216, 296n13

Adventure (slave ship), 177, 190, 191

Africa: about, 4, 11; Coromantees, 71, 103–5, 107–9, 118, 122; empire discourse and, 48–49; patriarchalism and, 94–95; pharmacopeia and, 101; rescue of Africans rationalization and, 70, 71; scientific data and, 16–17; slave trade rationalization as rescue from, 70, 166; transnational/trans-imperial

relations, 196; white authority as challenged, 107; "whitening" versus "blackening" of society and, 134, 150. *See also* slave trade

African Americans, 22, 40, 215–16, 235. *See also* enslaved children; enslaved men; enslaved men, and slave management practices; enslaved people; enslaved women; enslaved women, and reproduction issues; enslaved women, and slave management practices

Africanization ("blackening"), 101, 134, 148–50, 169, 172, 173, 282n16

Africans: Barbados plantation system and, 228; Eboe people, 88, 123; enslaved people as tradesmen, 80; fear of, 22; as field hands, 89, 112; jobbing gangs and, 80–81; relations with people of African descent, 79–80; slave system and, 228; state of nature and, 29; suicides, 83, 175; war capitalism and, 224. *See also* enslaved children; enslaved men; enslaved men, and slave management practices; enslaved people; enslaved women; enslaved women, and reproduction issues; enslaved women, and slave management practices; people of African descent; race and racial ideologies, and Africans

agency, and enslaved people, 15, 16, 19, 89, 94, 101, 254n41

age of revolution, 7, 19, 130, 237, 241, 248

agro-industrial system, 6, 8, 90–91, 227. *See also* plantation system

Akan (Coromantees, Amina, Mina, Ga) enslaved people, 71, 103–5, 107–9, 118, 122, 123

Alert (slave ship), 177, 182

Allen, Robert, 91

amelioration: birth rates, 94, 101, 106; health and welfare issues and, 74, 95, 99, 100; infant mortality, 101; mortality rates for enslaved people, 93; planters and, 281n94; slave management practices and, 15, 16, 76, 93, 99–101, 106, 281n94; standards of living and, 95, 99

Acknowledgements

My original intention was to present in a book numerous essays that I had published about Jamaica, some in obscure places. Over time, and with the encouragement of my exemplary editor Bob Lockhart, this project became an entirely new book, except for one chapter previously published in French to which most readers will not have had access, about Jamaica in the period of revolution, with the greatest concentration being on the years before, during, and after the American Revolution.

The introduction to this book shows that I am deeply influenced in the continuing evolution of my scholarship on eighteenth-century Jamaica by the efflorescence of outstanding new work in Caribbean history that has emerged in the last few years. I have presented versions of some of these chapters in workshops, seminars, and conferences throughout the world. I cannot remember all the occasions on which some aspects of this work have been discussed with colleagues, but they are numerous. My findings here thus rest on a whole substructure of helpful criticism and robust debate about Jamaica, the American Revolution, the Atlantic World, and eighteenth-century imperialism. Many of my most important talks have been with colleagues in Europe, and I am delighted to once again acknowledge my friends in my two most important scholarly associations, the British Group in Early American History and the European Early American Studies Association. Fellowships in Paris have been especially important in shaping the wider context of my work. I would like to thank Cécile Vidal of École des hautes études en sciences sociales, Marie-Jeanne Rossignol and Allan Potofsky of Paris VII (Diderot), and Bertrand Van Ruymbeke of Paris VIII (Vincennes-Saint-Denis) for facilitating productive visits to Paris since 2015 and in providing funds for me to enjoy all that Paris has to offer. My colleagues in Britain and America have continued to keep me buoyed up with interest in Jamaica even after I left Britain in 2011 after eleven happy years at Brunel, the University of Sussex, and the University of Warwick.

Much of the work for this book was developed when I taught at the University of Sussex (2004–7) and in animated conversation with Richard Follett, Paul Betts, Saul Dubow, and Naomi Tadmor. Sussex remains my favorite academic location. My research continued while I taught at the University of Warwick

(2007–11), where my colleagues in Caribbean and Atlantic history—Gad Heuman, Cecily Jones, David Dabydeen, Tim Lockley, Tony MacFarlane, Rebecca Earle, and David Lambert—helped continue Warwick's leading role as a center of Caribbean studies. Simon Newman has been friend, mentor, and fellow Atlanticist whose career has intertwined with mine over many years. I believe I was partly responsible for him becoming interested first in Barbados and now in Jamaica, but the pupil (if he was ever such a thing) outpaced the master (not that this is what I am) a long time ago. Peter Mancall is not resident in Europe and works on different areas and time periods in the Atlantic World, but our careers have been so connected, especially as we vie with each other to try and attend conferences together in as many countries as possible, notably in Europe, that I see him as compatriot in both Europe as well as America.

It would be invidious to name all my friends who have helped me from Britain and America, but I would like to note the specific contributions of Giorgio Riello, Aaron Graham, Sherrylynne Haggerty, Emma Hart, and Andrew O'Shaughnessy to collaborative work that has fed into the findings presented in this book. Kit Candlin, Deirdre Coleman, and Christine Walker in Australia and Singapore have also engaged in joint work with me that has proved very important for shaping how I understand Jamaica in the second half of the eighteenth century. Michael McDonnell, Andrew O'Shaughnessy, Steven Pincus, Lige Gould, Patrick Griffin, Kate Fullager, Lisa Ford, Frank Cogliano, Jennifer Milam, Ed Gray, Pat Griffin, Mark Peterson, and Jane Kamensky have been important in widening my grasp of the American Revolution as a global event in which Jamaica played an important role, including being at conferences in Ireland, Australia, and Chile in which I tried out some of my ideas.

I owe most debts, however, to the scholars whose work on Jamaica has preceded mine. Once again, I can name only a few of the more important influences. Richard Sheridan and Sid Mintz are now dead, but their assistance and example when I was just starting out in Caribbean history, in the late 1980s, was crucial to how I saw Jamaican history. We all remain indebted to Dick for his pioneering work in West Indian economic history. Sid's notion that the Caribbean was a place of "precocious modernity" is at the heart of my understanding of plantation culture and slave societies in Jamaica between the First and Second Maroon Wars (ca. 1730–96). Barry Higman, a fellow Antipodean, gave me my first job at the University of the West Indies at Mona, and you always remember fondly the person who gave you your start. My time at Mona introduced me to many of the giants of West Indian history, many of whom were in the History Department —Roy Augier and Douglas Hall from past times; Kamau Braithwaite, Hilary Beckles, Verene Shepherd, Swithin Wilmot, Patrick Bryan, Brian Moore, Kathleen Monteith, and Carl Campbell from when I was there; and James Robertson from after I left. Jonathan Dalby arrived in Jamaica the same day I did and has always been a friend and supporter. Barry's monumental studies of Jamaica and the West Indies immediately following the period in which I am most interested forms the foundation of the social history of eighteenth-century Jamaica that I have practiced since going to Kingston in October 1987. The person who encouraged me to move to Britain in January 2000 after a productive decade at the University of Canterbury in New Zealand was Ken Morgan of Brunel. His work on slavery, the slave trade, Jamaican commerce, and the eighteenth-century British Atlantic World was a touchstone, along with that by Andrew O'Shaughnessy, when I started to think about the American Revolution and the Caribbean in the early 2000s, and he was the ideal academic colleague in all respects. His body of work on Jamaica in this period is extraordinarily impressive. Another senior scholar in Britain who was an inspiration was Betty Wood. Her attention to diverse plantation experiences, to gender, and to an area where I pay less attention than I should—religion—continues to fascinate me and shape my historical practice. The third person in Britain I wish to acknowledge as having constantly challenged and inspired my work on Jamaica is Nuala Zahedieh. Nuala has unparalleled knowledge about the earliest period of English and British involvement in Jamaica and reading this work and having numerous discussions with her have enriched my scholarship greatly. In the United States, the innovative, erudite, and highly original work of John McNeill has fed into my own less adequate works in numerous ways that can only improve my tentative conclusions and help to put local matters in a global context.

I know it is customary now in acknowledgments to make effusive declarations about the support of one's family. It has always struck me as amazing how many writers praise their partners as if they were cowriters who participated in every step of the intellectual journey from idea to published book. My wife, Deborah, neither reads nor talks about what I do in my study when I avoid domestic chores. But she provides support in every other way, and I have been especially appreciative of her absolute support and confidence in me in very difficult work circumstances in late 2018 and into 2019. That support may not have added to the intellectual quality of this book, but what she has done to help me, along with my delightful children, Eleanor and Nicholas, and especially my very dear friend, Deirdre Coleman, and my utterly supportive colleagues, Joy Damousi, Marie Connolly, and Gillian Wigglesworth, in very tough times, have enabled me to push on with this project.

I would like to dedicate this book to two great historians of early America whose personal example, unstinting support, and remarkable intellectual achievements have been essential in giving meaning to what little insights I have had into Jamaican history. I first met Jack Greene as a naive and unsure New Zealander arriving in a boiling-hot Baltimore in September 1983. I met Richard Dunn when I had a fellowship at what later became the McNeil Center of Early American

Studies in an equally warm Philadelphia in September 1986. We have moved from being mentors and mentee to academic colleagues and friends. I realized soon after meeting both men that, along with their estimable personal qualities that made them ideal role models, they each had a profound interest in Jamaica. It says something about how fascinating Jamaica is as an object of study that two of the very best twentieth-century historians, whose work in the twenty-first century confirms their standing at the very top of the historical profession, have written outstanding works on Jamaica and on the British West Indies. I hope they find this offering of some value and interest.